MATHEMATICS OLD & NEW

MATHEMATICS OLD & NEW

SAUL STAHL &
PAUL E. JOHNSON

DOVER PUBLICATIONS, INC.
MINEOLA, NEW YORK

The authors are indebted to their colleagues James Church and Ben Cobb for several informative discussions regarding probability and statistics. They also acknowledge several helpful comments by Mark Huneke, Jennifer Wagner, and the book's anonymous reviewers.

Chapter 4 is a condensation of portions of Saul Stahl's *A Gentle Introduction to Game Theory*, American Mathematical Society, Mathematical World Volume 13, Copyright © 1991, Providence, Rhode Island. The authors are grateful to the Society for its generous permission to include these selections here.

The games described in Exercises 4.2. 12-4.2 16 and 4.51-4.5.44 are reprinted from J. D. Williams's *The Complete Strategyst*, Dover Publications, Inc., Mineola, New York, 1986, with the permission of the RAND Corporation.

Bibliographical Note

Mathematics Old and New, first published by Dover Publications, Inc., in 2017, is a revised and updated version of *Understanding Modern Mathematics*, originally published in 2007 by Jones and Bartlett Publishers, Sudbury, Massachusetts.

International Standard Book Number

ISBN-13: 978-0-486-80738-6
ISBN-10: 0-486-80738-X

Manufactured in the United States by LSC Communications
80738X01 2017
www.doverpublications.com

Contents

Preface

This book is aimed at the popular "Topics in Mathematics" course that is a requirement for the B.A. degree in many universities. This course has no prerequisites above and beyond a minimal amount of high school algebra and is meant to better acquaint the students with mathematics through an exposure to its applications and an exposition of some of its culture.

All the textbooks written for this course have adopted the survey approach and touch many topics lightly. We, however, feel that the course's goals can be best accomplished by focusing on a few important topics and explaining both their significance and evolution in some depth. This calls for locating good mathematical ideas that are presentable at the level of the class, convincing the students of their value, and allowing enough time to permit the concepts to take root in the students' minds through drill and development.

In our opinion, it is a serious pedagogical mistake to teach easy applications just because they show how mathematics can be used in an applied context. The superficiality of the mathematics will not escape the students and they will lose respect for the subject matter. With the respect will also go their willingness to learn. For this reason we have concentrated on a few topics whose centrality to mathematics is well known: probability, statistics, game theory, linear programming, and symmetry. The topic of social choice was added to these because its paradoxes and surprises make it a favorite with students.

While these topics have been expounded in many texts, our development of most of them is unique:

Probability and statistics: The exposition is focused on the normal curve. The evolution of this curve is set forth in detail together with explanations of the scientific needs that motivated the invention of some of the standard statistical procedures. This amounts to a comprehensible explanation of how and why the normal curve came to occupy the eminent position it does in our society - a tale in which Columbus, Galileo, the French Revolution, Gauss, and the Average Man all play their parts.

Voting systems: This chapter explores voting mechanisms that can be used to decide among a discrete set of alternatives. Although individual voters may be transitive and logical in their behavior, the aggregation of their opinions through a voting procedure can often produce problematic outcomes. The chapter includes a wide range of voting procedures, including the long-standing ones like plurality, majority, and rank-order voting, as well as some very recent proposals intended to address the weaknesses of the older methods. Several criteria for evaluation of voting procedures are considered.

Game theory: Enough game theory is taught so that the students gain an appreciation of the Nash equilibrium, a concept for which the economists voted to award the 1994 Nobel Prize to the mathematician John Nash. The popularity of the movie *A Beautiful Mind* makes this subject even more topical. Non-trivial applications to economics, population biology, anthropology, political science and other social situations are discussed.

Linear programming: The exposition is conventional. We included this material only because of its popularity amongst the instructors of this course and because of its importance as a tool of applied mathematics.

Symmetry: Contrary to common practice in books for this audience, we have added three-dimensional symmetry to the standard two-dimensional material. This makes possible visual/calculational problems that have just the right level of difficulty. It also provides a context for recounting to them the mysterious tale of Monstrous Moonshine - the story of the classification of the finite simple groups, MONSTER, and the mysterious connections with non-Euclidean geometry.

Map colorings: The study of map colorings passed on two surprisingly difficult problems. While both were solved a century later, both could be considered genuinely difficult problems since their proofs were long and subsequently no shorter proofs have emerged. The Four Color surely holds the distinction of producing more false proofs than any other. However, we have chosen to expound on the resolution of the Ringel-Youngs Theorem for no other reason than the exotic nature of its proof.

The evolution of the normal distribution: It is arguable that the normal distribution, otherwise known as the bell-shaped curve, has as many users as do Newton's Three Laws. An examination of its history reveals much about the way scientists go about their business.

Chapter 1

Probability

Animal astragali - bones used for gambling - have been found in graves that are thousands of years old. These archaeological artifacts indicate that the interest in the likelihood of events goes back to antiquity. Probability, which is the quantification of the informal notion of likelihood, dates back only as far as the Italian Renaissance. Subjective and objective notions of probability are discussed and some formal properties of the latter are stated. These are applied to the derivation of the probabilities of compound events. The normal curve is introduced as a computational aid.

1.1 The Quantification of Likelihood

It is often said that the only certain things in life are death and taxes. Most other aspects of life and the universe have some uncertainty attached to them: although the sun is practically certain to rise tomorrow, it is known that some day it will fail to do so; on the next July Fourth the temperature in Kansas City may top 85°F; the Dow Jones index may rise tomorrow; a woman may be elected as president in the next U.S. presidential election. All readers will agree that these events have been listed in decreasing order of likelihood and this consensus indicates that the ephemeral quality of likelihood should in fact be quantifiable. Such quantifications were first phrased in terms of one's willingness to risk stakes in betting on such uncertain events, i.e., as odds. This, however, was too subjective a procedure for scientific purposes and eventually an objective set of rules were laid down which serve as an objective definition of probability.

The quantification of uncertainty is sometimes easy and sometimes hard. When a coin is flipped, a probability of 1/2 is assigned to the eventuality that it will come up *heads*. The coin might, of course, be weighted so as to render this inaccurate, but we think of this possibility as an irrelevant aberration. If the coin is fair, as most coins are believed to be, then the probability of *heads* is 1/2 and only a mathematician might quibble. But how can the likelihood of a woman being elected president be quantified objectively?

In these pages it will first be shown how to assign probabilities in elementary, almost trivial, situations. These simple contexts will then be combined so as to cover more complex and realistic questions.

A *random experiment* is a future incident with a set of possible *outcomes*. The set of outcomes is called the *sample space*. Thus the flip of a coin is a random experiment with two outcomes, *heads* and *tails,* and the sample space {*heads, tails*}. The toss of a standard die is a random experiment with six outcomes: 1, 2, 3, 4, 5, 6. The associated sample space is {1, 2, 3, 4, 5, 6}. The selection of a shirt to wear is also a random experiment whose sample space consists of all the shirts in one's wardrobe.

Depending on one's interests, different sample spaces can be associated with the same random experiment. Suppose a pair of dice are tossed. If we are interested in the sum of the faces, then the sample space is {2, 3, 4, 5, 6, 7, 8, 9, 10, 11, 12}. If, on the other hand, we are interested in the number of ones, then the sample space is {0, 1, 2}.

Given a random experiment and a sample space, each of the outcomes has a probability that quantifies the likelihood of that outcome transpiring when the experiment is performed. This probability will not be defined explicitly but will instead be computed, wherever possible, on the basis of some elementary properties that are universally accepted. If a random experiment has a finite number of outcomes then each of the outcomes has a probability between 0 and 1 and these probabilities are so distributed that

P1. *The sum of the probabilities of all the outcomes is 1.*
P2. *Equally likely outcomes have equal probabilities.*

If a coin is flipped then each of the outcomes *heads* and *tails* has probability 1/2. If a die is tossed, and there is no reason to doubt its fairness, then each of the outcomes 1, 2, 3, 4, 5, 6 has probability 1/6. In general, if a random experiment has n outcomes and there is no reason to believe that any of these outcomes are more likely to take place than others, then each of the outcomes is assigned the probability

$$\frac{1}{n}.$$

Properties P1 and P2 above are themselves the consequences of a deeper assumption, which can almost be taken as a definition, that the probability of an outcome in some sense describes its frequency if the experiment is performed a large number of times. It is tempting to say that the probability of a flipped coin coming up *heads* is 1/2 because when a coin is tossed, say a thousand times, it will come up *heads* approximately 500 times. The reason this cannot be taken as a genuine definition is that the same coin will also come up *heads* approximately 499 times and so the very same repeated experiment could be used to justify assigning a probability of 0.499 to *heads*. A new experiment for deciding between 0.499 and 0.5 as the appropriate probability can of course be designed, but, amongst other difficulties, this would call for millions of flips and it has certainly not been carried out by either the author or any of his acquaintances. Nevertheless we are all willing to recognize 0.5 as the correct probability on logical grounds alone. The reason for this is that while the actual frequencies fail to serve as a definition of the probabilities, the expected (but rarely realized) frequencies do have the following analogs of the properties 1 and 2 above:

F1. *The actual frequencies of all the outcomes add up to the total number of iterations of the experiment.*

F2. *The frequencies of equally likely outcomes are expected to be equal.*

Properties F1 and F2 of frequencies then translate into properties P1 and P2 of probabilities.

When the outcomes are not equally likely some reasoning is needed to compute their probabilities. One method for doing this is to "break down" outcomes into "simpler" outcomes that are equally likely. Such a "complex" outcome is called an *event*. More formally, an *event* is a set of outcomes. For example, if a die is tossed, one may speak of the event that its outcome is *odd*, which is identified with the set of outcomes $\{1, 3, 5\}$. Similarly, the event that the outcome is greater than 4 can be thought of as the set $\{5, 6\}$. The following property can now be added to P1 and P2:

P3. *The probability of an event is the sum of the probabilities of its constituent outcomes.*

When all the outcomes of the random experiment in question are equally likely this is equivalent to:

P3′. $\text{Probability of event} = \dfrac{\text{number of constituent outcomes}}{\text{total number of outcomes}}$

Hence the probability of a tossed die coming up even is

$$\frac{3}{6} = \frac{1}{2}$$

and the probability of its coming up greater than 4 is

$$\frac{2}{6} = \frac{1}{3}.$$

Suppose two dice are tossed. The 36 equally likely outcomes are then

(1, 1) (1, 2) (1, 3) (1, 4) (1, 5) (1, 6)

(2, 1) (2, 2) (2, 3) (2, 4) (2, 5) (2, 6)

(3, 1) (3, 2) (3, 3) (3, 4) (3, 5) (3, 6)

(4, 1) (4, 2) (4, 3) (4, 4) (4, 5) (4, 6)

(5, 1) (5, 2) (5, 3) (5, 4) (5, 5) (5, 6)

(6, 1) (6, 2) (6, 3) (6, 4) (6, 5) (6, 6)

and each of these outcomes has probability $\frac{1}{36}$. The event that the two dice show the same face is the set

$$\{(1, 1), (2, 2), (3, 3), (4, 4), (5, 5), (6, 6)\}$$

and so has probability

$$\frac{6}{36} = \frac{1}{6}.$$

On the other hand, the event that one of the faces is more than double the other is the set

$$\{(1, 3), (1, 4), (1, 5), (1, 6), (2, 5), (2, 6), (3, 1), (4, 1), (5, 1), (5, 2), (6, 1), (6, 2)\}$$

and has probability

$$\frac{12}{36} = \frac{1}{3}.$$

As an example of a random experiment whose outcomes are not equally likely we gave the addition of the faces that show after a pair of dice is tossed. The sample space is $\{2, 3, 4, 5, 6, 7, 8, 9, 10, 11, 12\}$ and the probability of each of the outcomes is computed by expressing it as a set of the outcomes of the previous experiment:

Sum	The event as a set	Probability
2	$\{(1, 1)\}$	$\frac{1}{36}$
3	$\{(1, 2), (2, 1)\}$	$\frac{2}{36} = \frac{1}{18}$
4	$\{(1, 3), (2, 2), (3, 1)\}$	$\frac{3}{36} = \frac{1}{12}$
5	$\{(1, 4), (2, 3), (3, 2), (4, 1)\}$	$\frac{4}{36} = \frac{1}{9}$
6	$\{(1, 5), (2, 4), (3, 3), (4, 2), (5, 1)\}$	$\frac{5}{36}$
7	$\{(1, 6), (2, 5), (3, 4), (4, 3), (5, 2), (6, 1)\}$	$\frac{6}{36} = \frac{1}{6}$
8	$\{(2, 6), (3, 5), (4, 4), (5, 3), (6, 2)\}$	$\frac{5}{36}$
9	$\{(3, 6), (4, 5), (5, 4), (6, 3)\}$	$\frac{4}{36} = \frac{1}{9}$
10	$\{(4, 6), (5, 5), (6, 4)\}$	$\frac{3}{36} = \frac{1}{12}$
11	$\{(5, 6), (6, 5)\}$	$\frac{2}{36} = \frac{1}{18}$
12	$\{(6, 6)\}$	$\frac{1}{36}$

Two events are said to be *complementary* if between them they constitute the entire sample space of a random experiment and they share no outcomes. Such are *heads* and *tails* for the coin flip and *even* and *odd* for the toss of a die. Not getting a *two* upon tossing a die is complementary to getting it.

P4. *If A and B are complementary events then*

$$\textit{Probability of } A + \textit{Probability of } B = 1.$$

Another common method for expressing the outcomes of one experiment as a combination of outcomes of another experiment is that of conjunction. Outcome A is said to be a *conjunction* of B and C if its occurrence is logically equivalent to the occurrence of both B and C. Such, for example, is the case for the following events when two dice are thrown:

A: The sum of the faces showing is 12;
B: The first die shows a 6;
C: The second die shows a 6.

Similarly, if two coins are flipped then outcome A below is the conjunction of B and C:

A: No heads are showing;
B: The first coin came up *tails*;
C: The second coin came up *tails*.

Two random experiments are said to be *independent* if information about the eventual outcome of either sheds no light on the eventual outcome of the other. If two coins are flipped then knowing that the first coin came up *heads* tells us nothing about what happens to the second coin. These two coin flips are therefore independent. Similarly, the tossing of two dice are also independent experiments. On the other hand, if two coins are glued *heads* to *heads* and flipped, then knowing that the first coin landed tails down tells us that the second coin landed *tails* up. The two coin tosses are therefore not independent of each other. Similarly, the height and the weight of a randomly selected person are both random experiments. However, since tall people tend to be heavier than short ones, these are not independent experiments.

Suppose now that event A is the conjunction of events B and C which are respectively associated with independent random experiments. Then

P5. *Probability of A = (Probability of B)·(Probability of C).*

Accordingly, if two dice are tossed, then the probability of the first and the second coming up 5 and 6, respectively, is

$$\frac{1}{6} \cdot \frac{1}{6} = \frac{1}{36}.$$

Similarly, if a coin is flipped and a die is tossed, then the probability that the coin comes up *heads* and the die shows a 3 is

$$\frac{1}{2} \cdot \frac{1}{6} = \frac{1}{12}.$$

The multiplication principle P5 is not limited to just two experiments. The probability of a conjunction of events associated with any number of pairwise independent experiments is the product of the probabilities of the individual events. Thus, if a family with five children is selected at random then the probability of all of them being daughters is

$$\frac{1}{2} \cdot \frac{1}{2} \cdot \frac{1}{2} \cdot \frac{1}{2} \cdot \frac{1}{2} = \frac{1}{32}.$$

Similarly, if a die is tossed four times then the probability of its showing 1, 2, 3, 4 in succession is

$$\frac{1}{6} \cdot \frac{1}{6} \cdot \frac{1}{6} \cdot \frac{1}{6} = \frac{1}{1,296}.$$

Exercises 1.1

1. Two standard dice are tossed. Compute the probabilities of the following events:
   (a) The two faces showing differ by 1;
   (b) The two faces showing differ by no more than 1;
   (c) The two faces showing differ by 2;
   (d) The two faces showing differ by no more than 2;
   (e) The two faces showing are both even;
   (f) The two faces showing are both odd;
   (g) Exactly one of the faces showing is a 5;
   (h) At least one of the faces showing is a 5;
   (i) At most one of the faces showing is a 5.
2. If the eight triangular faces of the solid of Figure 1.1 are marked with 1, 2, ..., 8, respectively, then it can be used as a fair 8-sided die. When two such 8-sided dice are tossed the sample space is:

(1, 1)	(1, 2)	(1, 3)	(1, 4)	(1, 5)	(1, 6)	(1, 7)	(1, 8)
(2, 1)	(2, 2)	(2, 3)	(2, 4)	(2, 5)	(2, 6)	(2, 7)	(2, 8)
(3, 1)	(3, 2)	(3, 3)	(3, 4)	(3, 5)	(3, 6)	(3, 7)	(3, 8)
(4, 1)	(4, 2)	(4, 3)	(4, 4)	(4, 5)	(4, 6)	(4, 7)	(4, 8)
(5, 1)	(5, 2)	(5, 3)	(5, 4)	(5, 5)	(5, 6)	(5, 7)	(5, 8)
(6, 1)	(6, 2)	(6, 3)	(6, 4)	(6, 5)	(6, 6)	(6, 7)	(6, 8)
(7, 1)	(7, 2)	(7, 3)	(7, 4)	(7, 5)	(7, 6)	(7, 7)	(7, 8)
(8, 1)	(8, 2)	(8, 3)	(8, 4)	(8, 5)	(8, 6)	(8, 7)	(8, 8)

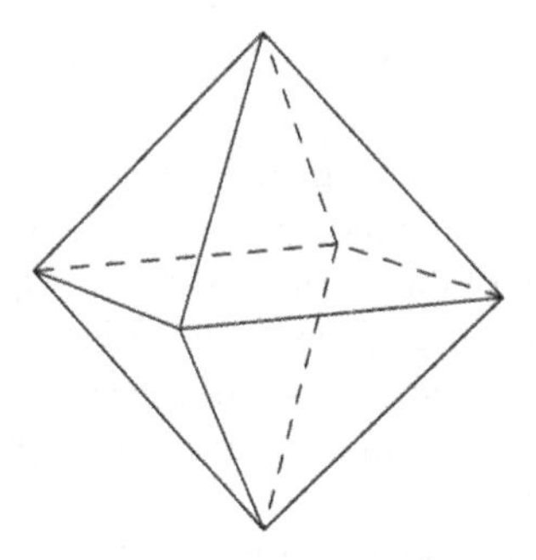

Figure 1.1: The octahedron, an eight-sided die.

Compute the probabilities of the following events when two 8-sided dice are tossed:

(a) The two faces showing differ by 1;
(b) The two faces showing differ by no more than 1;
(c) The two faces showing differ by 2;
(d) The two faces showing differ by no more than 2;
(e) The two faces showing are both even;
(f) The two faces showing are both odd;
(g) Exactly one of the faces showing is a 5;
(h) At least one of the faces showing is a 5;
(i) At most one of the faces showing is a 5.

3. Two 4-sided dice (see Figure 1.2) are tossed. Write down the sample space of this experiment.

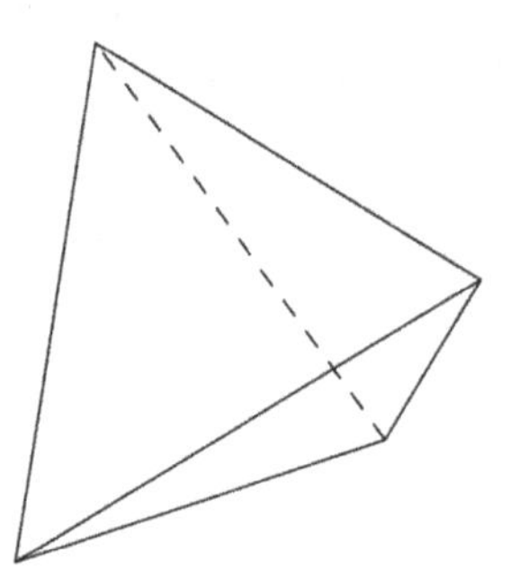

Figure 1.2: The tetrahedron, a four-sided die.

4. Two 4-sided dice are tossed. Compute the probabilities of the following events:
(a) The two faces showing differ by 1;
(b) The two faces showing differ by no more than 1;
(c) The two faces showing differ by 2;
(d) The two faces showing differ by no more than 2;
(e) The two faces showing are both even;
(f) The two faces showing are both odd;
(g) Exactly one of the faces showing is a 3;
(h) At least one of the faces showing is a 3;
(i) At most one of the faces showing is a 3.

5. A standard 6-sided die and a 4-sided die are tossed. Write down the sample space of this random experiment. Compute the probabilities of the following events:
(a) The two faces showing differ by 1;
(b) The two faces showing differ by no more than 1;

(c) The two faces showing differ by 2;
(d) The two faces showing differ by no more than 2;
(e) The two faces showing are both even;
(f) The two faces showing are both odd;
(g) Exactly one of the faces showing is a 3;
(h) At least one of the faces showing is a 3;
(i) At most one of the faces showing is a 3.

6. Two 4-sided dice (see Fig. 1.2) are tossed and the faces showing are added up. What is the corresponding sample space and what is the probability of each outcome?
7. Two 12-sided dice (see Figure 1.3) are tossed. Compute the probabilities of the following events:
   (a) The two faces showing differ by 1;
   (b) The two faces showing differ by no more than 1;
   (c) The two faces showing differ by 2;
   (d) The two faces showing differ by no more than 2;
   (e) The two faces showing are both even;
   (f) The two faces showing are both odd;
   (g) Exactly one of the faces showing is a 3;
   (h) At least one of the faces showing is a 3;
   (i) At most one of the faces showing is a 3.

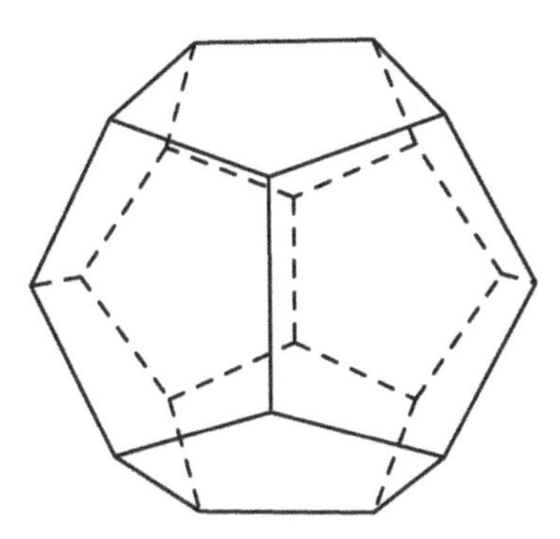

Figure 1.3: The dodecahedron, a twelve-sided die.

8. Two 8-sided dice are thrown. Compute the probability that the two faces showing add up to the following:

   a) 3 b) 4 c) 5 d) 6 e) 7 f) 8

   g) 9.

9. Two 12-sided dice are thrown. Compute the probability that the two faces showing add up to the following:

   a) 3 b) 4 c) 5 d) 6 e) 7 f) 8

   g) 9 h) 10 i) 11 j) 12 k) 13 l) 14

   m) 15 n) 16.

10. A card is selected at random from a standard deck of 52. Compute the probabilities of the following events:
    (a) The card is a diamond;
    (b) The card is red;
    (c) The card is a four;
    (d) The card is a face card (i.e., jack, queen, or king);
    (e) The card is a number between 5 and 7 (inclusive);
    (f) The card is a red ace.
11. Flip a coin on a table top and cover it with your hand before you see the outcome. Explain why the probability of seeing *heads* when the coin is uncovered is <u>not</u> 0.5. Can you evaluate this probability? If you had to bet on the outcome, what odds would be acceptable to you?

1.2 Galileo Galilei (A Historical Interlude)*

Around the year 1620 Galileo Galilei (1564–1643), by his own account, was ordered to consider the following question regarding the relative likelihoods of certain sums occurring when three dice are thrown:

The numbers 9, 10, 11, 12 can each be expressed as the sum of the faces of three dice in six ways, namely,

$$9:\quad 6+2+1=5+3+1=5+2+2=4+4+1=4+3+2=3+3+3$$

$$10:\quad 6+3+1=6+2+2=5+4+1=5+3+2=4+4+2=4+3+3$$

$$11:\quad 6+4+1=6+3+2=5+5+1=5+4+2=5+3+3=4+4+3$$

$$12:\quad 6+5+1=6+4+2=6+3+3=5+5+2=5+4+3=4+4+4$$

Why then does the experience of dice players indicate that 10 and 11 are more advantageous than 9 and 12?

By that time Galileo was one of the best known scientists in Europe and it is hard to imagine anyone *ordering* him to do anything except for his employer, the Grand Duke of Tuscany. In his response (Appendix A) Galileo explained to the Duke that the advantage (what is now called the likelihood or probability) of a number should be measured not by the frequency with which this number can be made up as a sum, but by the number of actual throws that realize that number as their sum.

In order to clarify this distinction, the sums on the previous page are referred to as *3-partitions* or simply *partitions*. It is clear from Galileo's letter that the Duke, assuming he was indeed the interlocutor, believed that the frequency of a number's 3-partitions should be the measure of its advantage. Not so, said Galileo, for this ignored the fact that 3-partitions can be produced in different ways. Of course the partition $3 = 1 + 1 + 1$ can only be produced by the single throw of three ones and the same goes for the partition $6 = 2 + 2 + 2$. However, the partition $4 = 2 + 1 + 1$ can be produced by three different throws: when the first die shows 2 and the second and third show the 1; or the second die a 2 and the first and third the 1; or the third die a 2 and the first and second the 1. This observation is summarized by the table below.

Partition:	$2 + 1 + 1$
	2, 1, 1
Throws:	1, 2, 1
	1, 1, 2

Similarly, the partition $8 = 3 + 3 + 2$ can also be produced by three different throws:

Partition:	$3 + 3 + 2$
	2, 3, 3
Throws:	3, 2, 3
	3, 3, 2

On the other hand, if a partition's three summands are all different, then it can be produced in six ways. For example, for $8 = 4 + 3 + 1$:

Partition:	$4 + 3 + 1$
	1, 3, 4
	1, 4, 3
Throws:	3, 1, 4
	4, 1, 3
	3, 4, 1
	4, 3, 1

Taking these frequencies into account, the following table displays the number of throws that produce the sums of 9 and 10, respectively:

Sum = 9		Sum = 10	
Partition	# of throws	Partition	# of throws
6 + 2 + 1	6	6 + 3 + 1	6
5 + 3 + 1	6	6 + 2 + 2	3
5 + 2 + 2	3	5 + 4 + 1	6
4 + 4 + 1	3	5 + 3 + 2	6
4 + 3 + 2	6	4 + 4 + 2	3
3 + 3 + 3	1	4 + 3 + 3	3
Total:	25	Total:	27

Thus there are 25 different throws that produce a sum of 9, and 27 that produce a 10. This difference, contended Galileo, accounted for the gamblers' aforementioned observation that 10 is more advantageous than 9. Having produced this explanation, Galileo also derived the number of throws that yield all the other possible sums and commented that the advantages of all the other configurations of the 3-dice game could be measured by these frequencies.

It is interesting that no explicit probabilities were computed in this letter which is one of the earliest documents wherein the likelihoods of unpredictable events are discussed in a mathematical manner. Galileo did compute the total number of distinct throws:

> *... since a die has six faces, and when thrown it can equally well fall on any one of these, only 6 throws can be made with it, each different from all the others. But if together with the first die we throw a second, which also has six faces, we can make 36 throws each different from all the others, since each face of the first die can be combined with each of the second, and in consequence can make 6 different throws, whence it is clear that such combinations are 6 times 6, i.e., 36. And if we add a third die, since each of its six faces can be combined with each of the 36 combinations of the other two dice, we shall find that the combinations of three dice are 6 times 36, i.e., 216, each different from the others.*

This done, however, he did not do what would be natural to us, which is to divide 25 by 216 and say that

$$\frac{25}{216}.$$

is the probability of 9 occurring as the sum of a throw of three dice. He only used the number 216 to check by adding that the frequencies derived for the various partitions did indeed account for all possible outcomes. The notion of the probability of an event was only implicit in his thoughts. The closest he came to such a concept in this context was in the term "advantage." Elsewhere (see Section 2.2) he also used "frequency" to denote likelihood. This bears testimony to the excruciating slowness with which fundamental mathematical concepts that seem natural to us actually evolved. Further examples of such resistance to new ideas will be given later.

It is interesting that the Duke's request was not for the solution of a practical problem but for a mathematical explanation of an observed phenomenon. His experience had already told him that 10 is more advantageous than 9 and he just wanted to know why this was so. In other words, he was motivated by the same belief that drove the classical Greek culture as well as the Renaissance era of which he was a prime mover: the belief that *Man is the Measure* and that the human mind is capable of imposing order on a seemingly inchoate Nature. This belief, that knowledge is worth having for its own sake and that the exercise of our logical faculties is as beneficial to our mental well being as physical exercise is to our bodies, is still a part of our own ethos today and it is in this spirit that the following variations on the Duke's games are offered here.

It was remarked above that the octahedron of Figure 1.1 can be used as a fair eight-sided die and Galileo's analysis will now be applied to the computation of the probability of the sum of the faces of a throw of three octahedral dice adding up to 9. The relevant partitions are

$$9 = 7 + 1 + 1 = 6 + 2 + 1 = 5 + 3 + 1 = 5 + 2 + 2$$

$$= 4 + 4 + 1 = 4 + 3 + 2 = 3 + 3 + 3.$$

Rather than list the throws that give rise to each partition we merely record their number:

Partition:	7 + 1 + 1	6 + 2 + 1	5 + 3 + 1	5 + 2 + 2
No. of throws:	3	6	6	3
		4 + 4 + 1	4 + 3 + 2	3 + 3 + 3
		3	6	1

for a total of $3 + 6 + 6 + 3 + 3 + 6 + 1 = 28$. Since the total number of possible throws is $8 \times 8 \times 8 = 512$ it follows that the requisite probability is

$$\frac{28}{512} = \frac{7}{128} \approx 5.5\%.$$

Other types of fair dice abound (see Figures 1.1–3 and Exercises 15–17). The method for the computation of the probability of any number n occurring as the sum of the faces of three d-faced dice is now clear:

(a) List the 3-partitions of n into three summands none of which exceed d. (Note that once you have stipulated that the first die shows some integer, say k, then the faces of the second and third must add up to $n - k$. Keep in mind that none of the faces exceed d and none exceeds the previous one.)
(b) Replace each partition by its corresponding count of 6 or 3 or 1.
(c) Add all the counts of part b and divide the sum by d^3.

Historical note. Galileo made no claim to originality in his letter and it can be inferred from the lack of argumentation in this document that these concepts were well understood by the experts of the time. In fact, the Italian mathematician Gerolamo Cardano (1501–1576) had already written his *Book on Games of Chance* more than fifty years earlier, which contained, amongst many others, an analysis of the very same 3-dice game discussed above. However, as this book was not published until 1663, Galileo was probably unaware of its existence. No other books on probability are known to have been written by that time and so it is possible that Galileo was simply basing his analysis on what he considered to be common sense.

Exercises 1.2

1. Three 6-sided dice are thrown. Compute the probability that the three faces showing add up to the following:

   a) 3 b) 4 c) 5 d) 6 e) 7 f) 8

   g) 11 h) 12 i) 13 j) 14 k) 15 l) 16

   m) 17 n) 18.

2. Three 8-sided dice are thrown (see Fig. 1.1). Compute the probability that the three faces showing add up to the following:

   a) 3 b) 4 c) 5 d) 6 e) 7 f) 8

   g) 9 h) 10 i) 11 j) 12 k) 13 l) 14

   m) 15 n) 16.

3. Three 4-sided dice are thrown (see Fig. 1.2). Compute the probability that the three faces showing add up to the following:

   a) 3 b) 4 c) 5 d) 6 e) 7 f) 8

   g) 9.

4. Three 12-sided dice are thrown (see Fig. 1.3). Compute the probability that the three faces showing add up to the following:

   a) 3 b) 4 c) 5 d) 6 e) 7 f) 8

   g) 9 h) 10 i) 11 j) 12 k) 13 l) 14

   m) 15 n) 16.

5. List all the 4-partitions of the following integers:

   a) 4 b) 5 c) 6 d) 7 e) 8 f) 9

   g) 10.

6. List all the 24 different throws that correspond to a 4-partition, all of whose summands are distinct. (Suggestion: work with the partition $4 + 3 + 2 + 1$.)
7. List all the 12 different throws that correspond to a 4-partition, two of whose summands are identical. (Suggestion: work with the partition $3 + 3 + 2 + 1$.)
8. List all the 6 different throws that correspond to a 4-partition that has two pairs of identical summands. (Suggestion: work with the partition $2 + 2 + 1 + 1$.)
9. List all the 4 different throws that correspond to a 4-partition that has three identical summands. (Suggestion: work with the partition $2 + 1 + 1 + 1$.)
10. Explain why there is only 1 throw that corresponds to a 4-partition, all of whose summands are identical.

In solving Exercises 11–16 you may make use of Exercises 6–10.

11. Four 4-sided dice are thrown. Compute the probability that the four faces showing add up to the following:

    a) 4 b) 5 c) 6 d) 7 e) 8 f) 9

    g) 10.

12. Four 6-sided dice are thrown. Compute the probability that the four faces showing add up to the following:

    a) 4 b) 5 c) 6 d) 7 e) 8 f) 9

    g) 10.

13. Four 8-sided dice are thrown. Compute the probability that the four faces showing add up to the following:

    a) 4 b) 5 c) 6 d) 7 e) 8 f) 9

    g) 10.

14. Four 12-sided dice are thrown. Compute the probability that the four faces showing add up to the following:

    a) 4 b) 5 c) 6 d) 7 e) 8 f) 9

    g) 10.
15. Design a fair die whose faces are numbered 1–10.
16. Design a fair die whose faces are numbered 1–5. (Hint: have each number appear twice.)
17. Design a fair die whose faces are numbered 1–7.

1.3 Binomial Processes

The 3-dice toss described in the previous section is actually a relatively complicated random phenomenon and does not constitute the traditional starting point for the formal study of probability. There are after all 16 possible outcomes (3 through 18) each of which can be broken down into several possible tosses. The simplest genuinely random event has only 2 possible outcomes and is called a *binomial, or Bernoulli, trial*, after the Swiss mathematical pioneer James Bernoulli (1654 - 1705) whose text *Ars Conjectandi* did much to establish the study of probabilities as a bona fide mathematical discipline. The flip of a coin, a baseball game, and the sex of an offspring are all examples of such bivalued binomial trials. While the person performing the binomial trial may have no preference for one outcome over the other, it is nevertheless customary to designate one of the outcomes as **success** and the other as **failure**. This nomenclature is not meant as a moral judgment of the two - its purpose is merely to provide us with words for discussing the situation. Probabilists could just as well have chosen to designate the two alternatives in the more neutral, though less colorful, terms of **Outcome I** and **Outcome II**. Happily, they did not. The decision as to which outcome should be chosen as the **success** is unimportant - all that is needed is some designation.

Given such a binomial trial, the probability of success, i.e., the probability that the outcome labeled **success** actually occurs, is denoted by p. The probability of failure is then $1 - p$. For example, suppose a coin is weighted so that upon being flipped it will come up *heads* 6 times out of 10. If *heads* is considered to be the success then $p = 0.6$; if *tails* is the successful outcome then $p = 0.4$. In the case of a fair coin $p = 0.5$ regardless of which outcome is deemed to be a success. If 15% of all cars of a certain model are defective to some extent, and one such car is selected at random and examined for defects, then $p = 0.15$. Note that here success is tantamount to selecting a defective car, an occurrence that most of us would consider regrettable, to say the least. Such are the vagaries of formal mathematical terminology. Anyone unhappy with this "misappelation" could of course designate the selection of a non-defective car as success, which would entail $p = 0.85$.

A *binomial process* consists of several repetitions of identical, though independent, binomial trials. Ten successive flips of a coin are an example of such a process, as are the births of several offspring. The iterations need not follow each other in time; they may be performed simultaneously. Such is the case when three dice are tossed or a handful of coins is thrown. It is reasonable when performing such a set of trials to ask for the likelihood of a certain number of successes: If a coin is flipped 100 times, what is the probability of its coming up heads 50 times? If a randomly selected family has 3 children, what is the probability of two of them being girls?

There is a formula for these probabilities. Specifically, if a process consists of n independent repetitions of the same binomial trial with success probability p, then the probability of exactly k successes is

$$b(k; n, p) = \binom{n}{k} p^k (1-p)^{n-k}$$

where $\binom{n}{k}$ (pronounced *n choose k*) is called *the choice number* and is to be computed by the formula

$$\binom{n}{k} = \frac{n(n-1)(n-2)...(n-k+1)}{1 \cdot 2 \cdot 3 \cdot ... \cdot k}.$$

The computation of the denominator of this expression is straightforward. The numerator is more easily understood if one realizes that it contains exactly as many factors as the denominator. Thus,

$$\binom{3}{2} = \frac{3 \cdot 2}{1 \cdot 2} = 3 \qquad \binom{4}{2} = \frac{4 \cdot 3}{1 \cdot 2} = 6 \qquad \binom{4}{3} = \frac{4 \cdot 3 \cdot 2}{1 \cdot 2 \cdot 3} = 4$$

$$\binom{5}{4} = \frac{5 \cdot 4 \cdot 3 \cdot 2}{1 \cdot 2 \cdot 3 \cdot 4} = 5 \qquad \binom{5}{3} = \frac{5 \cdot 4 \cdot 3}{1 \cdot 2 \cdot 3} = 10 \qquad \binom{5}{2} = \frac{5 \cdot 4}{1 \cdot 2} = 10$$

$$\binom{7}{4} = \frac{7 \cdot 6 \cdot 5 \cdot 4}{1 \cdot 2 \cdot 3 \cdot 4} = 35.$$

Example 1.3.1. *Assuming that every offspring has a 50-50 chance of being male (or female, for that matter), what is the probability that exactly 3 of a family's 5 children are male?*

Here $n = 5$, $k = 3$, and $p = 0.5$. Accordingly, the probability in question is

$$b(3; 5, 0.5) = \binom{5}{3} \cdot 0.5^3 \cdot 0.5^2 = 10 \cdot 0.125 \cdot 0.25 = 0.3125 = 31.25\%.$$

Example 1.3.2. *What is the probability of getting exactly 3 sixes in 7 throws of a single die?*

Strictly speaking, the toss of a die is not a binomial trial since it has 6 possible outcomes and not the required 2. However, this can be fixed by considering the outcome of a *6* as a success and the outcome of *not a 6* as a failure. This converts the toss of a die into a binomial trial with $p = \frac{1}{6}$ and, of course, $1 - p = 1 - \frac{1}{6} = \frac{5}{6}$. Here $n = 7$, $k = 3$, $n - k = 4$, and consequently the required probability is

$$b\left(3; 7, \frac{1}{6}\right) = \binom{7}{3}\left(\frac{1}{6}\right)^3\left(\frac{5}{6}\right)^4 = \frac{7 \cdot 6 \cdot 5}{1 \cdot 2 \cdot 3}\left(\frac{1}{6}\right)^3\left(\frac{5}{6}\right)^4 = 0.0781 \approx 7.8.\%.$$

Example 1.3.3. *Suppose 30% of the population of a certain city are fans of that city's pro basketball team. Suppose further that an advertising company's researcher addresses ten people at random while walking around the city and asks them whether they are indeed fans. What is the probability that exactly 3 of the respondents turn out to be fans?*

The choice of a person to interview is similar to the throw of a die. In each case we are looking at independent repetitions of a random event with two possible outcomes of which one is considered to be a success: finding a fan in one case and throwing a 6 in the other. Reasoning by analogy, we find that the probability of encountering exactly 3 fans is:

$$b(3; 10{,}0.3) = \binom{10}{3} \cdot 0.3^3 \cdot (1 - 0.3)^7 = \frac{10 \cdot 9 \cdot 8}{1 \cdot 2 \cdot 3} \cdot 0.3^3 \cdot 0.7^7 = 0.2668 \approx 26.7\%.$$

Similarly, if 45% of the population of this city is male and a surveyor phones 8 randomly selected numbers, then the probability of 3 of the people who answer the call being male is

$$b(3; 8, 0.45) = \binom{8}{3} \cdot 0.45^3 \cdot (1 - 0.45)^{8-3} = \frac{8 \cdot 7 \cdot 6}{1 \cdot 2 \cdot 3} \cdot 0.45^3 \cdot 0.55^5 = 0.2568 \approx 25.7\%.$$

These last examples, while seemingly practical, are in fact unrealistic as we are very unlikely to ever possess enough information to allow us to determine the precise proportion of fans or males in a town. In practice, one is more often interested in the reverse question of what information the results of the sampling provide regarding the proportion of fans in the entire population. This question will be dealt with in Section 2.5. At this point, however, it is possible to garner some information regarding the related issue of the likelihood of a sample being in some sense misleading.

Example 1.3.4. *One would expect a random sample of 10 inhabitants of this town to contain 3 fans and hence such a sample might give the surveyor the wrong impression if it contained no more than 1 fan. What is the probability of this happening?*

Since "no more than 1 fan" is tantamount to 0 or 1 fan, the probability of this happening is

$$b(0;10,0.3)+b(1;10,0.3)=\binom{10}{0}\cdot 0.3^0\cdot 0.7^{10}+\binom{10}{1}\cdot 0.3^1\cdot 0.7^9$$

$$\approx 0.0282 + 0.1211 = 0.1493 \approx 14.9\%.$$

Example 1.3.5. *A sample of 10 inhabitants could be considered to be misleading if it contains too many, say 5 or more, fans.*
The probability of this happening is

$$b(5;10,0.3)+b(6;10,0.3)+b(7;10,0.3)+b(8;10,0.3)+b(9;10,0.3)$$

$$+\,b(10;10,0.3)$$

$$=\binom{10}{5}\cdot 0.3^5\cdot 0.7^5+\binom{10}{6}\cdot 0.3^6\cdot 0.7^4+\binom{10}{7}\cdot 0.3^7\cdot 0.7^7+\binom{10}{8}\cdot 0.3^8\cdot 0.7^2$$

$$+\binom{10}{9}\cdot 0.3^9\cdot 0.7^1+\binom{10}{10}\cdot 0.3^{10}\cdot 0.7^0$$

$$\approx 0.1029 + 0.0367 + 0.0090 + 0.0038 + .00001 + 0.0000 = 0.150 = 15\%.$$

Put together, the above examples say that $14.9\% + 15.3\% \approx 30\%$ of the random samples could mislead the unwary surveyor, an uncomfortably high proportion.

Such probabilities are useful and need to be computed frequently. The above described method is unfortunately very unwieldy. For instance, suppose it is necessary to compute the probability of a sample of 20 inhabitants containing at least 10 fans (as opposed to the expected 6). The answer can be written as the sum

$$b(10;20,0.3)+b(11;20,0.3)+\ldots+b(19;20,0.3)+b(20;20,0.3).$$

Depending on the calculator at one's disposal as well as on one's skills, the evaluation of this sum could vary between a triviality and a nuisance. Before the advent of the electronic calculators, however, such problems constituted unpleasant and lengthy chores. They did arise often enough, however, to motivate the creation of appropriate tables. The use of these tables requires a new symbol. If a binomial process involves n repetitions of an elementary trial with a probability of success p, then the probability of at most k successes is denoted by

$$B(k; n, p) = b(0; n, p) + b(1; n, p) + \ldots + b(k; n, p). \tag{b}$$

Some of these values are tabulated in Table B. Accordingly, the probability that the sample of 20 inhabitants encounters at most 4 fans is

$$B(4; 20, 0.3) = 0.2375 = 23.75\%.$$

Similarly, the probability that a sampling of 16 inhabitants yields at most 7 fans is $B(7; 16, 0.3) = 0.9256 \approx 92.6\%$.

These tables can be used to derive other probabilities as well.

Example 1.3.6. *Suppose a random sample of 14 of the inhabitants of the town of the above examples are interviewed. Determine the probabilities of the following events:*

(a) *At least 7 are fans.* The complementary event is that at most 6 are fans and has probability $B(6; 14, 0.3)$. The required probability therefore is

$$1 - B(6; 14, 0.3) = 1 - 0.9067 = 0.0933 \approx 9.3\%.$$

(b) *The number of fans is between 4 and 7 (inclusive).* This probability is

$$B(7; 14, 0.3) - B(3; 14, 0.3) = 0.9685 - 0.3552 = 0.6133 \approx 61.3\%.$$

(c) The number of fans is exactly 5. This probability is

$$B(5; 14, 0.3) - B(4; 14, 0.3) = 0.7805 - 0.5842 = 0.1963 \approx 19.6\%.$$

Example 1.3.7. *In any match between the baseball teams NL and AL the first has the probability 0.6 of winning. Suppose the two teams meet in a World Series wherein a team must win four games to emerge victorious. What is the probability of team NL winning the series?*

The baseball World Series is a binomial process with $n = 7$, $k \geq 4$, and $p = 0.6$. Hence the desired probability is

$$1 - B(3; 7, 0.6) = 1 - 0.2898 = 0.7102 \approx 71\%.$$

Example 1.3.8. *In a certain baseball World Series the two teams are considered to be equally strong. Suppose the NL team wins the first two games of the series. What is the probability that this team will go on to win the entire series?*

In order to win the series, the team NL needs to win at least 2 of the remaining 5 games. It is therefore facing a binomial process with $n = 5$, $k \geq 2$, and $p = 0.5$ and the probability of its winning is

$$1 - B(1; 5, 0.5) = 1 - 0.1875 = 0.8125 = 81.25\%.$$

Exercises 1.3

Use the formula for $b(k; n, p)$ to compute the probabilities required in Exercises 1–7.

1. A fair die is tossed 5 times. What is the probability of obtaining (a) no 6's (b) exactly one 6 (c) exactly two 6's (d) exactly five 6's?
2. A fair die is tossed 6 times. What is the probability of obtaining (a) no 5's (b) exactly one 5 (c) exactly two 5's (d) exactly six 5's?
3. The probability that a patient recovers from a certain disease is 0.75. If a hospital has 7 patients with this disease, what is the probability of the following events: (a) none recover (b) exactly one recovers (c) exactly 5 recover (d) all recover?
4. In a certain city the need for money to buy drugs is stated as the reason for two thirds of all burglaries. Find the probability that among the next 8 burglaries the number of those due to this need is (a) 0 (b) exactly 6 (c) exactly 8.
5. It is known that one third of trucks will experience a tire blow out in a certain strenuous test. Suppose 5 trucks are tested. Find the probability that the number of trucks that experience a blow out is (a) 0 (b) exactly 2 (c) exactly 5.
6. It is estimated that a certain flu vaccine is 85% effective. If 13 people are inoculated find the probability that the number of successful vaccinations amongst them is (a) exactly 1 (b) exactly 13.
7. Of the cars passing through a toll booth, 15% are from outside the state. Find the probability that of the next 6 cars the number of out-of-staters is (a) 0 (b) exactly 1 (c) exactly 2.

   Use Formula (b) and Table B of the appendix to compute the probabilities required in Exercises 8–14.
8. A fair coin is tossed 15 times. What is the probability of obtaining (a) at most 9 heads (b) at least 9 heads (c) exactly 9 heads (d) between 8 and 12 heads inclusive?
9. A fair coin is tossed 20 times. What is the probability of obtaining (a) at most 12 heads (b) at least 12 heads (c) exactly 12 heads (d) between 8 and 12 heads inclusive?
10. The probability that a patient recovers from a certain disease is 0.7. If a hospital has 12 patients with this disease, what is the probability of the following events: (a) at most 5 recover (b) at least 5 recover (c) exactly 5 recover (d) between 5 and 10 (inclusive) recover?

11. In a certain city the need for money to buy drugs is stated as the reason for 80% of all burglaries. Find the probability that among the next 8 burglaries the number of those due to this need is (a) at most 6 (b) at least 6 (c) exactly 6 (d) between 4 and 6 inclusive.
12. It is known that 25% of trucks will experience a tire blow out in a certain strenuous test. Suppose 5 trucks are tested. Find the probability that the trucks that fail is (a) at most 2 (b) at least 2 (c) exactly 2 (d) between 1 and 3 inclusive.
13. It is estimated that a certain flu vaccine is 90% effective. If 12 people are inoculated, find the probability that the number of successful vaccinations amongst them is (a) at most 10 (b) at least 10 (c) exactly 10 (d) between 8 and 11 inclusive.
14. Of the cars passing through a toll booth, 25% are from outside the state. Find the probability that of the next 16 cars the number of out-of-staters is (a) at most 6 (b) at least 6 (c) exactly 6 (d) between 4 and 8 inclusive.

1.4 The Normal Approximation

Realistic polls may lead to such numbers as the probability that a random sample of 100 inhabitants contains at most 35 fans, which, according to the previous section, is $B(35;100, 0.4)$. There is no table for this at the end because the listing of the entries for n up to 100 is impractical, as is indicated by the rate at which the numbers of displayed entries for $n = 1, 2, ..., 20$ increase. On the other hand, Formula (b) yields the lengthy sum

$$\binom{100}{0} \cdot 0.4^0 \cdot 0.6^{100} + \binom{100}{1} \cdot 0.4^1 \cdot 0.6^{99} + \ldots + \binom{100}{34} \cdot 0.4^{34} \cdot 0.6^{66}$$
$$+ \binom{100}{35} \cdot 0.4^{35} \cdot 0.6^{65}$$

which is not easily calculated either by hand or with a handheld calculator. Of course, there are mathematical computer applications that will compute this sum in a jiffy, but even these will run into trouble with sums of binomial probabilities that arise in more technical circumstances. The mathematicians of the eighteenth century, who had no calculators available, resorted instead to their brains and in the process of overcoming this particular computational difficulty discovered a phenomenon that eventually became the cornerstone of the whole theory of statistics. While this discovery was first phrased algebraically it is more easily explained geometrically in terms of graphs.

Figures 1.4–7 display the probabilities associated with the flipping of a fair coin. These are binomial experiments with $p = 0.5$ and $n = 10, 20, 50, 100$, respectively. In these bar graphs the probability of each event of k successes is represented by the height of a rectangular strip of width 1 centered over k. Not surprisingly, all

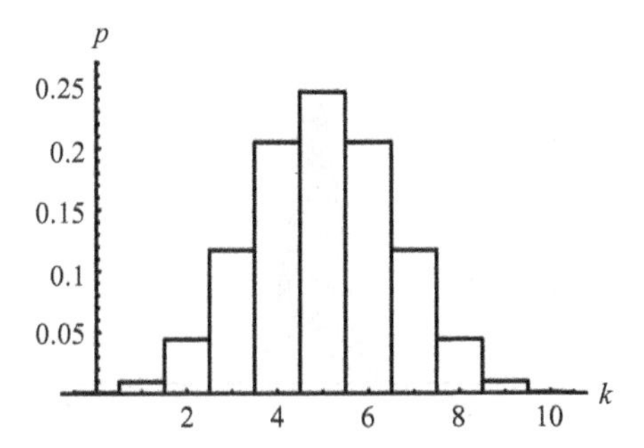

Figure 1.4: $n = 10, p = 0.5$

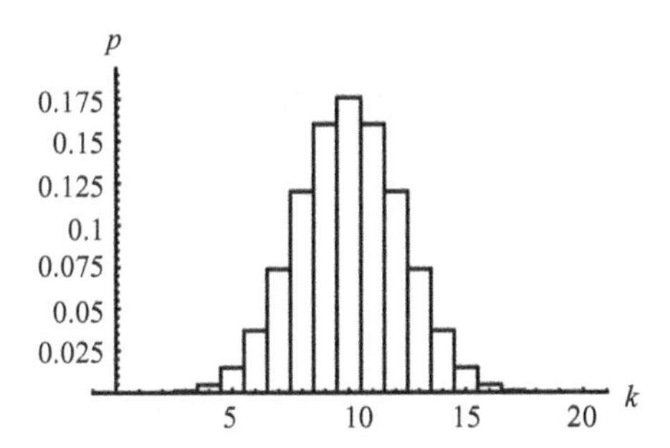

Figure 1.5: $n = 20, p = 0.5$

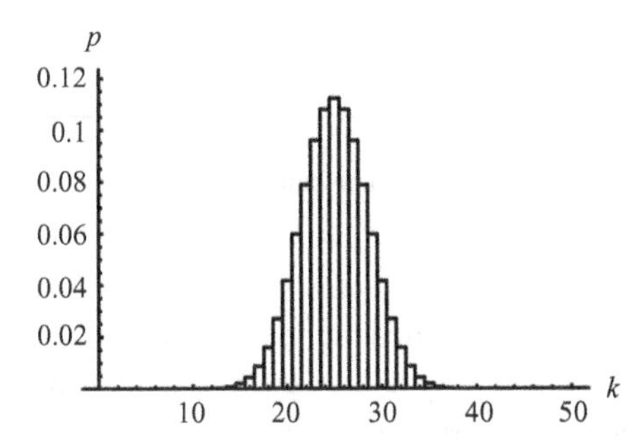

Figure 1.6: $n = 50, p = 0.5$

these graphs have the same general shape, which is popularly known as the bell shaped curve and is portrayed in Figure 1.8.

Figures 1.9–12 demonstrate how the bell-shaped curve can be fitted to the binomial probabilities of Figures 1.4–7 above. Figures 1.13–20 demonstrate that the symmetric

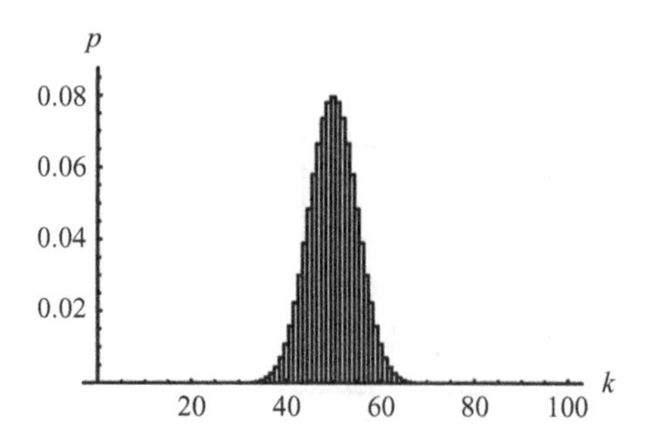

Figure 1.7: $n = 100, p = 0.5$

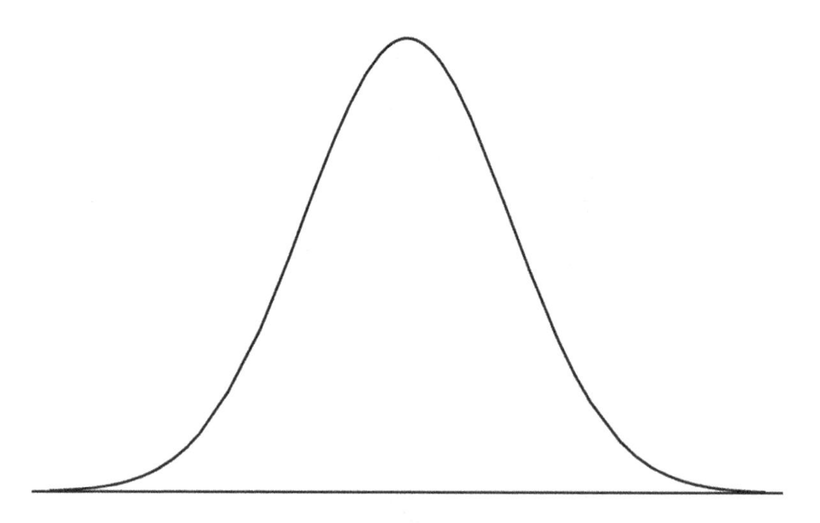

Figure 1.8: The bell-shaped curve.

bell-shaped curve can also be fitted to the asymmetric binomial probabilities associated with $p = 0.3$ and $p = \frac{1}{6}$. It is clear that for reasonably large n, i.e., when the experiment is repeated many times, the asymmetries are so small as to be negligible.

The approximating bell-shaped curves in Figures 1.8–20 are all different and are all called *normal curves*. They differ in their widths, and both the location and height of their peaks. Nevertheless, in 1735 Abraham de Moivre (1667–1754) demonstrated that all these curves were identical in the sense that any one could be transformed into any other by a rescaling of the axes and/or a horizontal shift. Moreover, this equivalence greatly facilitated the computation of binomial probabilities.

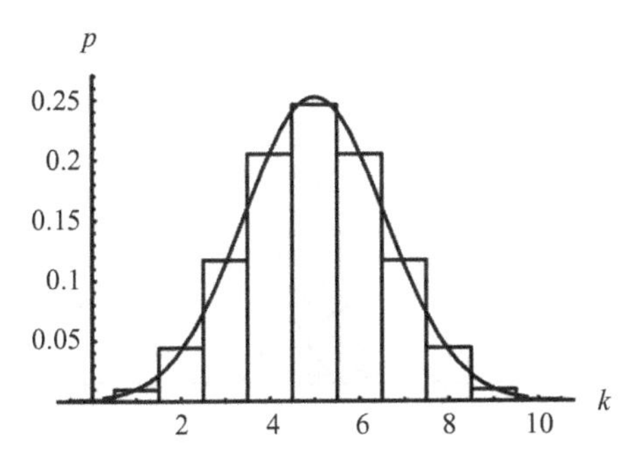

Figure 1.9: $n = 10, p = 0.5$

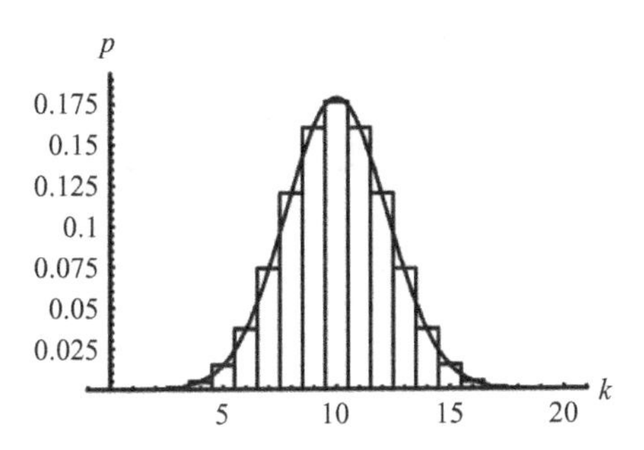

Figure 1.10: $n = 20, p = 0.5$

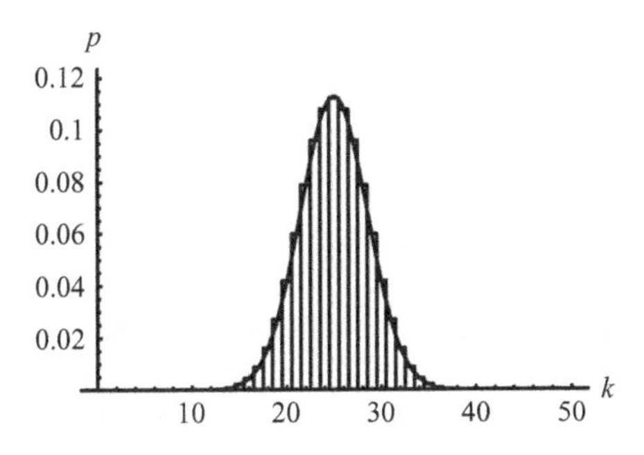

Figure 1.11: $n = 50, p = 0.5$

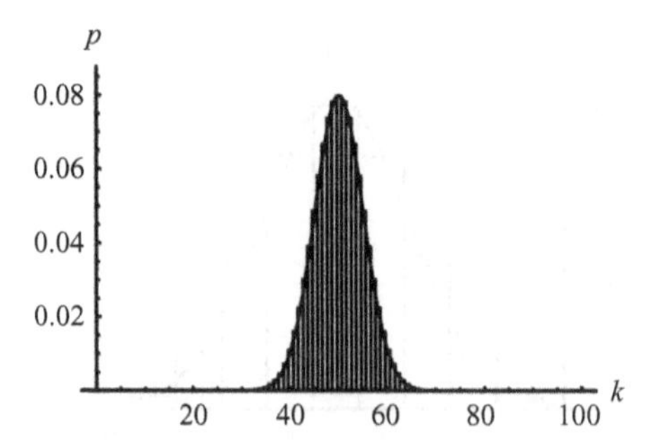

Figure 1.12: $n = 100, p = 0.5$

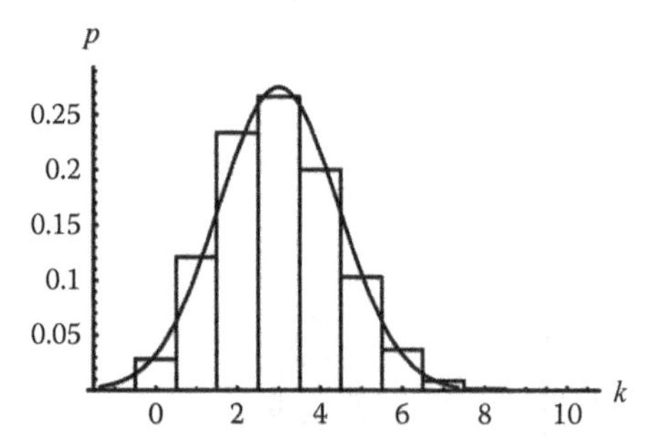

Figure 1.13: $n = 10, p = 0.3$

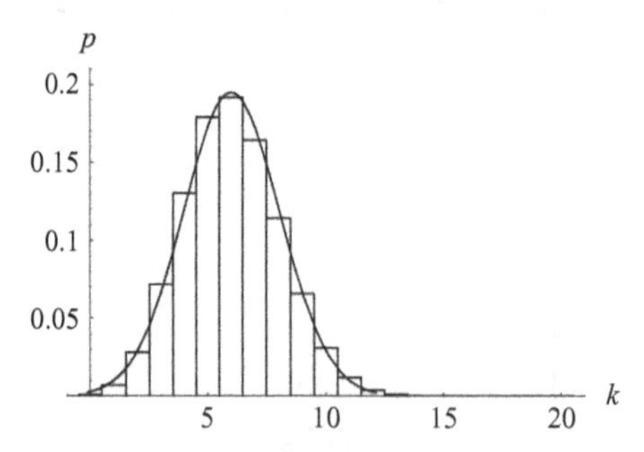

Figure 1.14: $n = 20, p = 0.3$

Consider, for the purpose of an illustration, the problem of computing $B(4; 10, 0.4)$ (for which the tables yield the value of 0.633). Because each of the strips in the graphical representation of the binomial probabilities has width 1, the probability represented by any strip actually equals its area. Hence, the probability $B(4; 10, 0.4)$ is represented by the shaded region in Figure 1.21.

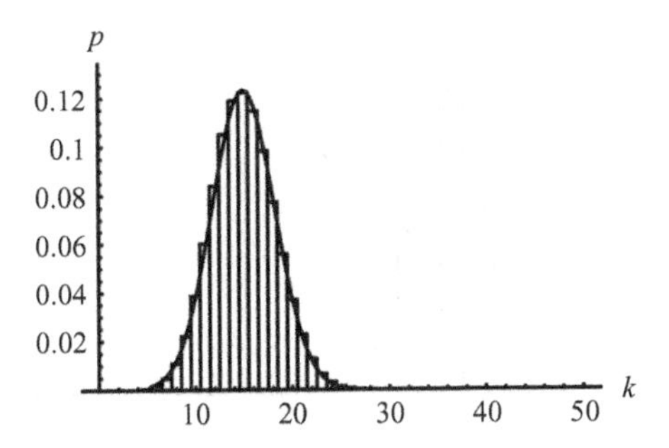

Figure 1.15: $n = 50, p = 0.3$

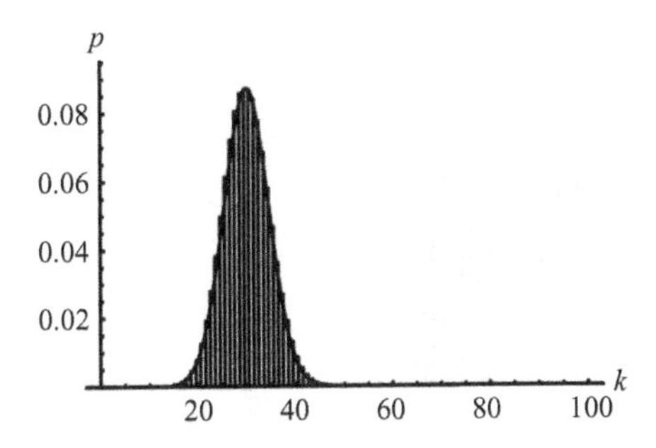

Figure 1.16: $n = 100, p = 0.3$

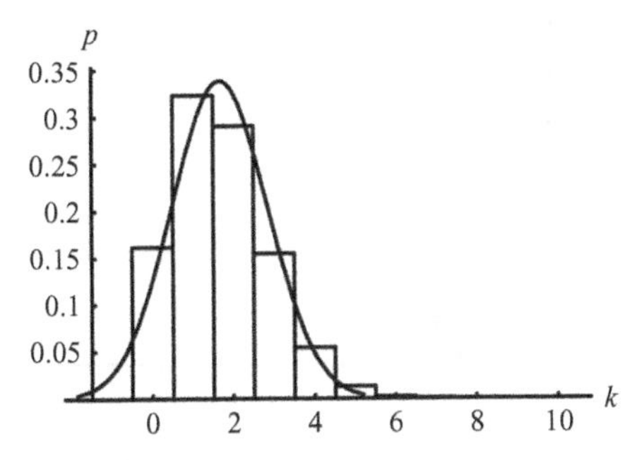

Figure 1.17: $n = 10, p = 1/6$

This area is approximately equal to the shaded region in Figure 1.22. Since all normal curves are essentially the same, it suffices to tabulate such areas for a single standard normal curve and then provide a method for transporting this information to arbitrary normal curves. The *standard* normal curve is the graph of the function

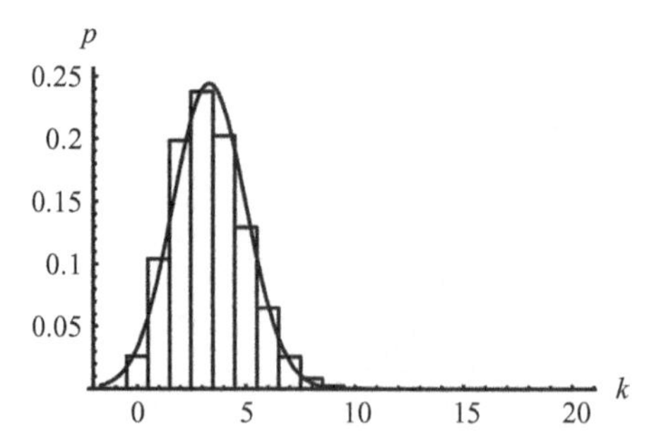

Figure 1.18: $n = 20, p = 1/6$

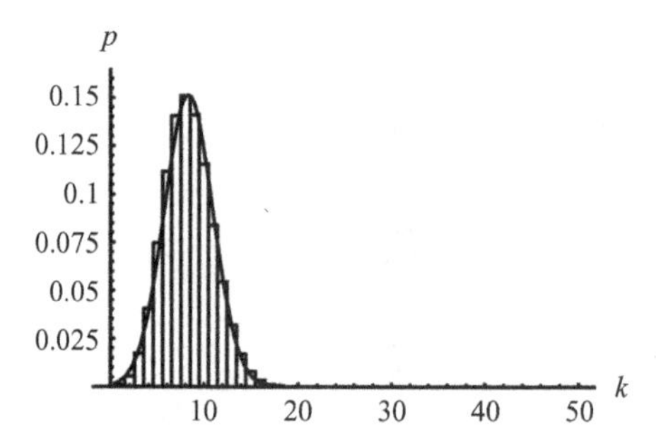

Figure 1.19: $n = 50, p = 1/6$

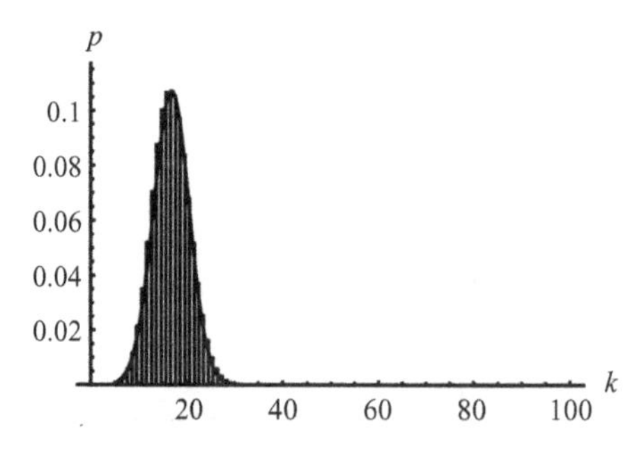

Figure 1.20: $n = 100, p = 1/6$

$$y = \frac{e^{-x^2/2}}{\sqrt{2\pi}}$$

which is displayed in Figure 1.23. Here π is the well-known ratio of the circumference of a circle to its diameter, with value 3.14159... whereas e is another important

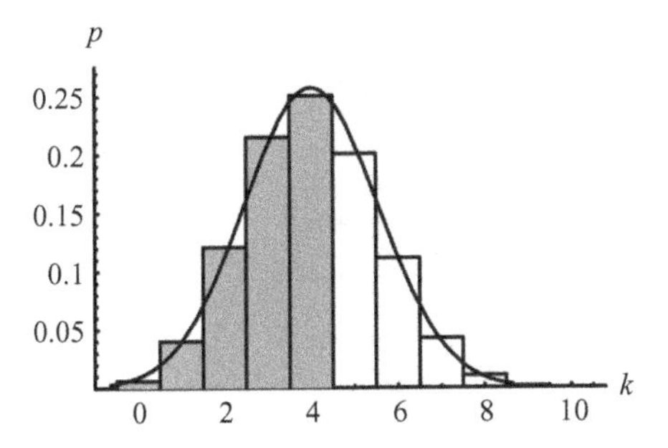

Figure 1.21:

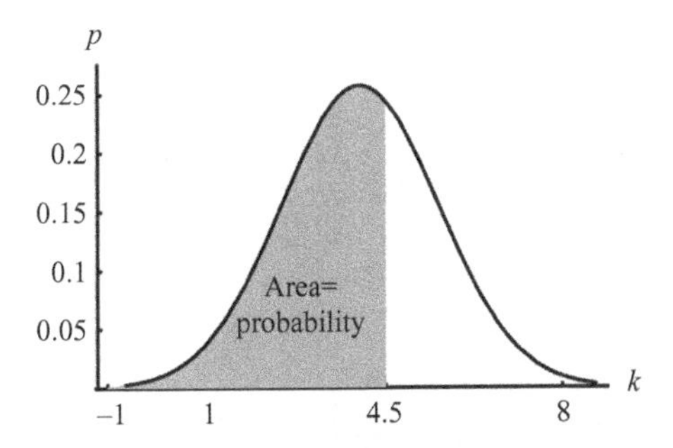

Figure 1.22:

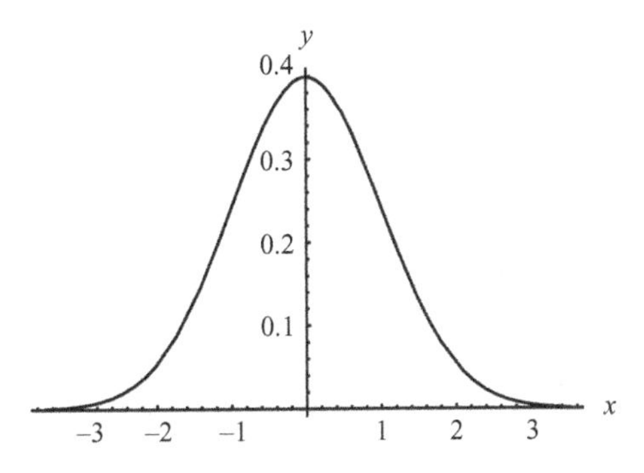

Figure 1.23:

mathematical constant with value 2.71828... . The area of the region underneath the standard normal curve and to the left of the number x is denoted by $A(x)$ and its values are tabulated in Table A of the appendix. This table shows that $A(1) = 0.8413$, $A(0.5) = 0.6915$, and $A(-1.17) = 0.1210$. The evaluation of the cumulative binomial probabilities is given by the formula

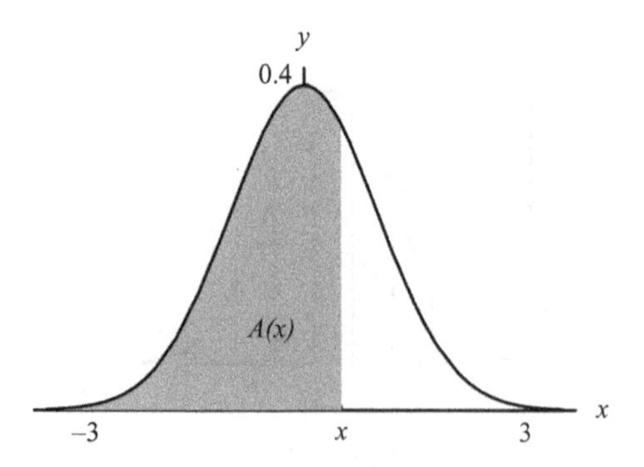

Figure 1.24:

$$B(k;n,p) = A\left(\frac{k+0.5-pn}{\sqrt{p(1-p)n}}\right).$$

Accordingly,

$$B(4;20,0.3) = A\left(\frac{4.5-0.3\cdot 20}{\sqrt{0.3\cdot 0.7\cdot 20}}\right) = A(-.73) = 0.2327 \approx 23.3\%$$

and

$$B(7;16,0.3) = A\left(\frac{7.5-0.3\cdot 16}{\sqrt{0.3\cdot 0.7\cdot 16}}\right) = A(1.47) = 0.9292 \approx 92.3\%,$$

both of which agree closely with the previously obtained results (see p. 1.27) of 23.75% and 92.6%, respectively. Moreover, the evaluation of untabulated binomial probabilities is now also feasible. For example,

$$B(35;100,0.4) = A\left(\frac{35.5-0.4\cdot 100}{\sqrt{0.4\cdot 0.6\cdot 100}}\right) = A(-0.92) = 0.1788 \approx 17.9\%.$$

Example 1.4.1. *It is known that the recovery rate for a certain rare disease is seven out of eight. Suppose 25 people contract this disease. What is the probability that (a) no more that 20 recover (b) at least 20 recover (c) from 20 to 23 recover (d) exactly 21 recover?*

(a) This is $B(20;25,\,7/8) = 1 - A\left(\frac{20.5-(7/8)\cdot 25}{\sqrt{(7/8)\cdot(1/8)\cdot 25}}\right) = A(-0.83) = 0.2033 \approx 20.3\%$.

(b) This is $1 - B(19; 25, 7/8) = 1 - A\left(\frac{19.5 - (7/8)\cdot 25}{\sqrt{(7/8)\cdot(1/8)\cdot 25}}\right) = 1 - A(-1.44)$
$= 1 - 0.0749 = 0.9251 \approx 92.5\%$.

(c) This is $B(23; 25, 7/8) - B(19; 25, 7/8) =$

$$A\left(\frac{23.5 - (7/8)\cdot 25}{\sqrt{(7/8)\cdot(1/8)\cdot 25}}\right) - A\left(\frac{19.5 - (7/8)\cdot 25}{\sqrt{(7/8)\cdot(1/8)\cdot 25}}\right) = A(0.98)$$
$$-A(-1.44) = 0.8365 - 0.0749 = 0.7616 \approx 76.2\%.$$

(d) This is $B(21; 25, 7/8) - B(20; 25, 7/8) =$
$$A\left(\frac{21.5 - (7/8)\cdot 25}{\sqrt{(7/8)\cdot(1/8)\cdot 25}}\right) - A\left(\frac{20.5 - (7/8)\cdot 25}{\sqrt{(7/8)\cdot(1/8)\cdot 25}}\right) = A(-0.23)$$
$$- A(-0.83) = 0.4090 - 0.2033 = 0.2057 \approx 20.6\%.$$

In the second half of the nineteenth century, the Englishman Sir Francis Galton did much to popularize the normal curve amongst the scientists of the day. As part of this effort he constructed in 1873 a device, called the *quincunx*, that strikingly demonstrated the ability of this curve to approximate binomial probabilities. In this device, shot was poured through an array of pins and since each time a shot hit a pin it would go to the right or to the left with an equal probability of 0.5, the paths taken by these shots simulated a binomial experiment (success = right, say, and failure = left). The shots then accumulated in compartments at the bottom of the quincunx where their outline would resemble a normal curve. Quincunxes are displayed in many science museums.

Exercises 1.4

1. A coin is tossed 200 times. What is the probability of obtaining (a) at most 90 heads (b) at least 90 heads (c) exactly 100 heads (d) between 90 and 110 heads inclusive?
2. A die is tossed 300 times. What is the probability of obtaining (a) at most 40 ones (b) at least 40 ones (c) exactly 50 ones (d) between 40 and 60 (inclusive) ones?
3. It was demonstrated in Section 1.1 that if three dice are rolled then the probability of the faces adding up to 9 is $\frac{25}{216}$. Suppose three such dice are rolled 200 times. Find the probability of obtaining this sum (a) at most 25 times (b) at least 25 times (c) between 20 and 30 times inclusive (d) exactly 25 times.
4. It was demonstrated in Section 1.1 that if three dice are rolled then the probability of the faces adding up to 10 is $\frac{27}{216}$. Suppose three such dice are rolled 200 times. Find the probability of obtaining this sum

(a) at most 27 times (b) at least 27 times (c) between 20 and 30 times (inclusive) (d) exactly 27 times.

5. Of the cars passing through a toll booth, 25% are from outside the state. Find the probability that of the next 1000 cars the number of out-of-staters is (a) at most 260 (b) at least 260 (c) exactly 260 (d) between 240 and 280 (inclusive).
6. A fair coin is tossed 150 times. What is the probability of obtaining (a) at most 60 heads (b) at least 60 heads (c) exactly 75 heads (d) between 60 and 80 heads inclusive?
7. A fair coin is tossed 30 times. What is the probability of obtaining (a) at most 12 heads (b) at least 12 heads (c) exactly 15 heads (d) between 8 and 12 heads (inclusive)?
8. The probability that a patient recovers from a certain disease is 0.7. If a hospital has 25 patients with this disease, what is the probability of the following events: (a) at most 12 recover (b) at least 12 recover (c) exactly 15 recover (d) between 12 and 18 recover (inclusive)?
9. In a certain city the need for money to buy drugs is stated as the reason for 80% of all burglaries. Find the probability that among the next 200 burglaries the number of those due to this need is (a) at most 150 (b) at least 150 (c) exactly 160 (d) between 150 and 170 (inclusive).
10. It is known that 25% of trucks will experience a tire blow out in a certain strenuous test. Suppose 35 trucks are tested. Find the probability that the number of trucks that experience a blow out is (a) at most 12 (b) at least 12 (c) exactly 10 (d) between 10 and 12 (inclusive).
11. It is estimated that a certain flu vaccine is 90% effective. If 1300 people are inoculated, find the probability that the number of successful vaccinations amongst them is (a) at most 1200 (b) at least 1200 (c) exactly 1200 (d) between 1150 and 1250 (inclusive).

1.5 Activities/Websites

1. Activate any quincunx. Let it run for a few minutes until the region below the normal curve is almost covered and then print the page. Does the outcome support the contention that the normal curve can be used to approximate binomial probabilities?
2. A contestant on a TV show faces three doors: one hides a car and the other two hide goats. The contestant selects a door, whereupon the host, who knows exactly which door hides what prize, opens one of the other two doors to reveal a goat. The contestant is then given the choice of staying with his or her original selection, or switching to the third door. Should the contestant switch or not?

(a) Go to the website http://www.shodor.org/interactivate/activities/monty3/ and obtain the answer by running some electronic simulations.

(b) Find a theoretical answer (on the web, if need be).

(c) Do the simulations and the theoretical answer agree?

3. Theoretical considerations indicate that in a group of seven people the probability of two of them being born on the same day of the month is slightly greater than 50%. Check this out by dividing your classmates into groups of seven and counting the groups with multiple birthdays. Use the phrase "birthday problem" to find some relevant information on the web.

Chapter 2

Statistics

All measurements, scientific or otherwise, are to be taken with a grain of salt because they involve some errors. While these inevitable errors were first regarded as merely a nuisance, they eventually became the object of scientific inquiry themselves. It was discovered that measurement errors follow probabilistic rules that are similar to those that govern binomial processes. This eventually led to myriads of applications of the normal curve both in the sciences and everyday life.

2.1 Summarizing Data

Modern science and technology are data-driven and all theories, no matter how elegant or "commonsensical," must be backed up by facts. Unfortunately, raw data often comes in bewildering quantities and in such cases it is easy to miss significant patterns because of the profusion of details. For this reason it is customary to summarize such data by means of quantitative parameters, groupings, graphs, or any combination thereof.

The quantitative parameters used to describe data are of two kinds: *measures of central tendencies* and *measures of variability*. Those of the former kind attempt to pinpoint the "center" of the data and those of the latter pertain to their dispersion.

Central Tendencies. The most commonly used measure of central tendency is the *average*, also known as the *mean* or the *arithmetic mean*. The average of the data $x_1, x_2, ..., x_n$ is denoted by $\bar{x}$ and equals

$$\bar{x} = \frac{x_1 + x_2 + \ldots + x_n}{n}.$$

As the computation of the average calls for a division, it will generally require more digits behind the decimal point than the data and we agree to stop after one additional digit has been obtained. Thus, the mean of the data 3, 7, 6 is 5.33... ≈ 5.3, that of the data 3, 7, 7 is 5.66... ≈ 5.7, and that of 1.3, 1.7, 1.7 is 1.566... ≈ 1.57.

The *median* of a set of data is almost as commonly computed as its mean. This measure is denoted by $\tilde{x}$ and its determination calls for first ordering the data in increasing order so that it may be assumed that $x_1 \leq x_2 \leq \ldots \leq x_n$ and then setting

$$\tilde{x} = \begin{cases} x_{(n+1)/2} & \text{if } n \text{ is odd} \\ (x_{n/2} + x_{n/2+1})/2 & \text{if } n \text{ is even} \end{cases}$$

For example, if the data are 15.7, 4.9, 3.2, 2.7, 4.1, 10.5, 8.9, then they need to be rearranged as 2.7, 3.2, 4.1, 4.9, 8.9, 10.5, 15.7. Since there are 7 data, the median is

$$\tilde{x} = x_{(7+1)/2} = x_{8/2} = x_4 = 4.9.$$

On the other hand, if there are 8 data: 15.7, 4.9, 3.2, 13.4, 2.7, 4.1, 10.5, 8.9, then they are rearranged as 2.7, 3.2, 4.1, 4.9, 8.9, 10.5, 13.4, 15.7 and

$$\tilde{x} = \frac{x_{8/2} + x_{(8/2)+1}}{2} = \frac{x_4 + x_5}{2} = \frac{4.9 + 8.9}{2} = 6.9.$$

Loosely speaking, the median is a number that splits the data by size into two groups of equal magnitudes. Unfortunately, this is all too often imprecise. The data 1, 2, 2, 3, 3, 3, 3, 5, 9, whose median is 3, has 7/9 of its members at 3 or below and 8/9 of its members at 3 or above, both of which fractions are considerably bigger than 1/2. More precisely, it can be asserted that for *at least* half of the data

$$x_i \leq \tilde{x}$$

and for *at least* half of the data

$$x_i \geq \tilde{x}.$$

The *mode* of a set of data is its most frequently occurring member. Thus, the mode of the data 1, 2, 2, 3, 3, 3, 3, 5, 9 is 3. Whereas every set of data has both a mean and a median, its mode need not be well defined. Thus, the data 1, 2, 2, 2, 3, 3, 4, 5, 5, 5 has no mode (it is sometimes said to be *bimodal*).

The mode is considered to be the least *sensitive* (or, equivalently most *robust*) of these three measures of centrality because there are many ways to modify a data set without affecting its mode. For example the data set

1, 2, 2, 2

and its extension

$$1, 2, 2, 2, 3, 3, 4, 4, 5, 5, 6, 6, 7, 7$$

clearly have the same mode. The median is also quite insensitive. For example any modification of all the data (except the median itself) that does not affect their relative magnitudes will not affect the median. Such, for example, is the case when the data

$$1, 2, 3, 4, 5, 6, 7$$

is modified to

$$0.001, 0.02, 0.3, 4, 50, 600, 7000.$$

The mean is the most sensitive of these measures. Any change in the value of some of the data is very likely to affect the value of their average.

Sensitivity and its opposite - robustness - have both advantages and disadvantages. Those who wish to protect against outliers - data that lie so far from the rest that one has good reason for discounting their validity or relevance - are advised to use the median since its value is mostly independent of such accidents. If, however, all the data are judged to be equally trustworthy, then the average is considered to be the better representative. The mode should be employed when it is crucial that the representative number should be an integer. Thus, the typical family in a certain town has two or three children, but never 2.6 children. Similarly, biologists, when describing the typical nest of a species of birds, use the mode for the number of eggs.

Variability. The second most important aspect of a set of data is its *variability*, or *spread*. One easy way of measuring this is simply to look for the difference between the largest and smallest data; this is called the *range* of the data. For example, the range of the data 4.7, 2.3, 1.6, 8.3, 5.2, 7.4, 2.5 is

$$8.3 - 1.6 = 6.7.$$

The range is not very sensitive and, for example, does not distinguish between the two data sets

$$1, 2, 3, 4, 5, 6, 7, 8, 9 \quad \text{and} \quad 1, 5, 5, 5, 5, 5, 5, 5, 9,$$

the first of which clearly has much more variability than the second. Sophisticated use of data, as will be illustrated below, calls for a more intricate measure. Our strategy is to define the variability of the data in terms of their distances from their common mean, i.e., in terms of $x_i - \bar{x}$. However, since the sum

$$(x_1 - \bar{x}) + (x_2 - \bar{x}) + \ldots + (x_n - \bar{x})$$

is always zero (see Exercises 15, 16), it is necessary to make these terms positive so that they do not cancel each other out. For reasons that will be explained later, it has become customary to square the individual differences and work with

$$(x_1 - \bar{x})^2 + (x_2 - \bar{x})^2 + \ldots + (x_n - \bar{x})^2.$$

This sum is then averaged to

$$\frac{(x_1-\bar{x})^2 + (x_2-\bar{x})^2 + \ldots + (x_n-\bar{x})^2}{n}$$

because it is necessary to make sure that data sets such as

$$1, 3 \quad \text{and} \quad 1, 1, 1, 1, 3, 3, 3, 3$$

have equal variability (see Exercises 1, 2). Finally, in order to offset the effect of the squaring, a square root is taken and the *standard deviation* of the data $x_1, x_2, \ldots, x_n$ is defined as

$$\sqrt{\frac{(x_1-\bar{x})^2 + (x_2-\bar{x})^2 + \ldots + (x_n-\bar{x})^2}{n}}.$$

For example, to compute the standard deviation of the data 5, 7, note that their mean $\bar{x}$ is 6 and so their standard deviation is

$$\sqrt{\frac{(5-6)^2 + (7-6)^2}{2}} = \sqrt{\frac{(-1)^2 + 1^2}{2}} = \sqrt{\frac{2}{2}} = \sqrt{1} = 1.$$

It is common to denote the standard deviation by the letter s. Thus, for the data 2, 3, 5, 10, we have

$$\bar{x} = \frac{2+3+5+10}{4} = \frac{20}{4} = 5$$

and

$$s^2 = \frac{(2-5)^2 + (3-5)^2 + (5-5)^2 + (10-5)^2}{4} = \frac{9+4+0+25}{4} = \frac{38}{4}$$

so that

$$s = \sqrt{\frac{38}{4}} \approx 3.1.$$

Similarly, for the data 4.7, 2.3, 1.6, 8.3, 5.2, 7.4, 2.5, we have

$$\bar{x} = \frac{4.7+2.3+1.6+8.3+5.2+7.4+2.5}{7} = \frac{32}{7} \approx 4.57$$

and

$$s^2 = \frac{1}{7}[(4.7 - 4.57)^2 + (2.3 - 4.57)^2 + (1.6 - 4.57)^2 + (8.3 - 4.57)^2 + (5.2 - 4.57)^2 + (7.4 - 4.57)^2 + (2.5 - 4.57)^2] \approx 5.7992$$

so that

$$s = \sqrt{5.7992} \approx 2.41.$$

Groupings and Histograms. Another way of coping with the obfuscation caused by the large size of a data set is to subdivide it into convenient groups and then represent those by means of bar graphs. The process of grouping has not been standardized and generally depends on the data as well as subjective considerations. Two procedures will described here.

Consider the following data that might describe the ages of the students in a certain university class:

{21, 20, 21, 22, 21, 18, 20, 19, 19, 21, 19, 21, 20, 21, 20, 20, 23, 20, 20, 22, 20, 21, 21, 19, 22, 22, 17, 21, 20, 21, 21, 23, 19, 22, 21, 18, 19, 19, 21}.

The *frequency table* of this data is obtained by recording the frequency f_i with which the ith age occurs rather than listing them all. It is customary to also include in such tables the *relative frequency*

$$r_i = \frac{f_i}{n}$$

where n denotes the total number of data.

Ages in a university class			
Group index i	Age	Frequency f_i	Relative frequency r_i
1	17	1	0.03
2	18	2	0.05
3	19	7	0.18
4	20	9	0.23
5	21	13	0.33
6	22	5	0.13
7	23	2	0.05
		39	1

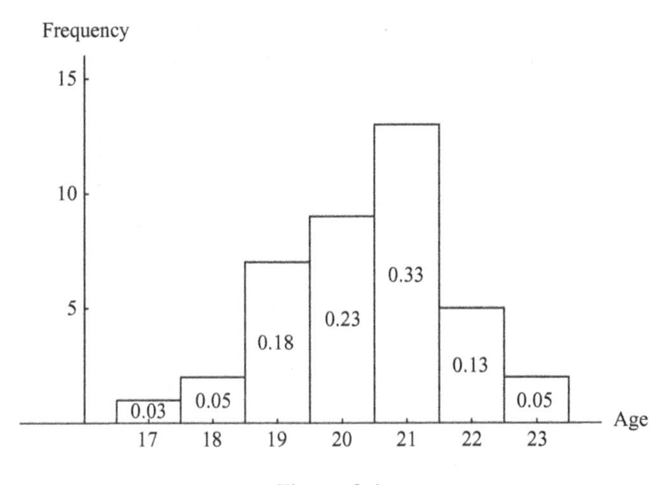

Figure 2.1:

The *histogram* of a set of data is a visual presentation of its frequencies as a bar graph.
The average and the standard deviation of a data set are easily computed from its frequency table:

$$\begin{aligned}\bar{x} &= r_1x_1 + r_2x_2 + \ldots + r_kx_k \\ s &= \sqrt{r_1(x_1-\bar{x})^2 + r_2(x_2-\bar{x})^2 + \ldots + r_k(x_k-\bar{x})^2}\end{aligned} \qquad \text{(d)}$$

Thus, for the data of the table above:

$$\begin{aligned}\bar{x} &= 0.03{\cdot}17+0.05{\cdot}18+0.18{\cdot}19+0.23{\cdot}20+0.33{\cdot}21+0.13{\cdot}22+0.05{\cdot}23 \approx 20.4 \\ s^2 &= (0.02(17-20.4)^2+0.05(18-20.4)^2+0.18(19-20.4)^2+0.23(20-20.4)^2 \\ &\quad + 33(21-20.4)^2+0.13(22-20.4)^2+0.05(23-20.4)^2] \approx 1.814\end{aligned}$$

so that

$$s \approx \sqrt{1.814} \approx 1.35.$$

When the data are spread out more finely each number is likely to occur only once and so it is better to group them into classes. Consider, for example, the following data which might represent the ages of the students attending an adult education class, where the students' ages are likely to vary much more than in a university class:

{26, 49, 43, 42, 46, 60, 34, 43, 41, 25, 20, 54, 46, 41, 56, 76, 33, 47, 56, 46, 34, 38, 45, 25, 33, 36, 38, 40, 65, 35, 49, 41, 25, 37, 43, 53, 51, 32, 18, 34, 39, 60, 19, 48, 46}.

Given that these numbers represent ages and that people tend to think of age in terms of decades, it is reasonable to group them as follows:

Ages in an adult education class			
Group index i	Age groups	Frequency f_i	Relative frequency r_i
1	10–19	2	0.04
2	20–29	5	0.11
3	30–39	12	0.27
4	40–49	17	0.38
5	50–59	5	0.11
6	60–69	3	0.07
7	70–79	1	0.02
		45	1

Approximations to the mean and standard deviation of a data set are easily derived from the histogram by means of formulas analogous to those of (d) above. The average of the limits of the i-th data group is called its *mark* and is denoted by m_i. These marks replace the actual data in (d) so that we obtain

$$\bar{x} = r_1 m_1 + r_2 m_2 + \ldots + r_k m_k$$

$$s = \sqrt{r_1 (m_1 - \bar{x})^2 + r_2 (m_2 - \bar{x})^2 + \ldots + r_k (m_k - \bar{x})^2}.$$

Thus, the above table of ages in an adult education class is augmented as follows:

Ages in an adult education class				
Group index i	Age	Frequency f_i	Relative frequency r_i	Mark m_i
1	10–19	2	0.04	14.5
2	20–29	5	0.11	24.5
3	30–39	12	0.27	34.5
4	40–49	17	0.38	44.5
5	50–59	5	0.11	54.5
6	60–69	3	0.07	64.5
7	70–79	1	0.02	74.5
		45	1	

These tables' grouped mean and standard deviation are

$$\bar{x} = 0.04 \cdot 14.5 + 0.11 \cdot 24.5 + 0.27 \cdot 34.5 + 0.38 \cdot 44.5 + 0.11 \cdot 54.5 \\ + 0.07 \cdot 64.5 + 0.02 \cdot 74.5 = 41.5$$

$$s^2 = 0.04(14.5-41.5)^2 + 0.11(24.5-41.5)^2 + 0.27(34.5-41.5)^2 \\ + .38(44.5-41.5)^2 + .11(54.5-41.5)^2 + 0.07(64.5-41.5)^2 \\ + 0.02(74.5-41.5)^2 \approx 155.00$$

$$s = \sqrt{1.55} \approx 12.4.$$

Exercises 2.1

1. Find the standard deviation of the data {1, 3}.
2. Find the standard deviation of the data {1, 1, 1, 1, 3, 3, 3, 3}.
3. Find the standard deviation of the data {10, 21,15, 14, 9, 12, 8, 10, 9, 11, 12}.
4. Find the standard deviation of the data {23, 20, 31, 38, 26, 19, 35, 32, 27, 32, 25}.
5. Find the standard deviation of the data {2.2, 1.5, 1.7, 2.0, 0.9, 2.3, 1.3, 1.7, 0.8, 1.2, 1.7}.
6. a) Organize the following data by creating a frequency table:

   2 1 3 0 1 0 2 5 1 2 0
   2 2 1 4 3 1 1 0 0 0 2
   2 1 0 3 1 1 3 4 2 1 3
   0 1

   b) Compute the mean and the standard deviation of the data of part a.
7. a) Organize the following data by creating a frequency table:

   66 61 66 61 63 68 64 67 60 63 69
   67 67 60 67 60 62 68 65 60 64 65
   60 64 62 67 62 69 63 65 64 69 63
   69 64 68 67 61 68 69

   b) Compute the mean and the standard deviation of the data of part a.
8. a) Organize the following data by creating a frequency table with groups 0.5–0.9, 1.0–1.4, etc.

   1.0 1.8 2.4 1.2 2.0 0.7 1.3 2.4 1.1 3.3 2.4
   0.8 2.3 1.7 1.0 2.8 1.4 3.0 0.8 1.2 1.4 2.2
   1.5 3.2 2.1 2.7 1.8 1.6 2.3 2.6 1.3 2.9 1.9
   1.2 0.5

   b) Compute the mean and the standard deviation of the data of part a.

9. a) Organize the following data by creating a frequency table with groups 5–9, 10–14, etc.

24 32 24 32 11 30 10 11 28 20 11
25 25 30 22 36 30 24 17 26 21 25
29 37 31 39 19 25 22 8 37 29 19
16 26 15 24 13 29 21 19 35 15 23
29 24 27 26 20 18

   b) Compute the grouped mean and the standard deviation of the data of part a.

10. a) Complete the table below to a frequency table and draw its histogram:

Group	Frequency
0–9	99
10–19	273
20–29	196
30–39	79
40–49	67
50–59	20

   b) Compute the grouped mean and the standard deviation of the data of part a.

11. a) Complete the table below to a frequency table and draw its histogram:

Group	Frequency
30–36	9
37–43	73
44–50	96
51–57	9
58–64	7
65–71	6

   b) Compute the grouped mean and the standard deviation of the data of part a.

12. a) Complete the table below to a frequency table and draw its histogram:

Group	Frequency
15.3–15.5	9
15.6–15.8	27
15.9–16.1	19
16.2–16.4	7
16.5–16.9	6

   b) Compute the grouped mean and the standard deviation of the data of part a.

13. a) Complete the table below to a frequency distribution and draw its histogram:

Group	Frequency
16–18	111
19–21	1,380

22–24	1,158
25–27	928
28–30	847
31–33	591
34–36	878

b) Compute the grouped mean and the standard deviation of the data of part a.

14. Toss a die one hundred times and organize the outcomes in the form of a frequency table. Compute the mean and the standard deviation of this data.
15. Verify that expression (cc) is indeed 0 for the data $\{1, 1, 1, 1, 3, 3, 3, 3\}$.
16. Verify that expression (cc) is indeed 0 for the data $\{1, 2, 3, 6, 10\}$.

2.2 Totaling Measurement Errors

The scientific custom of taking multiple observations of the same quantity and then selecting a single estimate that best represents it has its origin in the early part of the sixteenth century. By that time European mariners had stopped hugging the shore and taken to sailing for months in uncharted seas with no signposts or landmarks. The ship's location could be computed to some extent on the basis of the position of the moon relative to the fixed stars as well as the difference between the true north and the compass needle. All this called for observations that were unfortunately often beset with errors. These errors could be due to the motion of the ship, the imprecision of the instruments, the fallibility of the observers, atmospheric disturbances, and many other causes. It was in order to protect against such errors that the custom of repeating one's observations many times arose. This led to the natural question of how to decide on the best possible estimate. Should one use the "middle" observation, commonly known as the *median*? Should one *average* by adding all the measurements and dividing by their number? Should outlying observations that lie far from all the others be disregarded as patently irrelevant? As erroneous conclusions regarding the position of the ship could, and often did, lead to disastrous consequences this was an important question.

At first each observer had his own method for producing the best estimate which method was rarely, if ever, explicitly written down. A case in point is provided by Johannes Kepler (1571–1630), one of the founders of modern astronomy who at one time, in trying to pinpoint the location of a certain star, chose the numerical value 24.55 to represent the four observations: 23.65, 27.62, 23.30, and 29.80. Since the average of these four numbers is 26.09 and their median is 25.64, it is clear that he had neither of these popular estimates in mind. It is not known what he did have in mind.

There are many recorded instances of such arbitrary choices by astronomers. Sad to say, it seems that quite frequently observers would simply choose such a number as best confirmed their preconceived notions. Eventually, the need for a standardized procedure backed by a mathematical theory of observational errors was recognized by the scientific community. The first to make an attempt at an explicit theory of errors was none other than Galileo Galilei in his epoch making book *Dialog Concerning the Two Chief World Systems – Ptolemaic and Copernican*, which appeared in 1632 and was suppressed by the Roman Catholic Church.

Galileo's discussion of the errors inherent in astronomical observations has been summarized as follows [Hald]:

G1. *There is only one number which gives the distance of the star from the center of the earth, the true distance.*
G2. *All observations are encumbered with errors, due to the observer, the instruments, and the other observational conditions.*
G3. *The observations are distributed symmetrically about the true value; that is, errors are distributed symmetrically about zero.*
G4. *Small errors occur more frequently than large errors.*
G5. *The most probable hypothesis is the one which requires the smallest corrections of the observations.*

Points G1, G2, and G4 are self-explanatory. Point G3 is to be understood as referring to all potential measurements of a certain quantity rather than any actual set of such numbers. After all it is quite conceivable that two measurements of some length should both be overestimates. An alternative statement of this principle says:

G3′. *If e is any real number then the errors of magnitudes e and $-e$ are equally likely to occur.*

The discussion of Galileo's fifth, and most interesting, point requires a clarification of his terminology. Galileo thought of every measurement as being accompanied by an error and therefore needing, in principle, a correction. Multiple measurements resulted in multiple errors whose cumulative effect needed to be taken into consideration. But how should the errors be combined? It would clearly be misguided to simply add them and let overestimates cancel out underestimates, for the same reason that two hunters whose shots pass respectively one foot over and under the same duck must be conceded to have missed their mark. In order to prevent such fallacious cancellations, Galileo drew a slight distinction between an *error* and its *correction*. The latter is the positive part (technically, absolute value) of the former. Thus, both the errors $-0.1''$ and $0.1''$ require the same correction of $0.1''$.

Point G5 above will now be illustrated by means of an example. Suppose a repeated measurement conducted by scientists A and B yielded the estimates 4.01, 4.03 and 4.11 (the units don't matter here) of the same quantity. Suppose further that scientist A prefers to think of 4.11 as an outlier, decides to ignore it, and proposes

$$4.02 = \frac{4.01 + 4.03}{2}$$

as an estimate of the true value. Scientist B, however, does not wish to throw away any of the data and proposes

$$4.05 = \frac{4.01 + 4.03 + 4.11}{3}$$

as his candidate for the true value. Which of these two estimates is more likely to be the true value? The import of Galileo's point G5 is that estimates should be judged by the sum of the corrections they entail. For A's and B's estimates these sums are respectively

$\underline{4.02}$: $$|4.01-4.02| + |4.03-4.02| + |4.11-4.02| = 0.01+0.01+0.09 = 0.11$$

and

$\underline{4.05}$: $$|4.01-4.05| + |4.03-4.05| + |4.11-4.05| = 0.04+0.02+0.06 = 0.12.$$

Since $0.11 < 0.12$ if follows from Galileo's criterion that 4.02 is a better candidate for the unknown true value than 4.05. As it happens, this is the only sense in which Galileo used this total corrections criterion: to decide between subjectively selected candidates for the true value.

However, in the example under consideration there is a better candidate yet: the middle measurement 4.03. The sum of its corrections is

$\underline{4.03}$: $$|4.01-4.03| + |4.03-4.03| + |4.11-4.03| = 0.02+0+0.08 = 0.10.$$

The table below indicates that 4.03 is in fact the best candidate for the true value, at least in Galileo's sense of minimizing the sum of the corrections.

Candidate:	4.01	4.02	4.03	4.04	4.05	4.06	4.07	4.08
Total Correction:	0.12	0.11	0.10	0.11	0.12	0.13	0.14	0.15

The optimality of 4.03 can also be displayed visually by means of a graph. If x denotes any estimate of the measured quantity, then the total of the concomitant corrections is

$$y = |4.01-x| + |4.03-x| + |4.11-x|.$$

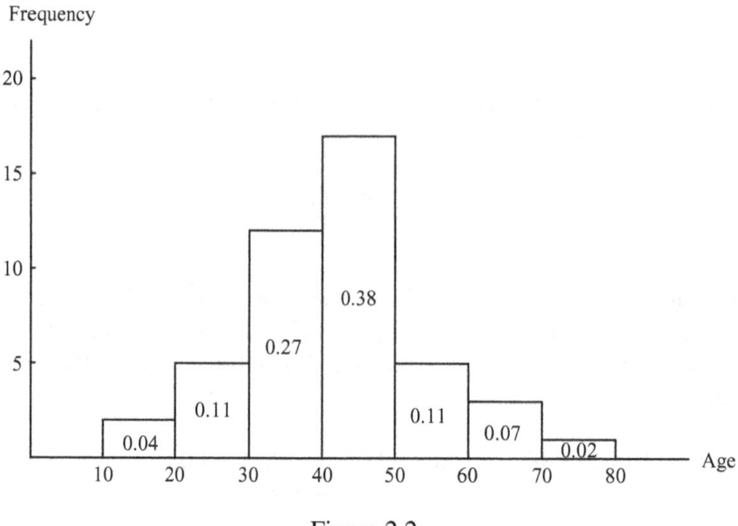

Figure 2.2:

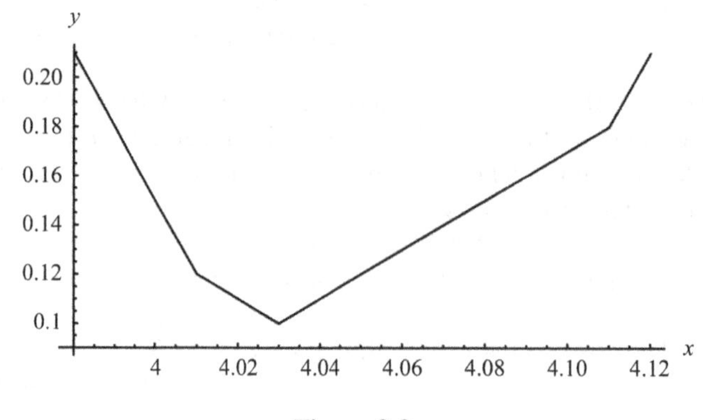

Figure 2.3:

A graph of this function, obtained by means of a calculating device, is displayed in Figure 2.3. The optimality of 4.03 manifests itself in the fact that the lowest point on this graph occurs at $x = 4.03$.

As noted above, this best candidate is the middle one of the three measurements. This turns out to be no coincidence. When the number of measurements is odd the middle one, otherwise known as the *median*, is invariably the optimal Galilean candidate. When the number of measurements is even, either of the two middle ones or any number in between them will serve, and it is customary to use the midpoint of these two middle values as the *median*. This is illustrated by the table below which lists the total corrections for a variety of candidates in the context of Kepler's four data mentioned above: 23.65, 27.62, 23.30, and 29.80.

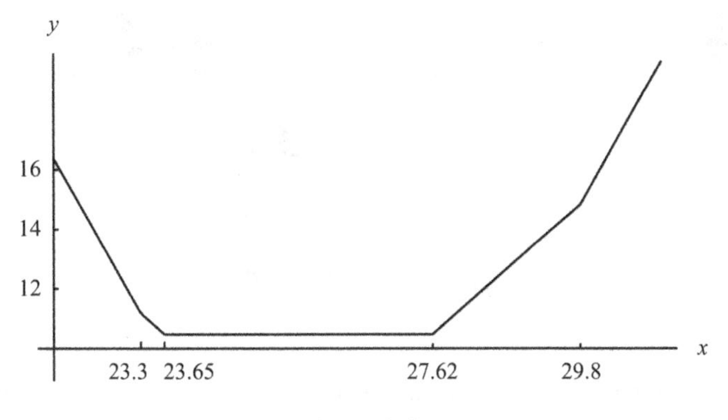

Figure 2.4:

Candidate:	23.50	24.00	25.00	26.00	27.00	28.00	29.00	29.50
Total correction:	10.77	10.47	10.47	10.47	10.47	11.23	13.23	14.23

Once again the optimality of the median is also demonstrated visually by means of the graph in Figure 2.4.

Since Galileo did not comment on this interesting coincidence it may be concluded that he did not notice it, probably because he expected the characterization of the best candidates to require a much more complicated resolution and so chose not to pursue it. He was, after all, preoccupied with the true nature of the solar system and at that time not much depended on the question of what to do with discrepant multiple observations.

Galileo's fifth point proposing the minimization of the total error can be viewed as an instance of *Occam's Razor* at work. This razor is an informal, though widely accepted, philosophical/scientific principle which stipulates that

> *All else remaining the same, the simplest possible hypothesis should be accepted as the most likely explanation.*

In the context of multiple errors, the simplest possible hypothesis could be interpreted as the estimate that calls for the least amount of correction.

Galileo's suggestion that the best candidate is that which minimizes the corrections was largely ignored by other astronomers and mathematicians. Instead, a variety of mostly undocumented ad hoc procedures were used. Eventually, sometime during the eighteenth century, the average of the measurements became widely accepted as the most likely true value. It is not known how this preference became established. The justification offered by Roger Cotes (1682–1716), one of Sir Isaac Newton's disciples, for this choice was that if the various measurements $m_1, m_2, \ldots, m_n$ were pictured on the number line and equal weights were placed at these points (see Figure 2.5), then the center of gravity of this system of weights would be located at exactly

Figure 2.5:

$$\overline{m} = \frac{m_1 + m_2 + \ldots + m_n}{n}$$

which is of course the average of the measurements. This explanation was not universally accepted and scholarly arguments against the use of the average can still be found as late as 1777. After all, measurements are not corporeal bodies and do not possess weight. Why then should this fictitious center of gravity of erroneous measurements have anything to do with the most likely true value?

At the beginning of the 19th century, two of the foremost mathematicians of the time, the German C. F. Gauss (1777–1855) and the Frenchman A. M. Legendre (1752–1833), proposed, independently of each other, an optimality criterion that bears the same relation to the average as Galileo's least total corrections principle bears to the median. This criterion, known as the *least squares method*, called for squaring the errors, rather than taking their absolute values, before they are totaled. For example, in reference to the data 4.01, 4.03, and 4.11 used above, the totaled squared corrections are:

$\underline{4.02}$: $(4.01-4.02)^2+(4.03-4.02)^2+(4.11-4.02)^2$
$= (-0.01)^2+0.01^2+0.09^2 = 0.0083$

$\underline{4.03}$: $(4.01-4.03)^2+(4.03-4.03)^2+(4.11-4.03)^2$
$= (-0.02)^2+0^2+0.08^2 = 0.0068$

$\underline{4.05}$: $(4.01-4.05)^2+(4.03-4.05)^2+(4.11-4.05)^2$
$= (-0.04)^2+(-0.02)^2+0.06^2 = 0.0056$

$\underline{4.06}$: $(4.01-4.06)^2+(4.03-4.06)^2+(4.11-4.06)^2$
$= (-0.05)^2+(-0.03)^2+0.05^2 = 0.0059.$

Of the above sums of squares, the least is 0.0056, given by 4.05, which also happens to be the average of 4.01, 4.03, and 4.11. The optimality of 4.05 can also be displayed visually by means of a graph. If x denotes any estimate of the measured quantity, then the total of the squares of the concomitant errors is

$$y = (4.01-x)^2+(4.03-x)^2+(4.11-x)^2.$$

A graph of this function, obtained by means of a calculating device, is displayed in Figure 2.6. This graph's lowest point lies right over $x = 4.05$.

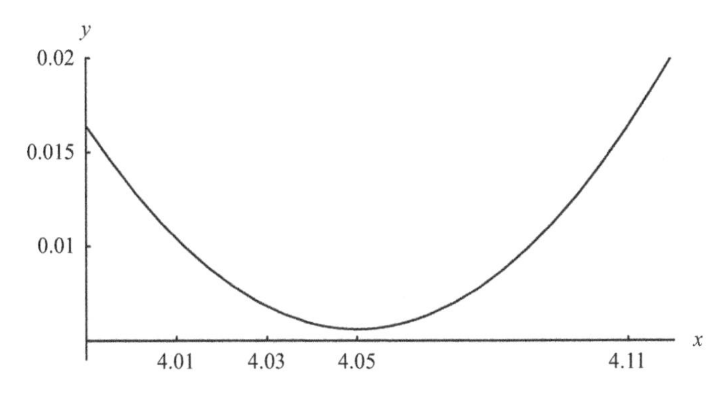

Figure 2.6:

Let us re-examine Kepler's measurements 23.65, 27.62, 23.30, 29.80 from this point of view. The table below gives the sums of the squares of the corrections for the listed candidates:

24: $(23.65-24)^2+(27.62-24)^2+(23.30-24)^2+(29.80-24)^2 = 47.36$

25: $(23.65-25)^2+(27.62-25)^2+(23.30-25)^2+(29.80-25)^2 = 34.62$

25.635 (median): $(23.65-25.635)^2+(27.62-25.635)^2+(23.30-25.635)^2 + (29.80-25.635)^2 = 30.68$

26.092 (average): $(23.65-26.092)^2+(27.62-26.092)^2+(23.30-26.092)^2 + (29.80-26.092)^2 = 29.84$

27: $(23.65-27)^2+(27.62-27)^2+(23.30-27)^2+(29.80-27)^2 = 33.14.$

Of these candidates, 25.635 is the median of Kepler's data and 26.092 is the average. Once again it is the average that yields the minimum sum of the squares of the corrections. The optimality of the average is also demonstrated by Figure 2.7, which displays the graph of the sum of the squared errors

$$y = (23.65-x)^2+(27.62-x)^2+(23.30-x)^2+(29.80-x)^2.$$

Here too the graph's lowest point lies over the average $x = 26.092$.

Exercises 2.2

For each of the sets of measurements in Exercises 1–8,

a) Compute the median and the average.

b) Compute the sum of the corrections associated with both the median and the average. Which is smaller?

c) Compute the sum of the squares of the errors associated with both the median and the average. Which is smaller?

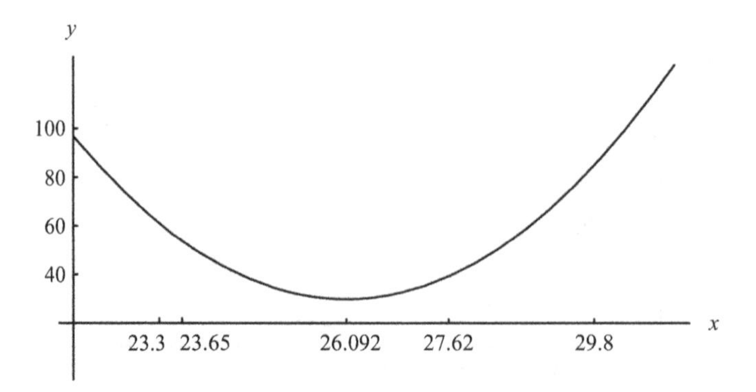

Figure 2.7:

1. {1, 2, 3, 4, 10}
2. {9, 9, 9, 8, 7, 6, 6, 6, 5, 5}
3. {5.1, 5.2, 5.3, 5.4, 6.0}
4. {10.2, 10.2, 10.3, 10.3, 10.5}
5. {9.1, 9.1, 9.1, 9.2, 9.2}
6. {1.5, 1.6, 1.8, 1.9, 2.0, 2.0}
7. {8.5, 8.5, 8.5, 8.7, 8.7, 8.7}
8. {8.5, 8.7, 8.7, 8.7, 8.7, 8.7}

2.3 Least Squares and Regression

The context for Legendre's proposal of the least squares was that of geodesy - the scientific discipline concerned with the exact shape of the earth. In the spirit of their country's recent revolution France's scientists had decided to abandon their antiquated measurement system and to adopt a new one, more scientific and efficient, whose unit of length, the *meter,* would equal

$$\frac{1}{40{,}000{,}000}$$

of the circumference of the earth. This necessitated an accurate determination of said circumference which, in turn, depended on the exact shape of the earth. They were aware that this shape deviated somewhat from the perfect sphere. This distortion is due to the centrifugal force produced by the earth's daily rotation around its axis, which slightly flattens its north-south silhouette into an ellipse, i.e., an oval shape whose horizontal axis is somewhat longer than its vertical axis. In order to determine the exact dimensions of these axes, observations were taken in 1799 along a nearly 700 mile long arc that, of course, lay completely within the French territories of the time. As indicated by Figure 2.8, ellipses come in many shapes and the question then arose of which ellipse best fitted these observations. It was in order to answer this question that the least squares approach was adopted.

A different procedure, one based on the Galilean approach of totaling the corrections and known as the Boscovich method, had been proposed in 1770. It never

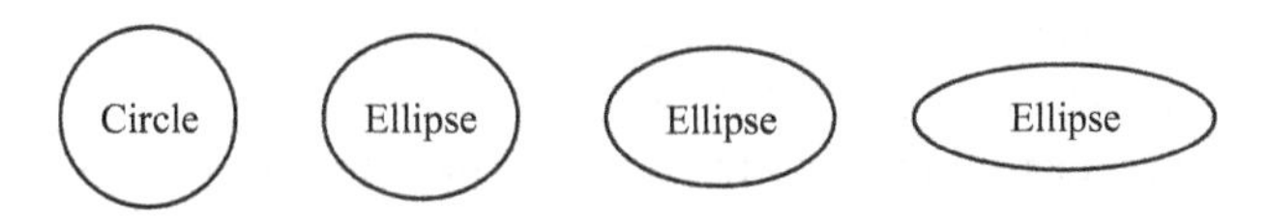

Figure 2.8:

gained popularity and sank into virtual oblivion once the least squares approach had been promulgated. The reasons for the scientists' preference for the latter method may seem strangely subjective and possibly irrelevant. The mathematics of Legendre's method was *easier* and more *satisfactory.* Note that the curves of Figures 2.3 and 2.4, which display the sum of the *corrections*, have corners, whereas those of Figures 2.6 and 2.7, which display the sums of the *squares of the errors*, are smooth, albeit curved. This difference means that the methods of calculus, an important mathematical discipline whose development was intimately tied to the evolution of modern science and technology, are applicable only to the latter but not to the former. It was this applicability of calculus to the method of least squares that made the scientific community prefer it over the Boscovich method. In other words, it was the tractability of the method of least squares to the tools of calculus rather than any objective demonstration of its superiority that drew mathematicians to it. Considering modern science's insistence on the objectivity of its theories and on the need to verify them by means of factual observation, this subjective criterion of convenience is rather surprising. Another reason for the alacrity with which the scientific community adopted the least squares approach may have been that it was so closely associated with the averaging procedure which, by some folk process whose details are lost, had already been adopted by the same group during the preceding half-century.

It will now be shown how the method of least squares is used to make predictions on the basis of experience and/or observations.

Example 2.3.1. *A retail merchant's experience indicates that there seems to be a fairly direct correlation between his sales and his advertising expenditures. In particular, the following data were recorded on four consecutive weeks:*

Advertising costs ($):	40	20	30	25	
Sales ($):	4,850	3,150	4,500	4,050	(c)

Using this data, the merchant wishes to describe the relationship between the expenditures and the sales as a function $y = f(x)$ where x denotes the expenditures and y the sales. Such a function would make it possible for him to project the sales resulting from any amount he might wish (or have available) to spend on advertising. As the relationship between advertising and sales is subject to random fluctuations, this projection would not be accurate, but still would be better than no projection at all. Such functions and equations, that represent the way real world variables affect each other, are called models.

Having no other information to go on, the merchant should assume that the function $f(x)$ is as simple as possible. This means that $f(x)$ is *linear* in the sense that its graph is a straight line, or more technically, there are constants m and b such that

$$f(x) = mx + b.$$

The specific values of m and b are then to be determined from the data of Table (c). Had there been only two observations (say the first two in the table), the procedure would have been straightforward. One would interpret these observations as the points (40, 4,850), (20, 3,150) of the Cartesian coordinate plane and then look for the equation of the straight line that passes through these two points. The two-point form of the equation of this straight line is:

$$y - 3{,}150 = \frac{4{,}850 - 3150}{40 - 20}(x - 20) = 85(x - 20) = 85x - 1700$$

or

$$y = 85x - 1{,}700 + 3{,}150 = 85x + 1{,}450.$$

This function is demonstrably inaccurate at $x = 25$ and $x = 30$. At these values it predicts sales of

$$f(25) = 85 \cdot 25 + 1{,}450 = 3{,}575 \text{ and } f(30) = 85 \cdot 30 + 1{,}450 = 4{,}000$$

respectively, which differ from the actual 4,050 and 4,500. These discrepancies are illustrated in Figure 2.9.

Such errors are endemic to the situation. Because of the inherent randomness of the universe, data points will rarely align themselves along a perfectly straight line and so any proposed straight line model will entail some errors. If the data points are $(x_1, y_1), (x_2, y_2), ..., (x_n, y_n)$ and the model is $f(x) = mx + b$, then the errors are defined as

$$e_i = y_i - (mx_i + b) \quad i = 1, 2, ..., n.$$

For example, for the data points of Table (c) and the model $y = 90x + 1{,}300$ the errors are:

$$e_1 = 4{,}850 - (90 \cdot 40 + 1{,}500) = -250 \qquad e_2 = 3{,}150 - (90 \cdot 20 + 1{,}500) = -150$$
$$e_3 = 4{,}500 - (90 \cdot 30 + 1{,}500) = 300 \qquad e_4 = 4{,}050 - (90 \cdot 25 + 1{,}500) = 300.$$

Since the errors are inevitable they can be at best minimized and it is commonplace to use the least squares method criterion proposed by Gauss and Legendre. In other words, we make the assumption that

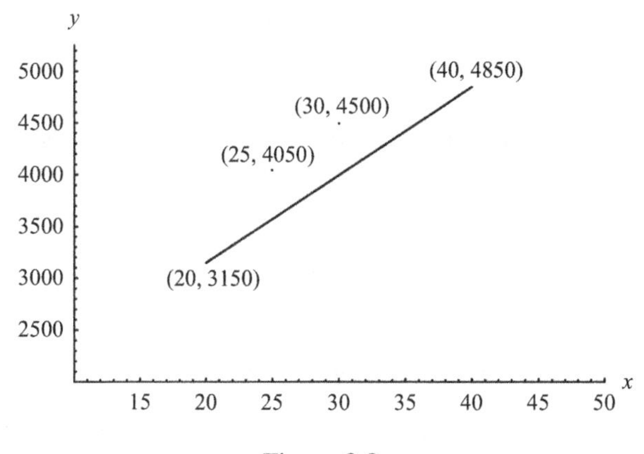

Figure 2.9:

That model is best which minimizes the sum of the squares of the errors.

The straight line $y = mx + b$ which minimizes the sum of the squares of the errors is called the *regression* line. Its parameters m and b are computed by the following procedure:

For the data points (x_1, y_1), (x_2, y_2), ..., (x_n, y_n) set:

$S_x = x_1 + x_2 + ... + x_n$ $\qquad S_y = y_1 + y_2 + ... + y_n$

$S_{x^2} = x_1^2 + x_2^2 + ... + x_n^2$ $\qquad S_{xy} = x_1y_1 + x_2y_2 + ... + x_ny_n.$

Then,

$$m = \frac{nS_{xy} - S_xS_y}{nS_{x^2} - (S_x)^2} \quad \text{and} \quad b = \frac{S_y - mS_x}{n}.$$

For example, using the four data points of Table (c), we obtain

$$S_x = 40 + 20 + 30 + 25 = 115$$

$$S_x = 40 + 20 + 30 + 25 = 115$$

$$S_y = 4{,}850 + 3{,}150 + 4{,}500 + 4{,}050 = 16{,}550$$

$$S_{x^2} = 40^2 + 20^2 + 30^2 + 25^2 = 3{,}525$$

$$S_{xy} = 40 \cdot 4{,}850 + 20 \cdot 3{,}150 + 30 \cdot 4{,}500 + 25 \cdot 4{,}050 = 493{,}250$$

$$m = \frac{4 \cdot 493{,}250 - 115 \cdot 16{,}550}{4 \cdot 3{,}525 - 115^2} = 79.7$$

$$b = \frac{16{,}550 - 79.7 \cdot 115}{4} = 1{,}845.7.$$

These calculations yield the regression line $y = 79.7x + 1{,}845.7$. Using this regression line one would project that if the merchant of Table (c) were to spend \$60 on advertising his sales would go up to

$$79.7 \cdot 60 + 1{,}845.7 = 6{,}627.7$$

dollars.

It might also be of interest to compute this model's errors:

$$e_1 = 4{,}850 - (79.7 \cdot 40 + 1{,}845.7) = -184 \qquad e_2 = 3{,}150 - (79.7 \cdot 20 + 1{,}845.7) = -290$$
$$e_3 = 4{,}500 - (79.7 \cdot 30 + 1{,}845.7) = 263 \qquad e_4 = 4{,}050 - (79.7 \cdot 25 + 1{,}845.7) = 211.$$

This regression line and its errors are illustrated in Figure 2.10.

The credibility of the average and the method of least squares were greatly enhanced by the *Ceres* incident. On January 1, 1801 the Italian astronomer Giuseppe Piazzi sighted a heavenly body that he strongly suspected to be a new planet. He announced his discovery and named it *Ceres.* Unfortunately, six weeks later, before enough observations had been taken to make possible an accurate determination of its orbit, so as to ascertain that it was indeed a planet, Ceres disappeared behind the sun and was not expected to re-emerge for nearly a year. Interest in this possibly new planet was widespread and astronomers throughout Europe prepared themselves by compu-guessing the location where Ceres was most likely to reappear. The young Gauss, who had already made a name for himself as an extraordinary mathematician, proposed that an area of the sky be searched that was quite different from those suggested by the other astronomers and turned out to be right. He became a celebrity and the theory of observational errors that helped

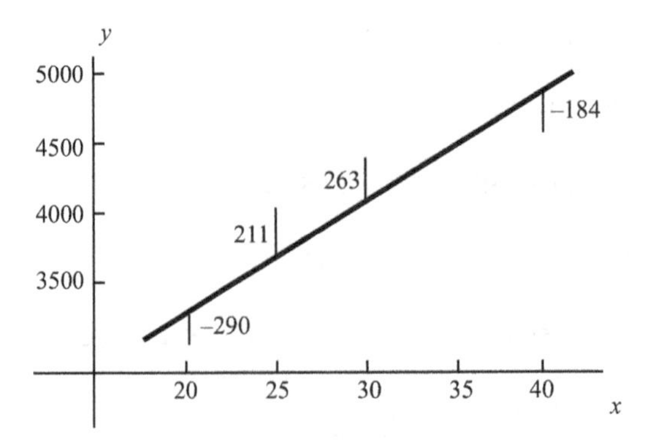

Figure 2.10:

him locate Ceres was immediately recognized as a milestone in the evolution of the scientific method.

Gauss' theory of errors was based on the following three assumptions:

N1. *Small errors are more likely than large errors.*
N2. *For any real number e the likelihoods of errors of magnitudes e and −e are equal.*
N3. *In the presence of several measurements of the same quantity, the most likely value of the quantity being measured is their average.*

On the basis of these principles and some clever mathematics Gauss then concluded that the mathematical curves that best described the distribution of observational errors were none other than the normal curves that were being used for approximating binomial probabilities. In other words, Gauss proved that the histograms of data obtained by taking multiple measurements of the same object will, in general, have the shape of the normal curve. Figure 2.11 displays some astronomical measurements pertaining to the right ascension of the Polar Star that originate in the publications of the Royal Observatory of Greenwich. As the unit of measurement is rather technical, it is omitted. After all, it is the shape of the data that is the main concern here.

Gauss's assumptions N1 and N2 stand to reason and had also been stated by Galileo. The third principle was revolutionary and inspired. It should be borne in mind that averaging had only become commonplace amongst astronomers during the previous half-century, and even then its adoption was in all likelihood a matter of subjective convenience rather than objective science. Despite the ubiquity of averaging, there was no proof, either theoretical or experimental, that the average constitutes a better candidate for the true value than the median or any other estimator. Nor has such a proof been offered since then.

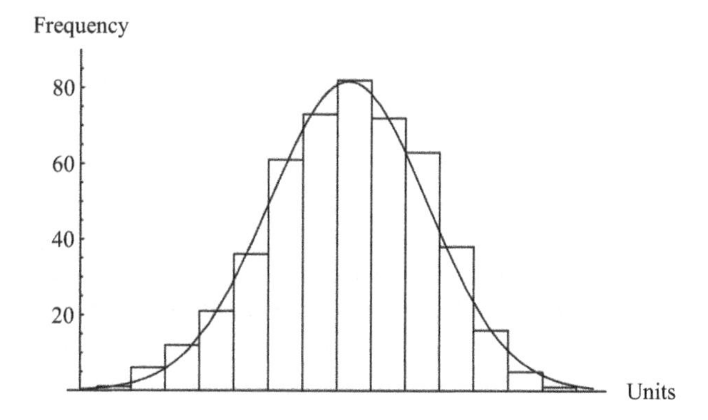

Figure 2.11: The histogram of some astronomical measurements.

This is a rather interesting issue. The normal curve, which was devised as a tool for approximating binomial probabilities and then was used by Gauss to describe the distribution of measurement errors, was later (see Sections 2.4–5) demonstrated to have a myriad of other applications to the real world. So much so that it is now considered to be the cornerstone of both mathematical and applied statistics. Nevertheless, the justification of the eminent position it holds in science boils down to the subjective criterion that averaging is the best way to handle the difficulties presented by random errors.

Gauss himself seems to have been uncomfortable with Assumption N3, though not with the conclusion that the error curves are normal. He subsequently justified the normality of the error curve by means of a variation of the least squares approach described above. However, as was noted before, the preference for least squares is also based on highly subjective grounds.

The reason for the applicability of De Moivre's approximation curve to the distribution of errors has been explained as follows: assume that just like matter, errors have an atomic structure so that each error is actually an aggregate of "atomic" or elementary errors, all of which have equal magnitude. If these elementary errors occur independently of each other and with equal probabilities, then one can think of the probability of occurrence of an actual error as a binomial probability whose binomial trials are the elementary errors. We hasten to add that this is hardly a scientific explanation. The existence of elementary errors was hypothesized solely for the purposes of this and other explanations. They have no factual basis.

This section is brought to a close with a restatement of the three distinct but equivalent scientific principles that can be used to explain the central role played by the normal curve in the handling of random errors:

R1. *The average is the best estimate for the true value.*
R2. *When aggregating errors, the individual errors should be squared.*
R3. *The distribution of measurement errors is best described by the normal curve.*

Exercises 2.3

For each of the sets of data points in Exercises 1-8 below.

a) *Find the line of regression.*
b) *Compute all the errors associated with the line of regression.*
c) *Use the line of regression to estimate the value of y that corresponds to x_0.*
d) *Graph the points and the regression line as well as its associated errors.*
e) *Find the equation of the straight line that contains the first and last points; compute the errors associated with this line and verify that the sum of their squares is at least as large as the corresponding sum for the regression line of part a.*

1. $\{(0, 2), (1, 1)\}, x_0 = 2$
2. $\{(0, 0), (1, 2)\}, x_0 = 2$
3. $\{(0, 0), (1, 1), (2, 2)\}, x_0 = 3$
4. $\{(0, 0), (1, 1), (2, 3)\}, x_0 = 5$
5. $\{(1, 2), (2, 3), (3, 2)\}, x_0 = 0$
6. $\{(0, 5), (1, 4), (3, 3), (4, 1)\}, x_0 = 2$
7. $\{(10, 26), (15, 20), (25, 10)\}, x_0 = 20$
8. $\{(12, 55), (18, 30), (20, 25)\}, x_0 = 10$

9. The following data are the selling prices y of a certain make and model of used car x years old:

x (years):	1	2	3	5
y (dollars):	6,250	5,725	5,350	4,945

Find the regression line for these data and use it to project the price of a 10-year-old car.

10. The grades of a class of 5 students on a midterm exam (x) and on the final exam (y) are as follows:

x (midterm grade):	77	50	71	81	99
y (final grade):	82	66	78	47	99

Find the regression line of these data and use it to predict the final exam grade of a student who scores 60 on the midterm.

11. To plan for contingencies, a municipal utility wishes to determine the relationship between the average temperature (in degrees Fahrenheit) and the average daily household consumption of water (in gallons). Given the following data:

Month:	January	April	July	October
Average daily temperature	26	46	71	61
Average daily consumption	180	193	245	238

Find the line of regression of these data and use it to project the average daily water consumption in a month whose average daily temperature is 50°F.

2.4 Normally Distributed Data

As the saying goes, nothing succeeds like success, and, subsequent to Gauss's spectacular prediction of the location Ceres, the least squares approach as well as the normality of the distribution of random errors soon became standard tools of all the physical sciences. It took another fifty years for scientists to come to the realization that they had in their possession a tool whose scope far exceeded the mere description of measurement errors. This next step was taken by the Belgian astronomer and sociologist Adolphe Quetelet (1796–1874) who was struck by the possibility that most data, and not just measurements and their accompanying errors, when grouped into classes, might display a normal shape. His first major example, given in 1846, was drawn from an 1817 article of the *Edinburgh Medical and Surgical Journal,* and listed the frequencies of chest measurements, in inches, of 5,738 Scottish soldiers.

Chest Measurements of Scottish soldiers	
Girth	Frequency
33	3
34	18
35	81
36	185
37	420
38	749
39	1,073
40	1,079
41	934
42	658
43	370
44	92
45	50
46	21
47	4
48	1
	5,738

These data are graphed in Figure 2.12 together with a normal curve that follows the general contours of the histogram. Quetelet did not justify his claim of the

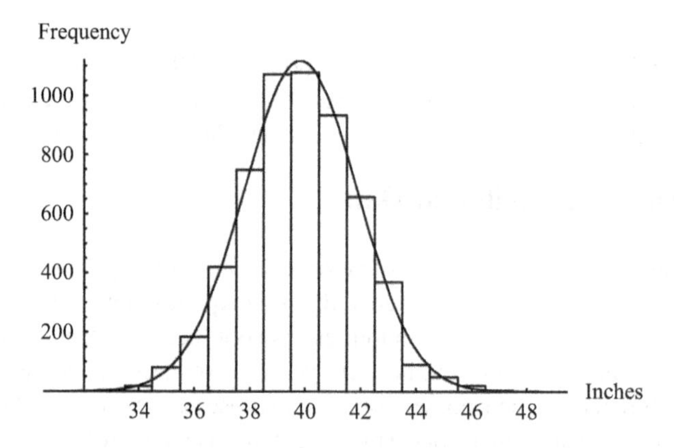

Figure 2.12: Chest measurements of Scottish soldiers.

normality of the data by means of such a drawing. Instead, he asserted that this table would be practically indistinguishable from a table produced by an inexperienced person taking 5,738 measurements of the chest of some statue. Given the subsequent widespread belief in the normality of all data, and the historical role that this example played in the creation of this tenet Quetelet's own words on this topic should be of interest:

> I now ask if it would be exaggerating, to make an even wager that a person little practiced in measuring the human body would make a mistake of an inch in measuring a chest of more than 40 inches in circumference? Well, admitting this probable error, 5,738 measurements made on one individual would certainly not group themselves with more regularity, as to the order of magnitude, than the 5,738 measurements made on the scotch soldiers; and if the two series were given to us without their being particularly designated, we should be much embarrassed to state which series was taken from 5,738 different soldiers, and which was obtained from one individual with less skill and ruder means of appreciation. [*Letters Addressed to H. R. H. the Grand Duke of Saxe Coburg and Gotha, on the Theory of Probabilities, As Applied to the Moral and Political Sciences,* Translated by O. G. Downes, C. & E. Leyton, London 1849, p. 92.]

In other words, Quetelet suggested that the deviations of the chest measurements should be viewed as errors committed by an imprecise natural process. And since measurement errors were assumed by astronomers (of which Quetelet was one) to be governed by the normal curve it would be reasonable to expect the chest measurements to also fall into the same pattern.

Quetelet's assertion that the chest measurements have a normal distribution should be taken with a healthy dose of skepticism, particularly so in view of the fact that he made several arithmetical mistakes in transcribing his data. The grounding of the belief in this supposedly universal "shape" of data is not unshakable and has been attacked by some specialists. Nevertheless, in the absence of compelling reasons for thinking otherwise, it is quite customary to assume the normality of most data. This assumption has very strong computational consequences that are very similar to those that proved so useful in the computation of binomial probabilities but have a much wider range of applications. These are the subject of this section's remainder.

It is common practice in statistics to use the term *population* to refer to any collection of data. If $\{x_1, x_2, ..., x_n\}$ is a population then F_T denotes the *cumulative frequency* (i.e., number) of the data that do not exceed the threshold T and

$$R_T = \frac{F_T}{n}$$

denotes their *cumulative relative frequency*. For example, for the chest measurements of the Scottish soldiers we have

$$F_{36} = 3 + 18 + 81 + 185 = 287$$

$$R_{36} = \frac{287}{5{,}738} \approx 0.05$$

and

$$F_{40} = 3 + 18 + 81 + 185 + 420 + 749 + 1{,}073 + 1{,}079 = 3{,}608$$

$$R40 = \frac{3{,}608}{5{,}738} \approx 0.63.$$

The assumption of the *normality of a population* is tantamount to the following statement.

NORMALITY ASSUMPTION 2.4.1. *If a population has average $\overline{x}$ and standard deviation s, then*

$$R_T \approx A\left(\frac{T-\overline{x}}{s}\right) \quad \text{and} \quad F_T \approx n \cdot A\left(\frac{T-\overline{x}}{s}\right)$$

where $A(t)$ denotes the entry of Table A of the appendix that corresponds to a t that is described in the left and top borders of the table.

As a first example we consider the ages of the last example of Section 2.1 for which the grouped mean and standard deviation were computed to be

$$\overline{x} = 41.5 \quad \text{and} \quad s = 12.4.$$

Accordingly, the number of students of age at most 49 is

$$F_{49} \approx 45A\left(\frac{49-41.5}{12.4}\right) = 45A(0.60) = 45.07257 \approx 33.$$

Note that the actual number of students in this age group is

$$2+5+12+17 = 36.$$

Similarly, the number of students of age at least 40 is

$$45 - 45 \cdot A\left(\frac{39-41.5}{12.4}\right) = 45 - 45 \cdot A(0-20) = 45 - 45 \cdot 0.4207 \approx 26.$$

This agrees with the actual number of

$$17+5+3+1 = 26.$$

It should also be of interest to see to what extent Quetelet's chest measurements fall into the normal pattern. For this table the mean and standard deviation are

$$\bar{x} = 39.8 \quad \text{and} \quad s = 2.0.$$

Consequently, in predicting the number of soldiers with chest measurements of 40 inches, the Normality Assumption 2.4.1 yields

$$f_{40} = F_{40} - F_{39} \approx 5{,}738 \cdot A\left(\frac{40-39.8}{2}\right) - 5738 \cdot A\left(\frac{39-39.8}{2}\right)$$
$$= 5738 \cdot A(0.10) - 5{,}738 \cdot A(-40) = 5{,}738 \cdot 0.5398 - 5{,}738 \cdot 0.3446 \approx 1{,}120$$

whereas the actual number is 1,079.

The following table augments the Scottish soldiers' chest measurements table with the theoretical frequencies, that were computed on the basis of the Normality Assumption 2.4.1.

Chest Measurements of Scottish Soldiers		
Girth	Actual frequency	Predicted frequency
33	3	4
34	18	18
35	81	67
36	185	193
37	420	434
38	749	762
39	1,073	1,047
40	1,079	1,127
41	934	950
42	658	626
43	370	323
44	92	131
45	50	41
46	21	10
47	4	2
48	1	0
	5,738	

There are obvious disagreements between the actual and the theoretical (normally based) frequencies. Nevertheless, the fit between the two is close enough to have convinced social scientists of the validity of Quetelet's assertion that these and all similar data are normally distributed. This validity is still widely accepted by default whenever there is no compelling reason for rejecting it. Two examples of non-normally distributed data are offered here as illustrations. Data obtained by rolling a die 1,000 times and recording the frequencies with which each face occurs will clearly not have a bell-shaped histogram and so is not normally distributed. Instead, the different frequencies will tend to remain equal to each other and their distribution is commonly described as uniform (see Figure 2.13).

Similarly, suppose each of the students in a large class keeps tossing a coin until a *heads* shows up, and records the number of tosses. One would then expect $\frac{1}{2}$ of the data to be 1, $\frac{1}{4}$ of the data to be 2, $\frac{1}{8}$ of the data to be 3, and so on. This is definitely a non-normal distribution. The histogram of this data would have the shape displayed in Figure 2.14.

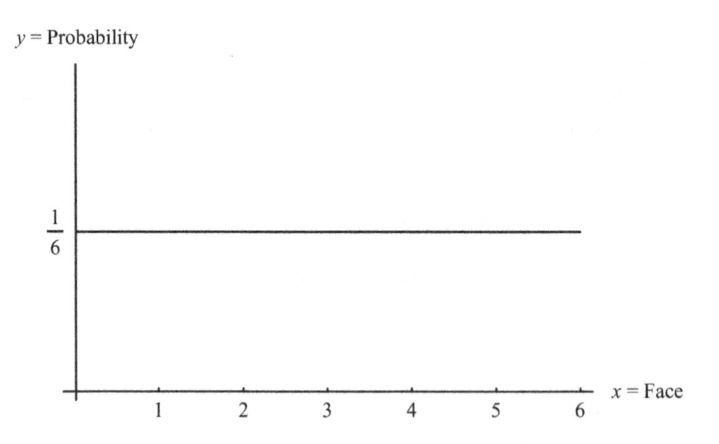

Figure 2.13: A non-normal distribution that describes data generated by a fair die.

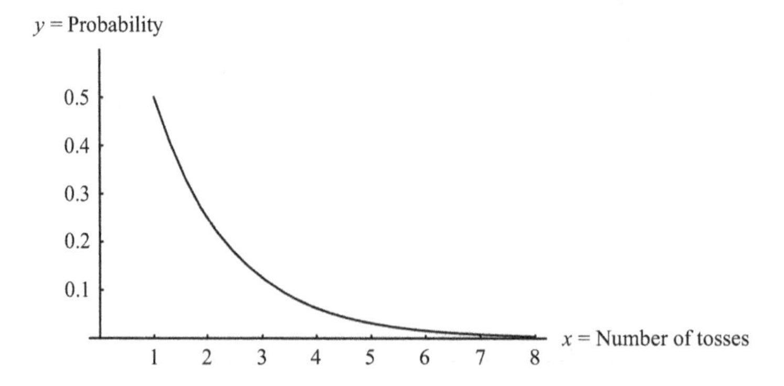

Figure 2.14: A non-normal distribution that describes the number of coin tosses required to obtain a *heads*.

Example 2.4.2. An airline reports that the mileage recorded by frequent flyers averages 22,987 with a standard deviation of 6,736. Assuming the normality of the data, estimate the proportion of flyers who have recorded the following mileages.

a) At most 20,000: $F_{20,000} \approx A\left(\frac{20,000-22,987}{6,736}\right) \approx A(-0.44) = 0.33 = 33\%.$

b) At least 30,000: $1-F_{30,000} \approx 1-A\left(\frac{29,999-22,987}{6,736}\right) \approx 1-A(1.04)$
$= 1-0.8508 = 0.1492 \approx 14.9\%.$

c) Between 15,000 and 25,000: $F_{25,000}-F_{15,000}$
$\approx A\left(\frac{25,000-22,987}{6,736}\right)-A\left(\frac{14,999-22,987}{6,736}\right) \approx A(0.30)-A(-1.19)$
$= 0.6179-0.1170 = 0.5009 \approx 50.1\%.$

Example 2.4.3. *Suppose next that the above airline wishes to award its top 5% customers a bonus advance of $100 towards their next ticket purchase and the next 10% an advance of $50. What are the cutoff mileages that qualify customers for these awards?*

Top 5%: We are looking for the threshold T below which 95% of the mileages lie (see Figure 2.15).

Hence

$$A\left(\frac{T-22,987}{6,736}\right) = 0.95.$$

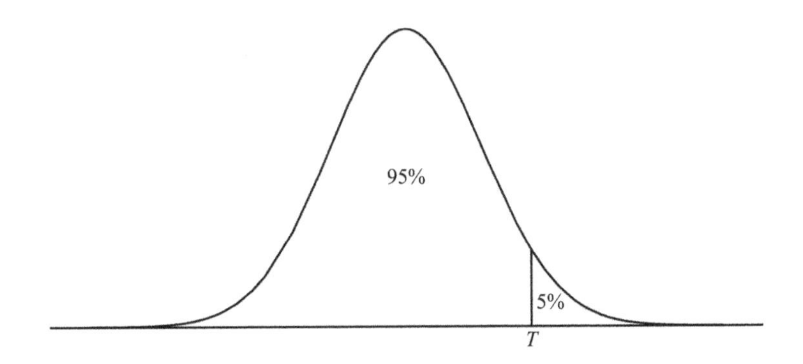

Figure 2.15:

The entries of Table A that are closest to 0.95 are 0.9495 and 0.9505 and the latter is selected arbitrarily. Hence

$$\frac{T-22{,}987}{6{,}736}=1.65.$$

By the rules of algebra we obtain the cutoff

$$T=6{,}736\cdot 1.65+22{,}987=34{,}101.4.$$

Thus, the company should award the \$100 bonus to any customer who has logged at least 34,101 miles.

Next 10%: By the previous calculations, the top cutoff of this group is 34,101 miles. The bottom cutoff is the threshold T below which 85% of the data lie (Figure 2.16).
Hence,

$$A\left(\frac{T-22{,}987}{6{,}736}\right)=0.85.$$

The entry of Table A closest to 0.85 is 0.8508 and so

$$\frac{T-22{,}987}{6{,}736}=1.04.$$

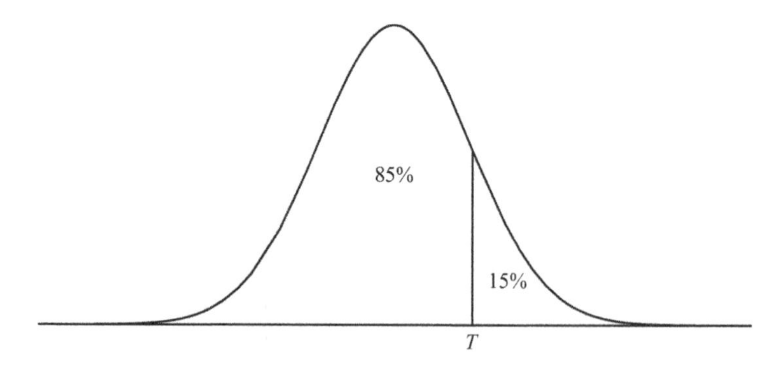

Figure 2.16:

By the rules of algebra we obtain the cutoff

$$T = 6{,}736 \cdot 1.04 + 22{,}987 = 29{,}992.44.$$

Thus, the $50 bonus should be awarded to all the customers with mileages between 29,993 and 34,101 inclusive.
For normal populations, it is a consequence of the Normality Assumption that the standard deviation provides a natural unit for describing the spread of the data. In particular, regardless of the values of the mean $\bar{x}$ and the standard deviation s,

34% of the data lie in each of the intervals $[\bar{x}, \bar{x} + s]$ *and* $[\bar{x} - s, \bar{x}]$,
14% of the data lie in each of the intervals $[\bar{x} + s, \bar{x} + 2s]$ *and* $[\bar{x} - 2s, \bar{x} - s]$,
2% of the data lie in each of the intervals $[\bar{x} + 2s, \bar{x} + 3s]$ *and* $[\bar{x} - 3s, \bar{x} - 2s]$

These percentages are illustrated in Figure 2.17.
For example, since the mean girth amongst the Scottish soldiers is 39.8″ and the standard deviation is 2″, one would expect 34% of 5,738 (i.e., 1,951) to have chests measuring either 40″ or 41″. The actual number is

$$f_{40} + f_{41} = 1{,}079 + 934 = 2{,}014.$$

Similarly, assume that the ages in the adult education class discussed in Section 2.1 are normally distributed. Since they are known to have mean $\bar{x} = 41.5$ and standard deviation $s = 12.4$, one would expect 14% of 45 = 6 of the students to have ages 54–66, which turns out to be exactly the right number.

A *random variable* is a phenomenon or a process that produces data. The daily high temperature in a certain city, the number of people in the next car to

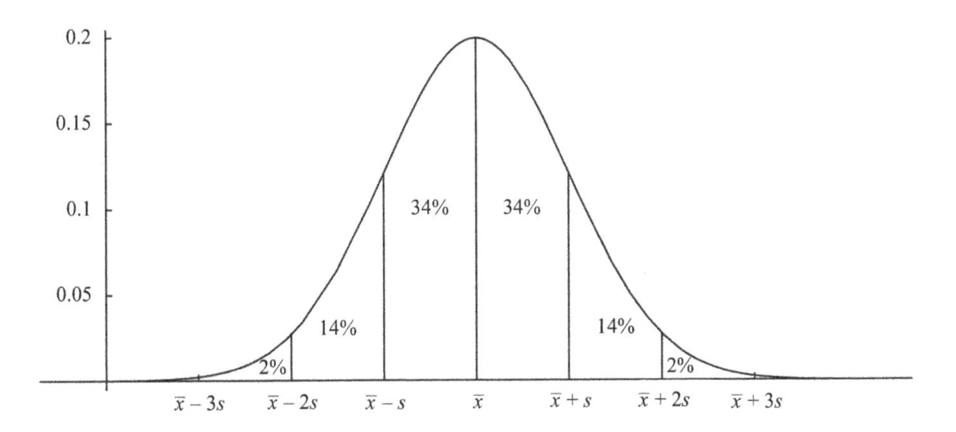

Figure 2.17:

arrive at a toll booth, the amount of coffee dispensed by a vending machine into a cup, the number of coin tosses required to accumulate 10 heads are all such random variables. Problems regarding the frequencies of certain types of data translate into questions regarding the probability of a random variable assuming corresponding values. Thus, the question of how many of the next 1,000 cars passing through a toll booth contain at most two riders can be rephrased as finding the probability that such a car contains two riders.

Random variables also have means and standard deviations which are idealizations of the averages and standard deviations of the populations that they produce. A random variable with mean 15 will produce populations with averages that are close to 15. A random variable with standard deviation 4 will produce populations with standard deviations that are close to 4. The Greek letters μ and σ are used to denote the mean and standard deviation of a random variable, respectively. The will be used much like the $\bar{x}$ and s of the data these variables generate.

A random variable is said to be normally distributed if the data it generates are normally distributed. If V denotes such a normally distributed random variable and T is any threshold, then the probability that V generates a number of value at most T is denoted by Prob[$V \leq T$] and is computed by the following analog of the Normality Assumption:

$$\text{Prob}[V \leq T] = A\left(\frac{T-\mu}{\sigma}\right). \qquad \text{(f)}$$

For example, if in a certain town the high temperature on July 4 is a normally distributed random variable with mean m = 94°F and standard deviation 5°F, then the probability that on the next July 4 the temperature will not exceed 100°F is

$$\text{Prob}[V \leq 100.5] = A\left(\frac{100-94}{5}\right) = A(1.2) = 0.8849 \approx 88.5\%.$$

Example 2.4.4. *A vending machine is set to dispense 7 ounces of coffee into its cups. Due to the inevitable imperfections in its machinery, the amount that is actually poured out is a normally distributed random variable with a standard deviation of 0.2 ounces. In order to prevent unsightly spills the machine is provided with cups that have volume 7.3 ounces.*

(a) *What percentage of the dispensations will result in a spill anyway?* This is the probability that the amount poured exceeds 7.3, which, by Formula (f) equals

$$1 - A\left(\frac{7.3-7}{0.2}\right) = 1 - A(1.5) = 1 - 0.9332 \approx 6.7\%.$$

(b) *If too large a portion of the cup remains unfilled the customer may feel shortchanged. Suppose the typical customer feels unhappy when half an ounce is left unfilled. What proportion of the customers will feel shortchanged?* Since $7.3-0.5=6.8$, the proportion of unhappy customers is

$$A\left(\frac{6.8-7}{0.2}\right)=A(-1)=0.1587\approx 15.9\%.$$

(c) *In order to antagonize fewer customers, the vending company will use smaller cups. What should be the size of the cup if the company is willing to allow for spills to occur in 10% of the sales?* This is a threshold determination whose diagram appears in Figure 2.18.
Hence

$$A\left(\frac{T-7}{0.2}\right)=1-0.1=0.9$$

and, using Table A in the reverse direction,

$$\frac{T-7}{0.2}=1.28$$

or

$$T=7+0.2\cdot 1.28=7.256.$$

Thus, a cup with capacity 7.256 ounces will reduce the number of unhappy customers while allowing for spills to occur no more than 10% of the time.

(d) *What proportions of the drinks dispensed will contain between 6.7 and 7.2 ounces?* This is

$$A\left(\frac{7.2-7}{0.2}\right)-A\left(\frac{6.7-7}{0.2}\right)=A(1)-A(-1.5)=0.8413-0.06668\approx 77.5\%.$$

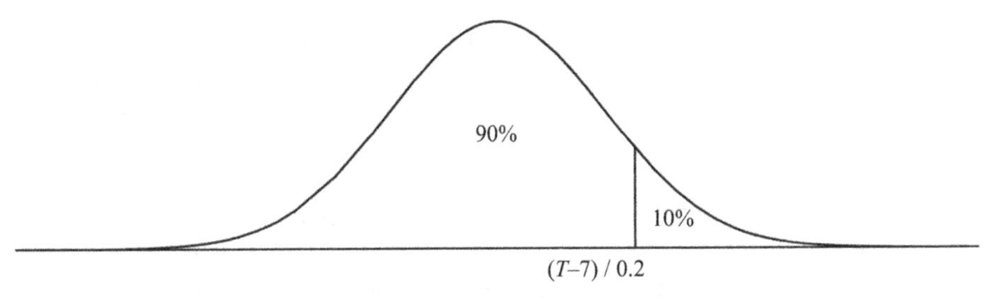

Figure 2.18:

Exercises 2.4

1. The yearly salaries of the employees of a certain large company are normally distributed with a mean of $28,500 and a standard deviation of $2,800. What percentage of the employees falls in the following salary ranges:
   a) between $20,000 and $30,000
   b) more than $30,000
   c) between $25,000 and $35,000
   d) less than $32,000
   e) a salary that deviates from the mean by more than $1,000
   f) a time that deviates from the mean by more than $4,000?

2. The yearly salaries of the employees of a certain large company are normally distributed with a mean of $28,500 and a standard deviation of $2,800. Find the cut-off salary for the following salary ranges:
   a) top 20%
   b) top 10%
   c) bottom 15%
   d) bottom 25%
   e) top 40%
   f) bottom 40%.

3. The scores of a certain national test are normally distributed with a mean of 72 and a standard deviation of 7. Find the cut-off scores for the following groups of scores:
   a) top 20%
   b) top 10%
   c) bottom 15%
   d) bottom 25%
   e) top 40%
   f) bottom 40%.

4. The scores of a certain national test are normally distributed with a mean of 72 and a standard deviation of 7. Find the percentage of scores that fall in the following ranges:
   a) between 70 and 80
   b) more than 85
   c) between 60 and 70
   d) less than 65
   e) between 65 and 85
   f) more than 72
   g) a score that deviates from the mean by more than 5
   h) a score that deviates from the mean by more than 10.

5. The heights of the 15,000 students of a certain university are normally distributed with a mean of 68 inches and a standard deviation of 4.7 inches. Determine the number of students that have the following heights:
   a) between 60 and 72 inches
   b) taller than 72 inches
   c) between 70 and 72 inches
   d) shorter than 72 inches
   e) shorter than 65 inches
   f) exactly 68 inches
   g) a height that deviates from the mean by more than 3 inches.

6. The heights of the 15,000 students of a certain university are normally distributed with a mean of 68 inches and a standard deviation of 4.7 inches. Determine the cutoffs for the following groups of heights:
   a) top 20% b) top 10% c) bottom 15%
   d) bottom 25% e) top 40% f) bottom 40%.

7. The mean weight of a box of cereal filled by a machine is 15.0 ounces, with a standard deviation of 0.6 ounces. Assuming a normal weight distribution, find the percentage of the boxes that have the following weights:
   a) less than 14 ounces b) between 14 and 16 ounces
   c) more than 16 ounces d) between 14.5 and 15.8 ounces
   e) that deviate from 15.0 by more than 1 ounce
   f) that deviate from 15.0 by more than 1.5 ounces.

8. The amount of time required to assemble a certain product is normally distributed with a mean of 3.4 minutes and a standard deviation of 0.5 minutes. Find the probability that a randomly selected employee will required the following amount of time to assemble the product:
   a) less than 2 minutes b) between 3 and 4 minutes
   c) between 2 and 3 minutes d) less than 4 minutes
   e) more than 4 minutes f) more than 5 minutes
   g) a time that deviates from 3.4 by more that 1 minute
   h) a time that deviates from 3.4 by more that 2 minutes.

9. The time it takes my favorite acrylic paint to dry is normally distributed with a mean of 2 hours and 45 minutes and a standard deviation of 30 minutes. Find the probability of the following drying times:
   a) between 2 and 3 hours b) less than 4 hours
   c) between 3 and 4 hours d) more than 3 hours
   e) between $1\frac{1}{2}$ and $2\frac{1}{2}$ hours f) more than 2 hours
   g) a time that deviates from the mean by more than 30 minutes
   h) a time that deviates from the mean by more than 40 minutes.

10. The amount of time between taking a pain reliever and getting relief is normally distributed with a mean of 23 minutes and a standard deviation of 5 minutes. Find the probability that the time between taking the medication and getting relief is as follows:
    a) between 20 and 25 minutes b) more than 20 minutes
    c) between 15 and 20 minutes d) more than 23 minutes
    e) between 18 and 28 minutes f) less than 20 minutes
    g) a time that deviates from the mean by more than 3 minutes
    h) a time that deviates from the mean by more than 7 minutes.

11. The mean weight of a box of cereal filled by a machine is 14.5 ounces, with a standard deviation of 0.5 ounces. Assuming a normal weight distribution, find the percentage of the boxes that will weigh the following amounts:
a) less than 14 ounces
b) between 14 and 16 ounces
c) more than 16 ounces
d) between 14.5 and 15.8 ounces
e) a weight that deviates from the mean by more than 0.5 ounces
f) a weight that deviates from the mean by more than 0.8 ounces.

12. The amount of time required to assemble a certain product is normally distributed with a mean of 3.6 minutes and a standard deviation of 0.6 minutes. Find the probability that a randomly selected employee will require the following amount of time to assemble the product:
a) less than 2 minutes
b) between 3 and 4 minutes
c) between 2 and 3 minutes
d) less than 4 minutes
e) more than 4 minutes
f) more than 5 minutes.

13. The amount of time between taking a pain reliever and getting relief is normally distributed with a mean of 24 minutes and a standard deviation of 4 minutes. Find the probability that the time between taking the medication and getting relief is as follows:
a) between 20 and 25 minutes
b) more than 20 minutes
c) between 15 and 20 minutes
d) more than 23 minutes
e) between 18 and 28 minutes
f) less than 20 minutes
g) a time that deviates from the mean by more than 3 minutes
h) a time that deviates from the mean by more than 4.5 minutes.

14. The amount of time required to assemble a certain product is normally distributed with a mean of 3.6 minutes and a standard deviation of 0.6 minutes. The company wishes to retrain its slower employees and reward the faster ones. Help the company find the cut-off assembly times that identify the following groups of employees:
a) fastest 20%
b) fastest 10%
c) slowest 15%
d) slowest 25%
e) fastest 40%
f) slowest 40%.

15. A population is normally distributed with mean 27.4 and standard deviation 2.4.
a) Find the intervals representing one, two, and three standard deviations from the mean.
b) What percentage of the data lies in each of the intervals in a)?

16. A population is normally distributed with mean 15.1 and standard deviation 3.1.

a) Find the intervals representing one, two, and three standard deviations from the mean.
b) What percentage of the data lies in each of the intervals in a)?

2.5 Statistical Inference

Two uses for the normal curve have been described thus far - the approximation of binomial probabilities and the abbreviation of normally distributed data. There are many other situations in which this curve supplies essential information and several of these, which are in very common use, will now be described.

Recall the town discussed in Examples 1.3.3–6, 30% of whose population were fans of its pro basketball team. In that context the question was asked of how likely a sample was to be misleading. Of course, the assumption that this proportion is known is unrealistic. Precise information of this sort is rarely available and, given the hypothetical nature of such numbers, the question of how misleading a sample can be is only of secondary interest. The reverse question:

> *If 30 out of 100 interviewees claimed to be fans, what inference can be made regarding the true proportion of fans in that town?*

is much more likely to be asked. Problems of this type were first proposed and worked out by James Bernoulli (1654–1705) in his posthumously published book *Ars Conjectandi* (1713), and it was he who suggested the format that the answer to such an a posteriori question should have, a format that is still very much in use today. Whereas it is impossible to determine the true proportion with absolute certainty on the basis of a sample, it is nevertheless possible to use the sample to trap this proportion within an interval so that one is *morally certain*, to use Bernoulli's phrase, that the proportion lies within the interval. Not only that, but Bernoulli showed how this moral certainty could be quantified in terms of a probability. Because he did not have the normal curve as a tool for estimating binomial probabilities, Bernoulli's mathematics turned out to be rather complex and impractical. Abraham De Moivre, who discovered the normal approximation to the binomial probabilities and incorporated it into the 1738 edition of his *Doctrine of Chances*, also showed how to express Bernoulli's intervals of moral certainty in terms of the normal curve. Today, the term *moral certainty* has been replaced by *confidence* and Bernoulli's and De Moivre's intervals are called *confidence intervals*.

Confidence Interval for Proportion. For any small positive number α, the $(1-\alpha)100\%$ *confidence interval* for the proportion p of a certain subpopulation, given by a sample of which the subpopulation accounts for proportion $\overline{p}$, has endpoints

$$\overline{p} \pm B\left(1-\frac{\alpha}{2}\right)\sqrt{\frac{\overline{p}(1-\overline{p})}{n}}$$

where $B(x)$ is obtained by reversing the reading of Table A. In other words,

$B(0.95) = 1.645$ because $A(1.645) = 0.95$
$B(0.975) = 1.96$ because $A(1.96) = 0.975$
$B(0.99) = 2.326$ because $A(2.326) = 0.99$
$B(0.995) = 2.575$ because $A(2.575) = 0.995$.

Example 2.5.1. *Suppose a sample of 500 inhabitants contains 160 fans of the local basketball team. Construct a 95% confidence interval for the true proportion of fans in the town.*

Since the specified confidence level is $(1-\alpha)100\% = 95\% = 0.95$, it follows that $\alpha = 0.05$ and so

$$B\left(1-\frac{\alpha}{2}\right) = B\left(1-\frac{0.05}{2}\right) = B(0.975) = 1.96.$$

In addition, $\overline{p} = 160/500 = 0.32$ and $1-\overline{p} = 1-0.32 = 0.68$. These numbers yield the endpoints

$$0.32 \pm 1.96 \cdot \sqrt{\frac{0.32 \cdot 0.68}{500}} \approx 0.32 \pm 0.04$$

of the confidence interval (0.28, 0.36). This means that we are 95% sure that the town's true proportion of fans lies between 28% and 36%. Depending on what stake one has in the issue under consideration, a confidence level of 95% may be insufficient. To obtain a 99% level of confidence it is necessary to set $\alpha = 1\% = 0.01$, so that

$$B\left(1-\frac{\alpha}{2}\right) = B\left(1-\frac{0.01}{2}\right) = B(0.995) = 2.575.$$

This yields the endpoints

$$0.32 \pm 2.575 \cdot \sqrt{\frac{0.32 \cdot 0.68}{500}} \approx 0.32 \pm 0.05$$

with a resulting confidence interval of (0.27, 0.37) which is unfortunately wider than the 95% interval of (0.28, 0.34) obtained above. Thus, the additional confidence comes at the cost of less accuracy. More accuracy can be obtained if one is willing to relinquish some of the certainty. For example, to obtain the 90% confidence interval we set $\alpha = 10\% = 0.1$, so that

$$B\left(1-\frac{\alpha}{2}\right)=B\left(1-\frac{0.1}{2}\right)=B(0.95)=1.645.$$

This yields the endpoints

$$0.32 \pm 1.645\sqrt{\frac{0.32\cdot 0.68}{500}} \approx 0.32 \pm 0.03$$

with a resulting confidence interval of (0.29, 0.35). The table below lists the above three intervals and clarifies the inverse relation between the confidence level and the interval's

Confidence level	Interval
99%	(0.27, 0.37)
95%	(0.28, 0.36)
90%	(0.29, 0.35)

width. The decision on which confidence level should be used must remain subjective. The 95% confidence level is the default level and also the one most commonly used by experimenters.

All of the above intervals may seem unsatisfactory since there is a considerable gap between the two endpoints. The difference between 29% and 35% is substantial, as is that between 27% and 37%. The term

$$B\left(1-\frac{\alpha}{2}\right)\sqrt{\frac{\overline{p}(1-\overline{p})}{n}} \tag{e}$$

is of course responsible for the width of the confidence interval. The variables α and n can be controlled by the pollster and their effect on the width will be discussed momentarily. The term $\overline{p}$, on the other hand, is unpredictable and so it is eliminated by reverting to the worst-case scenario, which happens to occur when $\overline{p} = 0.5$. In other words, since

$$B\left(1-\frac{\alpha}{2}\right)\sqrt{\frac{\overline{p}(1-\overline{p})}{n}} \leq B\left(1-\frac{\alpha}{2}\right)\sqrt{\frac{0.5(1-0.5)}{n}} \leq \frac{B\left(1-\frac{\alpha}{2}\right)}{2\sqrt{n}}$$

we define

$$MOE = \frac{B\left(1 - \frac{\alpha}{2}\right)}{2\sqrt{n}}$$

as the *margin of error* of the sample, knowing that if we succeed in keeping it small then the actual but unpredictable margin of (e) will also be small. In the previous example, in the case where $\alpha = 5\%$, the margin of error is

$$\frac{B(0.975)}{2\sqrt{500}} = \frac{1.96}{2\sqrt{500}} \approx 0.044$$

which is quite close to the actual error bound of 0.04.

In general, the margin of error should be thought of as the measure of accuracy of a statistically based estimation of some true proportion. The smaller the margin of error, the more accurate the test. It should be of interest to see how variations in the level of confidence affect the margin of error. The table below indicates the comparative widths of the confidence intervals associated with various standard confidence levels but identical sample sizes.

Confidence level:	99%	98%	95%	90%
α:	1%	2%	5%	10%
Width of C.I.	$\frac{2.58}{2\sqrt{n}}$	$\frac{2.33}{2\sqrt{n}}$	$\frac{1.96}{2\sqrt{n}}$	$\frac{1.65}{2\sqrt{n}}$
% Comparison of width	132%	119%	100%	84%

Thus, to increase the confidence level of the above test from 95% to 99% (without changing the sample) it would be necessary to increase the interval width by 32%. Unfortunately there are no guidelines for answering the trade-off question of whether the increase in confidence is justified by the accompanying decrease in the accuracy of the estimation. These decisions too must remain subjective.

It is also possible to narrow the width of the confidence interval by increasing the sample size. Because of the square root in the denominator of the MOE, doubling the accuracy (or halving the width) of the estimating interval calls for quadrupling the sample size, which can be quite expensive. In the $\alpha = 5\%$ case of the discussion above, where a sample of 500 resulted in an interval width of 8%, it would require sample sizes of 1000, 2000, and 4000 to obtain interval widths of 4%, 2%, and 1%, respectively.

It should be useful to have a method for computing the sample size that yields a prespecified accuracy. This is given by the formula

$$n = \left(\frac{B(1-\alpha/2)}{w}\right)^2$$

where w is the desired width of the confidence interval. Accordingly, if a confidence level of 98% and a width of 4% are desired, then the required sample size is

$$\left(\frac{B(0.99)}{0.04}\right)^2 = \left(\frac{2.326}{0.04}\right)^2 \approx 3382.$$

Once again, there are no general guidelines for resolving the tradeoff between the confidence level and the cost of the statistical test.

Confidence Interval for Mean. Confidence intervals can also be used to estimate unknown averages of large population. This is the sort of problem that arises in, say, estimating the average annual income of a large group of people on the basis of a sample. Here, too, no sample can be guaranteed to yield the true mean of a large population and it is necessary to make do with confidence intervals. Suppose a sample of n (in practice, n must be at least 30) data from a large population with unknown average m has average $\bar{x}$ and standard deviation s. Then the numbers

$$\bar{x} \pm B\left(1-\frac{\alpha}{2}\right)\frac{s}{\sqrt{n}}$$

are the endpoints of a $(1-\alpha)$ 100% confidence interval for the population average m. For example, if a random sample of 300 families in a certain town report an average income of \$34,355 with a standard deviation of \$9,811, then the numbers

$$34{,}511 \pm B\left(1-\frac{0.05}{2}\right)\cdot\frac{9{,}811}{\sqrt{300}} = 34{,}355 \pm 1.96\cdot\frac{9{,}811}{\sqrt{300}} \approx 4{,}355 \pm 1{,}110$$

constitute a 95% confidence interval for the average family income over the entire population. In other words, we are 95% confident that the average family income for the town lies between \$33,245 and \$35,465. The author, for one, considers this to be a much more satisfactory answer than the 95% confidence interval of (28%, 36%) for the proportion of a team's fans in a different town. The difference between 28% and 36% seems much larger than the difference between \$33,245 and \$35,465. Perhaps this is because 35,465 is only 7% greater than 33,245 whereas 36 is 29% greater than 28.

Hypothesis Testing For Mean. Another use of the normal distribution is that of hypothesis testing. Consider, for example, the case of a manufacturer's can of peas whose net weight is 16 ounces. Such a company is likely to check on the net content of its product. In doing so, attention must be paid to two different potential problems. The cans could in reality contain less than 16 ounces in which case the manufacturer could face charges of false advertising. On the other hand, the cans could contain more than 16 ounces in which case the company is placing itself at a disadvantage relative to its competitors. Thus, the company has an interest in making sure that the net contents of its can are as close to the advertised 16 ounces as possible. Due to the intrinsic randomness inherent in our universe in general, and in all manufacturing processes in particular, no two cans can have exactly the same weight and the tester must be satisfied with verifying that the average contents are as stated rather than the contents of each can. Thus, the tester will use samples to test the so-called *(null) hypothesis*:

H_0: *The mean weight of a can is 16 ounces.*

The idea that underlies hypothesis testing is quite straightforward. A sample of the product in question is examined. If there is a considerable difference between the average weight of the sampled cans and 16 ounces, then the sample is considered as strong evidence that the true average weight is not 16 ounces. Of course, it is necessary to quantify the notion of *considerable difference*, and this is where the normal distribution comes in. Given a population with standard deviation s and any number d, the quantity

$$2\left(1-A\left(\frac{d}{s/\sqrt{n}}\right)\right)$$

turns out to be the probability that an n-sample will have a mean that differs from the population's *mean* of m by at least as much as d. Consequently, we define

$$p\text{-value} \quad = 2\left(1-A\left(\frac{\bar{x}-m}{s/\sqrt{n}}\right)\right).$$

Suppose that in the above context 100 cans of peas are examined and their net weights average 15.87 ounces with a standard deviation of 0.42 ounces. Then,

$$p\text{-value} = 2\left(1-\mathrm{A}\left(\frac{15.87-16}{0.42/10}\right)\right) = 2(1-\mathrm{A}(-3.10)) = 2(1-0.999)$$
$$= 0.002 = 0.2\%.$$

To understand the significance of the p-value note that the low sample mean of 15.87 ounces could be the result of two causes:

(a) The average weight of the boxes is 16 ounces and the sample is a fluke;
(b) The average weight of the boxes really is less than 16 ounces.

The p-value is essentially the probability of the first of these two alternatives. As such it measures the test's *significance*. The smaller the p-value of a statistical test, the less likely it is to lead to an erroneous conclusion and hence the more significant, or meaningful, is the result of the test. The above obtained 0.2% means that if we conclude on the basis of this sample that the average weight really is less than 16 ounces, then there is only a chance of 0.2% of making a mistake. This fraction of a percent is so small that most people would be willing to accept that kind of a risk. In practice, most experimenters will accept a p-value (i.e., risk of drawing the wrong conclusion) that is less than 0.05 = 5% as a sufficiently strong grounds for rejecting a null hypothesis.

Example 2.5.2. *Whole milk is supposed to contain 3% fat. In order to test whether this is true for the milk sold by the How Now Brown Cow dairy, 30 of their containers were opened and it was found that their fat content average was 3.1% with a standard deviation of 0.3%. Find the p-value of this test and decide whether this test indicates the fat content is not 3%.*

The p-value of this test is

$$2\left(1 - A\left(\frac{3.1-3}{0.3/\sqrt{30}}\right)\right) = 2\left(1 - A(1.83)\right) = 2(1-0.9664) \approx 6.7\%$$

which exceeds 5%. Hence the sample cannot be understood as contradicting the null hypothesis that the fat content of the whole milk sold by the How Now Brown Cow dairy is 3%.

Example 2.5.3. *Suppose the sample of Example 2.5.2 had an average fat content of 3.12% instead of 3.1%. How would that affect the conclusion to be drawn from the test?*

The p-value of the test would be

$$2\left(1 - A\left(\frac{3.1-3}{0.3/\sqrt{30}}\right)\right) = 2\left(1 - A(2.19)\right) = 2(1-0.9857) \approx 2.9\%$$

which is less than 5%. Under these circumstances it would be proper to conclude that the average fat content of the dairy's whole milk is not 3%. This makes sense as, all other things being equal, the greater the difference between the hypothesized average and the sample average, the more likely the hypothesized average is to be wrong.

Example 2.5.4. *Suppose the sample of Example 2.5.3 had a standard deviation of 0.5% instead of 0.3%. How would that affect the conclusion to be drawn from the test?*

The p-value of the test would be

$$2\left(1-A\left(\frac{3.1-3}{0.5/\sqrt{30}}\right)\right)=2(1-A(1.10))=2(1-0.8642)\approx 27.1\%$$

which is more than 5%. Consequently, the test does not cast doubt on the claim that the fat content is indeed 3%. Notice that in the transition from the previous example to this one, the p-value, which is the likelihood of this test to be a fluke, has increased. Since the evidence of the original example was deemed too weak to reject the 3% hypothesis, so must be every test with an even greater p-value, all else being equal.

Example 2.5.5. *Suppose Example 2.5.2 had a sample of size 40 instead of 30. How would that have affected the conclusion to be drawn from the test?*

The p-value would have been

$$2\left(1-A\left(\frac{3.1-3}{0.3/\sqrt{40}}\right)\right) = 2\left(1-A(2.11)\right) = 2(1-0.9826) \approx 3.5\%$$

which is less than 5%. Under these circumstances it would have been proper to conclude that the average fat content of the dairy's whole milk is not 3%. This illustrates the beneficial effect of increasing the sample size. The same average and standard deviation become more discriminating when they come from a larger sample.

Hypothesis Testing For Proportion. The process of hypothesis testing can be applied to the proportion of subpopulations as well. For example, it might be desired to *test* the widely held belief that the proportion of fans in the aforementioned town (see Section 1.3) is indeed 0.3 = 30%. To do this, one might interview a random sample of inhabitants and compare the proportion of fans in the sample to the putative proportion of 0.3. Once again there is a p-value to help gauge strength of the evidence. Specifically, to test the null hypothesis

H_0: The proportion of a certain subpopulation is p

one takes a random sample (of size n) and computes the proportion $\overline{p}$ of the subpopulation present in it. In this context the likelihood of an error if the difference between p and $\overline{p}$ is used to discredit the null hypothesis is given by

$$p\text{-value} = 2\left(1 - A\left(\frac{\overline{p} - p}{\sqrt{\overline{p}(1-\overline{p})/n}}\right)\right).$$

Example 2.5.6. *Two hundred inhabitants of the aforementioned town are randomly selected and interviewed. Seventy of them claim to be fans. Is this strong enough evidence to conclude that the widely held belief that 30% of the townspeople are fans is wrong?*

Here $\overline{p} = 70/200 = 0.35$ so that $1 - \overline{p} = 0.65$. Consequently

$$\begin{aligned} p\text{-value} &= 2\left(1 - A\left(\frac{0.35 - 0.30}{\sqrt{0.35 \cdot 0.65/200}}\right)\right) = 2\big(1 - A(1.48)\big) \\ &= 2(1 - 0.9306) \approx 13.8\%. \end{aligned}$$

Since 13.8% exceeds the 5% cutoff level mentioned above, most statisticians would consider it a rash act to reject the 30% hypothesis on the basis of this evidence alone.

Example 2.5.7. *Suppose the sample of the previous example had a size of 400 rather than 200. How would that affect the conclusion to be drawn?*

In this case the p-value is

$$\begin{aligned} 2\left(1 - A\left(\frac{0.35 - 0.30}{\sqrt{0.35 \cdot 0.65/400}}\right)\right) &= 2\big(1 - A(2.10)\big) = 2(1 - 0.9821) \\ &= 0.0358 \approx 3.6\%. \end{aligned}$$

Since 3.6% is less than 5%, most statisticians would accept the result of this poll as indicating that the true proportion of fans is no longer 30%.

Exercises 2.5

1. Of a random sample of 120 shoppers in a certain town, 86 used coupons. Construct a $(1-\alpha)100\%$ confidence interval for the true proportion of coupon users where α is (a) 1% (b) 5% (c) 10%. At each of these levels of confidence determine the margin of error as well as the smallest sample size that guarantees an interval of width 0.04.

2. An examination of 250 randomly selected police files found that 156 involved firearms. Construct a α confidence interval for the true proportion of crimes that involve firearms where α is (a) 1% (b) 5% (c) 10%. At each of these levels of confidence determine the margin of error as well as the smallest sample size that guarantees an interval of width 0.03.
3. A certain medical screening test was found to misdiagnose 37 out of 300 randomly selected test cases. Construct a $(1-\alpha)100\%$ confidence interval for the true proportion of test failures where α is (a) 1% (b) 5% (c) 10%. At each of these levels of confidence determine the margin of error as well as the smallest sample size that guarantees an interval of width 0.02.
4. A new drug is tested on 400 randomly selected patients and is found to be effective in 343 of the cases. Construct a $(1-\alpha)100\%$ confidence interval for the true effectiveness of the drug where α is (a) 1% (b) 5% (c) 10%. At each of these levels of confidence determine the margin of error as well as the smallest sample size that guarantees an interval of width 0.05.
5. A survey of 150 randomly selected homeowners in a certain town finds that 10 of them do not have home insurance. Construct a $(1-\alpha)100\%$ confidence interval for the true proportion of uninsured homeowners in that town where α is (a) 1% (b) 5% (c) 10%. At each of these levels of confidence determine the margin of error as well as the smallest sample size that guarantees an interval of width 0.02.
6. Twenty-five boxes of a certain kind of cereal are opened and it is determined that their average net weight is 14.8 ounces with a standard deviation of 0.15 oz. Assuming that these weights are normally distributed, form a $(1-\alpha)100\%$ confidence interval for the true mean weight where α is (a) 1% (b) 5% (c) 10%.
7. A study of 45 randomly selected workers finds that the assembly of a certain product requires, on the average, 3.6 minutes, with a standard deviation of 1.5 minutes. Construct a $(1-\alpha)100\%$ confidence interval for the true proportion of coupon users where α is (a) 1% (b) 5% (c) 10%.
8. Seventy-five cereal boxes of a certain brand are found to average a net weight of 15.7 ounces with a standard deviation of 0.08 oz. Construct a $(1-\alpha)100\%$ confidence interval for the true average net weight where α is (a) 1% (b) 5% (c) 10%.
9. A survey of 250 randomly selected moviegoers has found that, on the average, each spent \$242 on movies during the past year, with a standard deviation of \$21. Construct a $(1-\alpha)100\%$ confidence interval for the true mean annual expenditure of the typical moviegoer where α is (a) 1% (b) 5% (c) 10%.
10. The average salary of random sample of 200 employees of a certain large company is \$29,342 with a standard deviation of \$2,531.

Construct a $(1-\alpha)100\%$ confidence interval for the true average net weight where α is (a) 1% (b) 5% (c) 10%.

11. In order to test the belief that 60% of shoppers in a certain town use coupons, 120 randomly selected shoppers were observed and 86 of those were found to use coupons. What is the p-value of this test? What conclusion should be drawn at the level of significance of (a) $\alpha = 0.5\%$ (b) $\alpha = 1\%$ (c) $\alpha = 10\%$?
12. In previous years the percentage of crimes that involve firearms has held steady at 50%. An examination of 250 recent police files found that 156 of them involved firearms. Does this indicate that the percentage in question changed? Draw a separate conclusion at each of the significance levels of (a) $\alpha = 0.5\%$ (b) $\alpha = 1\%$ (c) $\alpha = 10\%$.
13. A certain medical screening test is known to be 90% effective. A new procedure is found to be correct on 190 of 200 people. Does this indicate that the new procedure has a different effectiveness rate? Draw a separate conclusion at each of the significance levels of (a) $\alpha = 0.5\%$ (b) $\alpha = 1\%$ (c) $\alpha = 10\%$.
14. A survey found that of 185 voters who came to a certain booth, 99 were women. Does this indicate that the proportion of females amongst voters is different from 50%? Draw a separate conclusion at each of the significance levels of (a) $\alpha = 0.5\%$ (b) $\alpha = 1\%$ (c) $\alpha = 10\%$.
15. For years the proportion of coffee drinkers amongst American adults has been 75%. A study finds that of 300 randomly selected adult Americans 212 drink coffee. Has a change occurred? Draw a separate conclusion at each of the significance levels of (a) $\alpha = 0.5\%$ (b) $\alpha = 1\%$ (c) $\alpha = 10\%$.
16. The assembly of a certain product requires, on the average, 3.6 minutes. A new method is tried out on 50 employees and is found to require, on the average, 3.1 minutes with a standard deviation of 1.5 minutes. Is there enough evidence to indicate that the new method is substantially different? Draw a separate conclusion at each of the significance levels of (a) $\alpha = 0.5\%$ (b) $\alpha = 1\%$ (c) $\alpha = 10\%$.
17. I am a painter and the amount of time it takes my favorite acrylic paint to dry on a fair day is, on the average, two and a half hours. I test a random sample of thirty cans of a different kind of paint and find that the average drying time is two hours and ten minutes with a standard deviation of 45 minutes. Does the evidence indicate that the drying times are different? Draw a separate conclusion at each of the significance levels of (a) $\alpha = 0.5\%$ (b) $\alpha = 1\%$ (c) $\alpha = 10\%$.
18. A company claims that its cereal boxes contain, on the average, 16 ounces of cereal. Twenty-five boxes are examined by that company's test product division and found to contain, on the average, 15.9 ounces with a standard deviation of 0.2 ounces. Does the evidence contradict the claim? Draw a separate conclusion at each of the significance levels of (a) $\alpha = 0.5\%$ (b) $\alpha = 1\%$ (c) $\alpha = 10\%$.

19. In order to determine whether the average expenditure per person on movies has changed, a random sample of 30 people was asked to compare their expenditures in 1997 versus those of 1996. It was found that the average expenditure increased by $20 with a standard deviation of $55 in the reported differences. Does the sample indicate a change in people's spending habits? Draw a separate conclusion at each of the significance levels of (a) $\alpha = 0.5\%$ (b) $\alpha = 1\%$ (c) $\alpha = 10\%$.
20. A manufacturer of small electric motors asserts that on the average they will draw 1.8 amperes under normal load conditions. A sample of 25 of the motors was tested and it was found that the mean current was 1.96 amperes with a standard deviation of 0.32 amperes. Does the evidence bear out the manufacturer's claim? Draw a separate conclusion at each of the significance levels of (a) $\alpha = 0.5\%$ (b) $\alpha = 1\%$ (c) $\alpha = 10\%$.

2.6 The Central Limit Theorem

One of the reasons that the normal curve is such a popular tool in statistical inference is that even when a population is not normally distributed, its sample averages do tend to be so distributed. This fact, which is known as the Central Limit Theorem and was proved in 1810 by P.-S. Laplace (1749 – 1827), implies the near universality of the testing methods described in the previous section. It is now illustrated with several examples.

The data $\{1, 2, 3, 4, 5, 6\}$ are not normally distributed - there is neither a central aggregation not a tapering off of the data towards the ends. Visually, the histogram of this data is flat (Figure 2.19). Statistical inference, however, is concerned with samples and their averages. The simplest possible non-trivial sampling that can be done calls for selecting 2 data at random. There are fifteen such potential samples and they, and their sample averages, are listed in the table below.

2-Samples and their averages

2-Sample:	$\{1, 2\}$	$\{1, 3\}$	$\{1, 4\}$	$\{1, 5\}$	$\{1, 6\}$	$\{2, 3\}$	$\{2, 4\}$	$\{2, 5\}$	$\{2, 6\}$
Sample average:	1.5	2	2.5	3	3.5	2.5	3	3.5	4

2-Sample:	$\{3, 4\}$	$\{3, 5\}$	$\{3, 6\}$	$\{4, 5\}$	$\{4, 6\}$	$\{5, 6\}$
Sample average:	3.5	4	4.5	4.5	5	5.5

These sample averages are regarded as data themselves and as such they have a histogram that appears in Figure 2.20. Note that whereas this graph is still not bell-shaped, it is closer to being so than the completely flat histogram of Figure 2.19. This is one of the implications of the Central Limit Theorem. Regardless of the shape (of the histogram) of the original data, histograms of sample averages tend

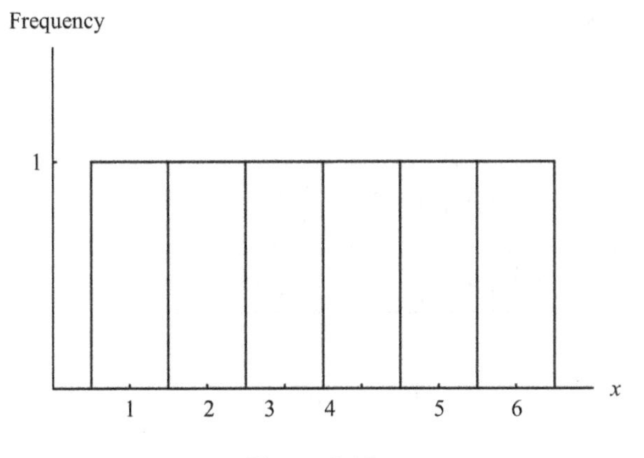

Figure 2.19:

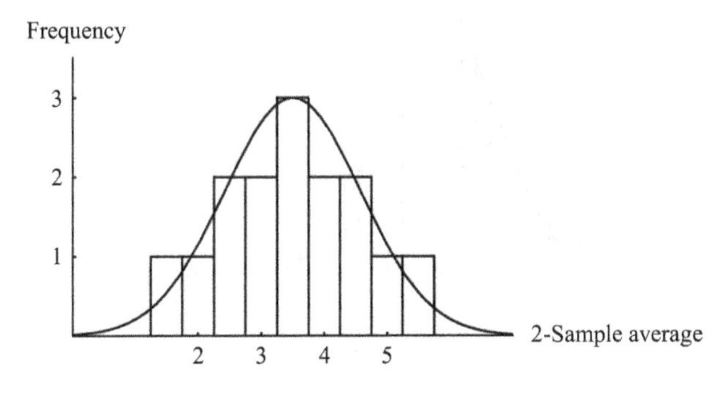

Figure 2.20: n = 6

to assume the bell shape. This assertion is also demonstrated by Figure 2.21 which tabulates the 2-sample averages of the data {1, 2, 3, 4, 5, 6, 7, 8}. The fit in both of these examples is somewhat forced. The histograms of the 2-sample averages are actually triangular, a shape that is still closer to the ideal bell shape than to the rectangle we started out with in Figure 2.20. However, the characteristic tapering off of the bell's rim becomes evident when averages of 3-samples are examined. Figure 2.22 displays the histogram of the 3-sample averages of the data {1, 2, 3, 4, 5, 6, 7, 8, 9} (see the abbreviated table below).

3-Samples and their averages

3-Sample:	{1, 2, 3}	{1, 2, 4}	...	{7, 8, 9}
Sample average:	2	$\frac{7}{3}$	...	8

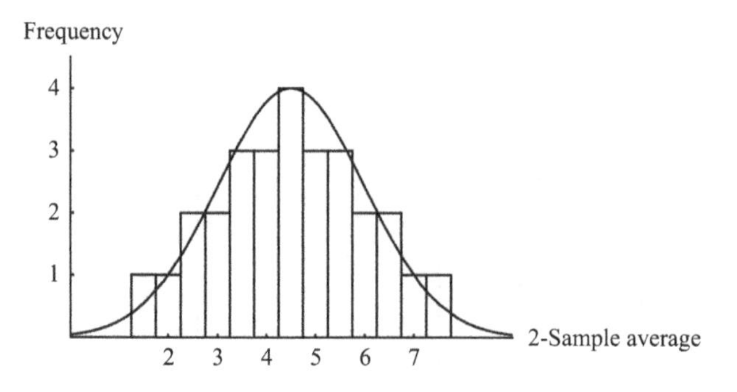

Figure 2.21: n = 8

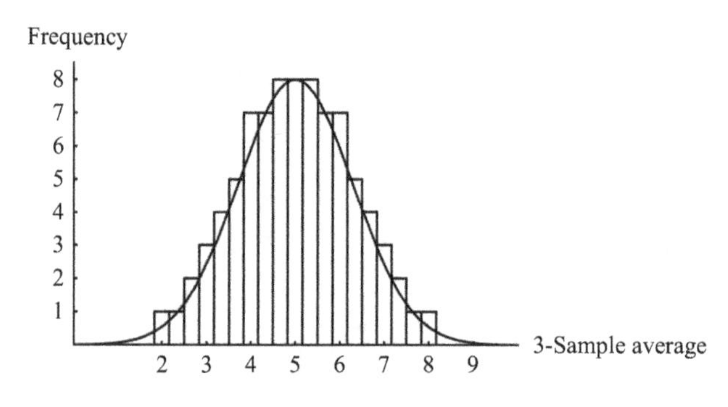

Figure 2.22: n = 9

This trend toward normality is also evident in Figure 2.23 which displays the histogram of the 4-sample averages of the data {1, 2, 3, 4, 5, 6, 7, 8, 9, 10}.

It is impossible to give an accurate statement of the Central Limit Theorem in this context as it is too technical. Informally, this theorem, which is the foundation of much of statistical practice, says that

For any population with average m and standard deviation s, and for any integer $n \geq 30$, the averages of the n-samples are (approximately) normally distributed with average m and standard deviation $s/\sqrt{n}$.

Exercises 2.6

For each of the populations in Exercises 1-13,

a) Construct the frequency table of the population and draw its histogram;
b) Compute the average and the standard deviation of the population;
c) Construct a frequency table for the 2-sample averages and draw its histogram;
d) Compute the average and standard deviation of the 2-sample averages;

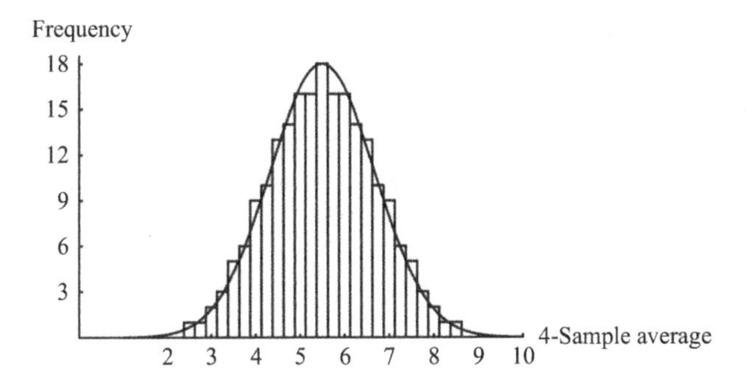

Figure 2.23: n = 10

e) Construct a frequency table for the 3-sample averages and draw its histogram;
f) Compute the average and standard deviation of the 3-sample averages.

1. $\{1, 2, 3, 4\}$
2. $\{2, 3, 4, 5\}$
3. $\{1, 2, 5, 6\}$
4. $\{1, 2, 2, 3\}$
5. $\{1, 2, 3, 3\}$
6. $\{1, 2, 2, 2\}$
7. $\{2, 2, 2, 2\}$
8. $\{1, 2, 3, 4, 5\}$
9. $\{1, 2, 2, 3, 4\}$
10. $\{1, 2, 2, 3, 3\}$
11. $\{1, 2, 2, 2, 3\}$
12. $\{1, 2, 2, 2, 2\}$
13. $\{3, 3, 3, 3, 3\}$

2.7 Activities/Websites

1. Collect and tabulate the heights (in inches) of the students in your class. Compute the mean and the standard deviation. Use Figure 2.12 to check whether these data are normally distributed.
2. Collect and tabulate the weights (in pounds) of the students in your class. Compute the mean and the standard deviation. Use Figure 2.12 to check whether these data are normally distributed.
3. Let each student in class flip a coin and keep on flipping, if necessary, until the first *heads* is obtained. Record the number of flips done by each student and have the students repeat the experiment until at least 30 data are obtained. Compute the average and the standard deviation of the data. (Theoretically, they should be 2 and $\sqrt{2}$, respectively.) Repeat the experiment with 3 times as many data.
4. Let each student in class toss a die and keep on tossing, if necessary, until the first 6 is obtained. Record the number of tosses done by each student and have the students repeat the experiment until at least 30 data are obtained. Compute the average and the standard deviation of the data.

(Theoretically, they should be 6 and $\sqrt{30}$, respectively.) Repeat the experiment with 3 times as many data.

5. It would seem that if a coin is spun then the probability of its coming up *heads* is 50%. On the other hand, it could be argued that since the faces are different, perhaps the 50% is incorrect. Test the 50% hypothesis at $\alpha = 5\%$ by spinning the same coin 100 times.

Chapter 3

Voting Systems

3.1 Introduction

We rank everything. We rank movies, bands, songs, racing cars, politicians, professors, sports teams, the plays of the day. We want to know "Who's number one?" and who isn't. As a local sports writer put it, "Rankings don't mean anything. Coaches continually stress that fact. They're right, of course, but nobody listens. You know why, don't you? People love polls. They absolutely love 'em" (*Woodling 2004*). Some rankings are just for fun, but people at the top sometimes stand to make a lot of money. A study of films released in the late 1970s and 1980s found that, if a film is one of the five finalists for the Best Picture Oscar at the Academy Awards, the publicity generated (on average) about \$5.5 million in additional box office revenue. Winners make, on average, \$14.7 million in additional revenue (*Nelson 2001*). In today's inflated dollars, the figures would no doubt be higher. Actors and directors who are nominated (and who win) should expect to reap rewards as well, since the producers of new films are eager to hire Oscar winners.

This chapter is about the procedures that are used to decide who wins and who loses. Developing a ranking can be a tricky business. Political scientists have, for centuries, wrestled with the problem of collecting votes and molding an overall ranking. To the surprise of undergraduate students in both mathematics and political science, mathematical concepts are at the forefront in the political analysis of voting procedures. Mathematical tools are important in two ways. First, mathematical principles are put to use in the scoring process that ranks the alternatives. Second, mathematical principles are used to describe the desired properties of a voting procedure and to measure the strengths and weaknesses of the procedures. Mathematical concepts allow us to translate important, but vague, ideas like "logical" and "fair" into sharp, formally defined expressions that can be applied to voting procedures.

In this chapter, we consider a number of example voting procedures and we discuss their strengths and weaknesses.

3.2 Mathematical Concepts

Transitivity. Real numbers are transitive. Every school child knows that transitivity means that

$$\textit{If } x > y \textit{ and } y > z, \textit{ then } x > z. \tag{1}$$

The symbol $>$ means "greater than" and $\geq$ means "greater than or equal to." Both of these are transitive binary relations. "Binary" means only two numbers are compared against each other, and transitivity means that many number can be linked together by the binary chain. For example, we can work our way through the alphabet. If $a > b$, and $b > c$, and ... $x > y$, then $a > z$. If a, b, and c represent policy proposals or football teams being evaluated, we expect that voters are able to make binary comparisons, declaring that one alternative is preferred to another, or that they are equally appealing. The parallel between "greater than" and "preferred to" is so strong, in fact, that we use the symbol $\succ$ to represent "preferred to." The subscript $\succ_i$ indicates that we are talking about a particular voter, so $x \succ_i y$ means that voter i prefers x to y and $x \succeq_i y$ means that x is as good as y. If i is indifferent, we write $x \approx_i y$. Using the binary relations $\succ_i$ and $\succeq_i$, it is now possible to write down the idea of transitive preferences. If a person's preferences are transitive, the following holds:

$$\textit{If } x \succ_i \mathrm{y} \textit{ and } y \succ_i \mathrm{z}, \textit{ then } x \succ_i \mathrm{z}. \tag{2}$$

This can be written more concisely as

$$\mathrm{x} \succ_i \mathrm{y} \succ_i \mathrm{z}. \tag{3}$$

According to William Riker, a pioneer in the modern mathematical theory of voting, transitivity is a basic aspect of reasonable human behavior (*Riker 1982*). Someone who prefers green beans to carrots, and also prefers carrots to squash, is expected to prefer green beans to squash. One of the truly surprising–even paradoxical–problems that we explore in this chapter is that social preferences, as expressed through voting procedures, may not be transitive. The procedure of majority rule, represented by the letter M, can compare alternatives so that $x \succ_M \mathrm{y}$, $y \succ_M \mathrm{z}$, and yet x does not defeat z in a majority election. The fact that individuals may be transitive, but social decision procedures are not, has been a driving force in voting research.

Ordinal Versus Cardinal Preferences. If $x \succ_i y$, we know that x is preferred to y, but we don't know by "how much." These are called ordinal preferences because they contain information only about ordering, and not magnitude. Can the magnitude of the difference be measured? The proponents of an alternative model, called cardinal preferences, have developed ingenious mathematical techniques to measure preferences.

The proponents of ordinal preferences argue that all of the information about the tastes of voters that is worth using is encapsulated in a statement like $x \succ_i y$. Even if we could measure the gap in the desirability between x and y in the eyes of the voter I, it would still not solve the problem that the gaps observed by voters i and j would not be comparable. There is no rigorous, mathematically valid way to express the idea that i likes policy x "twice as much" as voter j, even though such an interpersonal comparison is tempting. As a result, the ordinalists argue that no voting procedure should be designed with the intention of trying to measure "how much" more attractive x is than y. Only the ranking should be taken into account.

The approaches based on cardinal preferences offer the possibility of very fine-grained and, at least on the surface, precise social comparisons. If we had a reliable measure of preferences on a cardinal scale, perhaps it would be possible to say that x is 0.3 units more appealing than y. The most widely used method of deriving these preferences is the Von Neumann-Morgenstern approach, which is used in game theory (see Chapter 4). The theory that justifies the measurement techniques is somewhat abstract and complicated. Even in a laboratory setting, these preferences have proven difficult to measure reliably.

For whatever reasons, almost all voting procedures in use in the world today employ ordinal methods. We ask voters to declare their favorite or to rank the alternatives first, second, and third. As we discuss the weaknesses of various voting methods, readers will often be tempted to wonder if the use of cardinal preference information might lead to an easy fix. We believe that such quick fixes are unlikely to be workable, but they are often very interesting. The fact that one cannot compare preference scores across individual voters should never be forgotten, because many proposals implicitly assume that such comparisons are meaningful.

Exercises 3.2

1. Consider three lunch items, *steak, fish*, and *eggplant*. If Joe prefers steak to fish, we write $steak \succ_{Joe} fish$. Joe also prefers *fish* to *eggplant*. If Joe's preferences are transitive, should we conclude $steak \succ_{Joe} eggplant$ or $eggplant \succ_{Joe} steak$?
2. Suppose Jennifer tells us $steak \succ_{Jennifer} fish$ and $steak \succ_{Jennifer} eggplant$. Can we say whether $steak \succ_{Jennifer} fish$ or $fish \succ_{Jennifer} steak$?
3. If $pink \succ_{Mary} red$ and $red \succ_{Mary} green$, and then Mary says she prefers *green* to *pink*, would her choices seem reasonable to you? Why or why not?

3.3 The Plurality Problem

By far the most commonly used election procedure in the United States is simple plurality rule. Each eligible voter casts one vote and the winner is the candidate that receives the most votes. In terms borrowed from horse racing, it is sometimes called a "first past the post" procedure. The winner need only have more support than the second-placed competitor. This method is used in elections in other countries (e.g., Great Britain) and it is the method of balloting used in the selection of winners of the prestigious Oscar Awards, which are offered by the Academy of Motion Picture Arts and Sciences.

Two Candidates? No Problem! If there are only two candidates, then the plurality rule is a good method of choice. It is, in fact, equivalent to majority rule. Suppose the alternatives are x and y and the voters are $N = \{1,2,3, \ldots, n\}$. Voters for whom $x \succ_i y$ will vote for x. In **majority rule**, which we label M, an alternative with the support of more than one half of the voters is the winner. Using the notation

$$|\{i \in N : condition\}|$$

to indicate "number of elements in N for which *condition* is true," majority rule is stated as:

$$\frac{|\{i \in N : x \succ_i y\}|}{n} > \frac{1}{2} \textit{ implies } x \succ_M y. \tag{4}$$

The voters who oppose x, the ones for whom $y \succ_i x$, or are indifferent, $x \approx_M y$, are viewed as the opposition.

In contrast, under **plurality rule**, which we label P, the candidate that is preferred by a greater number of voters wins, even if that candidate has less than one half. Plurality rule is formally defined as

$$|\{i \in N : x \succ_i y\}| > |\{i \in N : y \succ_i x\}| \textit{ implies } x \succ_P y. \tag{5}$$

Note this ignores indifferent voters. In contrast, in a majority system, the indifference is treated as opposition. If indifferent voters exist, but they abstain from voting, then plurality and majority again coincide. Only if the indifferent voters are given a way to register their indifference, such as casting a "tie" vote, will the two methods diverge.

There is a mathematical proof of the superiority of majority rule known as May's Theorem (*May 1952*). Many people have been confused by the fact that May is actually discussing the system that we have defined as plurality rule. In the system that May calls majority rule, votes in favor of two alternatives are collected and the one with the most favorable votes wins (the indifferent voters are not counted as opposition). May's theorem states that when there are two alternatives, majority (what we call plurality) rule is the only decisive procedure that is consistent with these three elementary properties:

- **Anonymity:** Each person's vote is given the same weight.
- **Neutrality:** Relabeling the alternatives (switching the titles x and y) in voter preferences causes an equivalent relabeling of the outcome (the alternatives are undifferentiated by their labels sometimes called *undifferentiatedness* of alternatives).
- The electoral system is positively responsive in the following two senses:
    - **Strong monotonicity:** If there is a tie and then one voter elevates x in his preferences (that is, changes from $y \succ_i x$ to xI_iy or $x \succ_i y$, or from xI_iy to $x \succ_y y$), then x must be the winner.
    - **Weak monotonicity:** If x is the winner and one voter elevates x in his preferences, then x must remain the winner.

The formalization and proof of May's theorem is discussed in the exercises.

May's theorem is one of the most encouraging results in the study of voting procedures. It points our attention at a particular method of decision making and it formally spells out the virtuous properties of the procedure.

Shortcomings of Plurality Rule with More than Two Candidates. Because majority rule has so much appeal with two candidates, there is a natural tendency to try to stretch it to apply in an election with more candidates. The two most important methods through which this has been tried are **pure plurality rule** and the **majority/runoff election** system. In pure plurality, the winner is the one with the most votes, even if the vote share is only 5 or 10 percent. In the majority/runoff system, all of the candidates run against each other and the winner is required to earn more than one half of all votes cast. If no candidate wins a majority, then the top two vote getters are paired off against each other in a runoff election. This system does not typically include "none of the above" as an alternative in either stage, and so there is no method for voters to register indifference (and hence, majority and plurality rule with two candidates imply the same results).

The two stage majority/runoff system is used in some American state and local elections and in French national elections. This system has obvious flaws. The candidates who have the widest following might "knock each other out" by dividing the vote, allowing little known candidates to sneak through. The system puts voters in a difficult position of deciding whether they should vote for their favorite, who might be unlikely to win, or for one of the likely winners. The runoff system is also expensive; it requires the government to hold two full elections (ballots aren't free, after all).

It is almost too easy to point out the flaws in pure plurality rule. The most obvious weakness of the plurality rule is that the winner need not have widespread support. Suppose there are four candidates, $\{w, x, y, z\}$, running for mayor. The voters can be divided into groups according to their preferences. In Table 3.1,

	Voters			
Groups	1	2	3	4
Number of Members	20	24	26	30
Ranking				
First	z	y	x	w
Second	x	z	y	z
Third	y	x	z	x
Fourth	w	w	w	y

Table 3.1: Plurality election with 4 candidates.

the preference orderings of the four groups are illustrated. Note that the largest group, group 4, has 30 members and their favorite is w. This example is designed to emphasize the fundamental problem that a minority, 30 percent of the population, is able to force its will on the rest, and all of the other voters think that w is the *worst possible choice*! One of the most troublesome aspects of this example is that w would lose a head-to-head election against x (because x is preferred by groups 1, 2, and 3), y, or z.

Holding a runoff between the top two vote getters would help somewhat, since x would win in a head-to-head competition against w. However, the voters who support z would certainly protest. Note that z is **unanimously** preferred to x! How in the world can such a universally unlikeable character as x win an election?

The plurality problem is a "real life" problem with serious consequences. When a candidate wins an election with less than 50% of the vote, there is always concern that the wrong candidate won. Consider, for example, the fact that, in 1998, voters in the state of Minnesota elected former professional wrestler Jesse "The Body" Ventura as their governor. Ventura won with 37% of the vote, outdistancing the Republican candidate Norm Coleman (34%) and the Democrat Hubert Humphrey (28%). There's no way to know for sure what would have happened if either Coleman or Humphrey had faced Ventura in a one-on-one contest, but there's considerable speculation that either would have defeated him.

Even if you don't care much about who is governor of Minnesota, you probably do care who wins the coveted Grammy Awards for the best songs and albums. The winners are selected by a plurality vote with five nominees on the ballot. There is often concern that the best performers do not win. Music reporter Robert Hilburn suggested they consider changing their voting procedure, noting, "Looking over previous Grammy contests, it's easy to see where strong albums may have drawn enough votes from each other to let a compromise choice win. In 1985, two of the great albums of the decade – Bruce Springsteen's "Born in the U.S.A." and Prince's "Purple Rain" – went head to head in the best album category, allowing Lionel Richie's far less memorable "Can't Slow Down" to get more votes" (*Hilburn 2002*). Clever voters might try to outsmart the system, voting for their second-ranked alternative in order to stop a weak candidate from winning. Dishonest voting is euphemistically called strategic voting in the literature.

The plurality problem has been known for centuries. It was a focus of concern in French academic circles in the late 1700s. Two extremely interesting characters are the philosopher/scientists Jean-Charles de Borda and the Marquis de Condorcet. Recall that it was a time of revolution, both in America and in France. The philosophy of democracy was becoming well accepted. The Marquis de Condorcet (1743-1794), was an extremely influential philosopher of the Enlightenment, studying not only voting (*Condorcet 1785*), but also publishing on the rights of women, slavery and free markets. He was highly placed in academic circles, an eager proponent of revolution against the King and, eventually, elected to the legislature (and then later imprisoned by an opposing faction). He is credited with a comment which translates as, "The apparent will of the plurality may in fact be the complete opposite of their true will" (cited *Mackenzie 2000a*). Condorcet proposed a system of voting in which alternatives were paired off for comparison. Traveling in the same circles as Condorcet was Borda (1733-1799), an explorer, soldier, and scholar whose study of physics and mathematics had wide-ranging impact on science. He agreed with Condorcet on the shortcomings of the plurality system and he also proposed a new election procedure based on rank order voting. His manuscript, which was written sometime between 1781 and 1784 (see *Saari 1994*), indicates that he had previously presented his results on June 16, 1770 (*Borda 1781*). It appears that Borda had the best of it in the eyes of the French because, after the French Revolution (and until Napoleon took over) the Borda procedure was used in French elections. The Borda count fell into disuse after that, only to be revived in the modern era for rankings in American sports. Variants are used in legislative elections in the Pacific Island countries of Kiribati and Nauru (*Reilly 2002*). The debate between Borda and Condorcet has framed research on elections for two centuries. In the next section we consider Borda's method.

Exercises 3.3

1. Is there any way to win a majority of the vote without also winning a plurality of the vote?
2. Suppose there are 50 shares of stock in a company and the stockholders are allowed to vote on company policy. Each stockholder is allowed to cast one vote per share owned in a private ballot box. No voter's ballots have his/her name on it. Does this system violate the criterion of neutrality or anonymity?

Exercise 3.3.3	Voters		
Groups	1	2	3
Number of Members	24	16	10
Ranking			
First	x	y	x
Second	y	z	z
Third	z	x	y

3. This example has 50 voters and 3 alternatives.
   (a) Find the majority rule winner, if there is one.
   (b) Calculate the plurality vote totals for the candidates, and find the winner.
   (c) Does the plurality winner have a majority of the votes?
   (d) Confirm that a runoff election will choose the same winner.
   (e) If the winner of each type of election is compared one-on-one against each of the other candidates, is it a winner?

Exercise 3.3.4	Voters		
Groups	1	2	3
Number of Members	24	16	10
Ranking			
First	x	y	z
Second	y	z	y
Third	z	x	x

4. This example has 50 voters and 3 alternatives.
   (a) Which candidate would win a plurality election?
   (b) Does the plurality winner have a majority of the votes?
   (c) If there is a runoff election between the top two vote getters, which candidate will be the winner?
   (d) If the winner of each type of election is compared one-on-one against each of the other candidates, is it a winner?

Exercise 3.3.5	Voters			
Groups	1	2	3	4
Number of Members	24	19	21	36
Ranking				
First	x	y	z	w
Second	w	z	x	x
Third	z	x	y	y
Fourth	y	w	x	z

5. This example has 100 voters and 4 alternatives.
   (a) Find the plurality winner.
   (b) Does the plurality winner have a majority of the votes?
   (c) If there is a runoff election between the top two vote getters, which candidate will be the winner?
   (d) Suppose that, after the plurality vote, only one candidate is eliminated, and then another three-way plurality election is held. Will any of the candidates win a majority? If not, which candidate will win a runoff election between the top two vote getters?
6. In 2003, the voters of California were presented with an interesting choice. First, do you want to remove Governor Gray Davis? Second, if Davis is to be replaced, which of the following 150 candidates would

you most prefer? California law states that if the number who answer yes to the first question is more than 50% of the votes cast, then the candidate who receives the most votes will become governor. About 54% of the voters wanted to remove Davis, and the plurality winner of the second vote was body builder & movie star Arnold Schwarzenegger. Consider these questions about the 2003 California election.

(a) Can we be sure that a majority (or even a plurality) prefer Schwarzenegger to Davis as governor?

(b) The results indicated that 46% of the people voted to keep Davis, but they lost. What percentage of the vote must the replacement receive in order to take office? Does that seem sensible?

(c) What incentives did voters have to vote strategically (untruthfully) in the first and second parts of the ballot?

7. In the American criminal justice system, the members of a jury must agree unanimously in order to reach a verdict. If the result is not unanimous when deliberations end, there is a "hung jury" and a mistrial is declared. Explain why this system does not satisfy the principle of strong monotonicity.

8. Kenneth May showed that a system based on majority rule (in which indifferent voters abstain) is equivalent to a system in which the voters are anonymous, the alternatives are undifferentiated, and the procedure is strongly and weakly monotonic. In May's treatment, the voters consider two alternatives x and y and then cast votes by selecting a value from the set $\{-1, 0, 1\}$. A vote of 0 indicates indifference, while 1 indicates a preference for x and -1 indicates a preference for y.

(a) Create a mathematical definition of majority rule, assuming that the indifferent voters are ignored.

(b) Use your definition to reconsider the following: the winner is x and then one voter who favored y decides she is indifferent. If you apply your procedure for majority rule to the new voter preference data, does the result change?

(c) Suppose there is a tie and one voter changes her mind to prefer x. Apply your procedure to find out who the winner will be.

(d) On the basis of your calculations in b and c of this question, do you think your definition of majority rule is consistent with the monotonicity principle?

3.4 Rank-Order Voting: The Borda Count

If you are a fan of college football, you may be aware of the fact that (until 2006, at least) there has been no national tournament to answer the question "which team is the best." Instead, 100 (or so) college football teams are ranked at the end of the season, and eight of the top teams are selected to play in four spotlighted games that are known as the Bowl Championship Series. The top two ranked teams play in the national championship game, and after that game, there is another series of votes intended to rank all of the teams from top to bottom.

The formula used by the BCS to decide which teams should play in the national championship game has been a source of frustration and controversy. In 2001, the University of Nebraska was soundly defeated by the University of Colorado in the final regular season game. That loss prevented Nebraska from winning the Northern division of the Big 12 conference and it was not in the Conference championship game (which Colorado won). Nevertheless, the BCS scores had Nebraska ahead of Colorado and the Nebraska Cornhuskers were selected to face the University of Miami in the championship game. Nebraska was soundly defeated, giving grounds to the contention that the BCS had chosen the wrong team. At the end of the 2003 season, the University of Southern California was not selected for the championship game, even though the polls of writers and coaches placed USC in the top two. Decisions like that have led to a chorus of complaints over the years, summarized aptly by a student sports writer at the University of Kansas: division 1A college football uses "what may be the most complicated monstrosity on the planet" (*Bauer 2004*, p. 2) to rank the teams.

The people who run the BCS are not oblivious to the criticism of their methods of ranking teams, and every few years, they make repairs in their voting system. They raise or lower the weight put on the votes cast by coaches, sports writers, and, strangely enough, *computers*! So far, at least, they have not strayed far from the same basic system. Individuals (coaches, writers, and computers) are asked to provide rankings of the top 25 teams. The many individual rankings are summed to build an overall ranking. This is a variant of the method that was proposed by Jean-Charles de Borda (*Borda 1781*).

The Borda Count. The Borda count, which we label *BC*, is a method of combining ranking of many individual voters. Suppose there are k alternatives. The voters rank the alternatives and assign integer scores to them, beginning with 0 (for the worst) stepping up in integer units to $k-1$ (for the best). The rankings assigned by the voters are summed and the alternative with the largest Borda count is the winner. The Borda count generates a complete, transitive social ordering of the alternatives.

In mathematical terms, each voter submits a vector of scores that are keyed to the list of candidates. If the candidates are {x, y, z} and voter i submits the **Borda vector** $v_i = (0, 2, 1)$, that means that the voter thinks y is the best, z is second, and x is the worst. The Borda count earned by a candidate j is the sum of rankings assigned by the voters, $\sum_{i=1}^{n} v_{ij}$. The notation $x \succ_{BC} y$ indicates that x defeats y in the Borda count. Although we do not concern ourselves with ties in this chapter, we should mention that voters are allowed to register indifference by assigning two alternatives the same value. For example, (0.5,0.5,2) indicates that x and y are equally desirable (one vote is split between them), and both are less appealing than z.

The Borda count works sensibly in many situations. For example, if all of the voters agree about the ranking for all of the alternatives, it is easy to see that the

Voter preference profile			
	Voter Rankings		
Voter:	1	2	3
Ranking			
First	x	x	x
Second	y	y	y
Third	z	z	z

Borda count table				
	Borda Votes			Borda Count
Voter:	1	2	3	
Alternatives				
x	2	2	2	6
y	1	1	1	3
z	0	0	0	0

Table 3.2: Borda count with homogenous voters.

Borda count gives a meaningful reflection of their rankings. The rankings assigned by the three voters are presented in the top part of Table 3.2. The **Borda table**, shown in the bottom half of the figure, illustrates the numerical vectors that represent the Borda votes. This demonstrates that the "sum of ranks" procedure used in the BCS and other poll-style ranking systems is the same as the Borda count. It is, of course, painfully obvious that x should win. It is the most-preferred alternative for all of the voters and it has the highest Borda count.

A system of this sort has been the basis of many ranking systems used in sports and entertainment. Although there are bitter squabbles about the outcomes of these ranking procedures, the commentary in the public media seldom considers the possibility that there is a fundamental problem in the basic idea of the Borda count itself. Very peculiar – even alarming – things can happen with the Borda count.

A Paradox in the Borda Procedure. The problem that we now introduce has sometimes been called the "Inverted-Order Paradox" or the "Winner Turns Loser Paradox." It is a famous example that has been widely investigated (*Riker 1982,* p. 82, *Fishburn 1974).* There are seven voters, $\{1, 2, 3, 4, 5, 6, 7\}$, and four alternatives, $\{w, x, y, z\}$. In Table 3.3, the columns represent the Borda voting vectors assigned by the voters.

The winner, the candidate with the largest Borda count, is y. The overall ranking, from top to bottom, is $y \succ_{BC} x \succ_{BC} w \succ_{BC} z$. This is a strength of the Borda count. It generates a transitive ordering. Nevertheless, in the eyes of its critics, the ordering is complete nonsense.

	Borda Votes							Borda Count
Voter:	1	2	3	4	5	6	7	
Alternatives								
w	3	3	0	0	1	1	3	11
x	2	2	3	3	0	0	2	12
y	1	1	2	2	3	3	1	13
z	0	0	1	1	2	2	0	6

Table 3.3: Borda count with 4 alternatives.

	Borda Votes							Borda Count
Voter:	1	2	3	4	5	6	7	
Alternatives								
w	3	3	1	1	2	2	3	15
x	2	2	3	3	1	1	2	14
y	1	1	2	2	3	3	1	13
z	0	0	0	0	0	0	0	0

Table 3.4: Borda count after making z ineligible.

To bring some life into this discussion, suppose that the alternatives w, x, y, and z are bands competing for the prestigious Grammy award. After the votes have been collected, an investigative reporter exposes the fact that group z is composed of frauds who lip-sync their performances. (A group later shown to have been lip-syncing, *Milli Vanilli*, actually did win a Grammy award for Best New Artist after their success in 1989.[1]) After z has been exposed, the Grammy managers declare z ineligible. It would seem logical to expect that, if the ordering was $y \succ_{BC} x \succ_{BC} w \succ_{BC} z$, and we rule the last-place candidate ineligible, then the new ordering among the remaining alternatives should simply be $y \succ_{BC} x \succ_{BC} w$. Band z was a loser, anyway. The winner should still be y, shouldn't it?

Sadly enough for the proponents of the Borda count, the answer is no. Assuming that y would still win was a terrible mistake caused by a fundamental flaw in the Borda count. We can re-calculate the votes, supposing either that z is completely eliminated from consideration or that it is forced into last place on all of the ballots (the result is the same, either way). Let's begin by simply moving z to the bottom of each voter's ranking. This will assure that z cannot be the winner because it will finish in last place. The alternatives that were ranked below z are shifted up one notch, as indicated in Table 3.4.

[1] That award was revoked and Milli Vanilli was removed form the official list of award-winning performers.

After moving z to the bottom of each ballot, the Borda count indicates that w should be the winner. The ranking from top to bottom is $w \succ_{BC} x \succ_{BC} y \succ_{BC} z$. Compare that with the previous result and a truly astonishing outcome is revealed: *the ordering of the top three alternatives is exactly reversed.* Instead of winning, y now finishes in third place.

The befuddling aspect of this result is that the Borda ranking of w and x is influenced by the placement of the alternative z. If we are completely focused on the comparison of w and x, then z is an **irrelevant alternative**. When the irrelevant alternative affects the outcome, then the decision-making procedure is said to violate the principle of **Independence from Irrelevant Alternatives**, which holds that the social ranking of two alternatives should not be influenced by the placement of other alternatives in the ballots of the voters.

A story might persuade readers that a voting procedure ought to respect the principle of **Independence from Irrelevant Alternatives**. Suppose your friend Daffy and his family have received a gift certificate for a new car, either a Toyota or a Honda. When the Daffy family is focused on those two alternatives, the desirability of all other makes should be irrelevant. Suppose you are discussing this in the car dealer's parking lot and Daffy says out loud, "We prefer the Toyota to Honda. We will go in to pick that one." You figure all the work is done, so you go home. Later, Daffy drives up to your house in a shiny new Honda. You say, "What happened? You preferred Toyota!" Daffy says, "While we were in line, my son heard that Cadillacs have great durability. So we changed from Toyota to Honda!" Could the outcome be more ridiculous? It cannot possibly make sense to have the choice between a Honda and a Toyota depend on the road performance of a Cadillac. And yet, that can happen with the Borda count.

In the example illustrated in Tables 3.3 and 3.4, the terminology says z is an irrelevant alternative, but it is not, in fact, irrelevant when the Borda count is used. That is the sense in which the Borda count violates the principle of Independence from Irrelevant Alternatives. Although commentators on amateur figure skating competitions do not know the terminology, they have certainly puzzled over it many times because skating uses a Borda-style ranking system. In the 1995 World Figure Skating Championships, Michelle Kwan's surprisingly good performance put her in fourth place, but caused a reversal in the ranking of the second and third ranked skaters who had already finished before Kwan took the ice (*Mackenzie 2000b*).

Another variant of this story begins with three alternatives. In Table 3.5, the three alternatives are w, x, and y. The Borda winner is w. Now, suppose that someone who is in charge of the election desperately wants y to win. That someone is so desperate, in fact, that he is willing to sponsor the entry of a new candidate into the race. The new candidate, z, is chosen very carefully. When z is in the race, the preferences of the voters are the same ones we considered in Table 3.3. After adding in the losing candidate z, the Borda count selects y, just as the election manager wants. This displays the truly insidious nature of the violation of the principle of irrelevant alternatives. Not only is the decision illogical, but now it is prone to electoral manipulation. Without having precise information about that tastes of the voters,

	Borda votes							Borda Count
Voter	1	2	3	4	5	6	7	
Alternatives								
w	2	2	0	0	1	1	2	8
x	1	1	2	2	0	0	1	7
y	0	0	1	1	2	2	0	4

Table 3.5:

	Borda Votes					Borda Count
Voter:	1	2	3	4	5	
Alternatives						
w	3	3	1	1	1	9
x	2	2	0	3	3	10
y	1	0	2	0	2	5
z	0	1	3	2	0	6

Table 3.6: Condorcet winner (w) rejected by Borda count.

it might be difficult to slide in just the right candidate to bring about the desired outcome. One might mistakenly make the outcome even less desirable.

Another Paradox: The Borda Winner is a Loser? There is another closely-related (equally serious) problem with the Borda count. The Borda count can produce a winner that would lose in a head-to-head election against one of the other alternatives. In Table 3.3, the Borda winner is y, but in a head-to-head contest between y and x, x would win because it is preferred to y by voters $\{1,2,3,4,7\}$. This result is not necessarily a sign of trouble in the eyes of the advocates of the Borda count. The Borda method, they argue, averages out the extreme opinions, so that the winner is somehow "more representative" of the overall opinions of the voters.

The problem is actually more serious. When he proposed his voting system, Jean-Charles de Borda drew the attention of a fellow French academic, the Marquis de Condorcet. Condorcet pointed out what he thought was a crippling weakness in Borda's method. Consider Table 3.6, an example in which there are five voters and four alternatives. The Borda winner is x, but x is defeated in a head-to-head contest with w. In fact, w would defeat each of the other alternatives in a head-to-head contest. Alternative w is preferred to x by voters $\{1, 2, 3\}$, and to y by voters $\{1, 2, 4\}$, and to z by voters $\{1, 2, 5\}$. In Condorcet's honor, we call an alternative that can win a pairwise (head-to-head) contest against each of the other alternatives the **Condorcet winner**. A voting procedure which does not select the Condorcet winner is said to violate the **Condorcet criterion**.

Inside the Guts of the Borda Count. The ranking created by the Borda count is vulnerable to manipulation by the addition or subtraction of candidates. By working on a few examples, one can gain some good working knowledge of what kinds of changes will make a difference. One of the clearest examples is found by comparing an election with two candidates against an election with three candidates. Suppose there are 100 voters and they are divided into two groups. There are 60 voters in group 1 and they prefer x to y. Formally speaking,

$$Group\ 1: x \succ_i y \ for\ i \in \{1, 2, \ldots, 60\}.$$

There are 40 voters on the other side of the issue,

$$Group\ 2: y \succ_i x \ for\ i \in \{61, 62, \ldots, 100\}.$$

The Borda count is summarized in Table 3.7. Since group 1 has the most voters, its favorite is the winner.

Next, suppose that a third alternative is added, and it is very similar to the loser, y. We want to insert this new alternative into the preferences of the voters so that z is always right next to y. Because z is always grouped together with y, it is sometimes called a clone. (One standard for voting procedures is that their outcomes should not be changed by the addition of redundant alternatives (clones). This example is intended to show that the Borda count is not "cloneproof.") There are four possible orderings. The first two new orderings are found by placing z about y in the tastes of group 1:

- Group 1a: $x \succ_i y \succ_i z$
- Group 1b: $x \succ_i z \succ_i y$

The last two are obtained by inserting z into the preferences of group 2:

- Group 2a: $z \succ_i y \succ_i x$
- Group 2b: $y \succ_i z \succ_i x$

By inserting z, which is very similar to y in the eyes of the voters, we thus obtain preferences for four groups. The relative sizes of these groups are referred to as n_{1a}, n_{1b}, n_{2a}, n_{2b} in Table 3.8.

	Borda Votes		Borda Count
Voter:	1	2	
Number of Members	60	40	
Alternatives			
x	1	0	60*1 = 60
y	0	1	40*1 = 40

Table 3.7: Borda count with two alternatives.

	Borda Votes				Borda Count
Voter:	1a	1b	2a	2b	
Number of Members	n_{1a}	n_{1b}	n_{2a}	n_{2b}	
Alternatives					
x	2	2	0	0	$2 * n_{1a} + 2 * n_{1b}$
y	1	0	1	2	$1 * n_{1a} + 1 * n_{2a} + 2 * n_{2b}$
z	0	1	2	1	$1 * n_{1b} + 2 * n_{2a} + 1 * n_{2b}$

Table 3.8: Add one alternative to Table 3.7.

Suppose that the original groups are split exactly in half, so $n_{1a} = 30$, $n_{1b} = 30$, $n_{2a} = 20$, $n_{2b} = 20$. With that even division within the two groups, the Borda count is 120, 90, 90 for x, y, and z, respectively. Since y is similar to z in the eyes of the voters, the result seems plausible because y and z are tied in the Borda count. The original winner, the most favored alternative of the majority group, still wins.

We can make some magic by fiddling with the sizes of the subgroups. Keep in mind that there are always 60 voters for whom x is the most attractive alternative and there are always 40 for whom x is the least attractive. The only manipulation that we consider is the subdivision of the original two groups. (If you use a computer spreadsheet, it is pretty easy to consider a lot of conjectures.) For most of the examples that you try, the winner will be x, but there are some exceptions. The exceptions are found when there are many voters who have y preferred to z. That is, if you increase the values of n_{1a} and n_{2b} enough, then the Borda winner will change from x to y. If the groups are divided $n_{1a} = 50$, $n_{1b} = 10$, $n_{2a} = 5$, and $n_{2b} = 35$, then the Borda counts for x, y, and z are 120, 125, and 105. Even though x is still the first-ranked alternative for 60 of 100 voters, the alternative y is the Borda winner.

The algebra of the situation is enlightening. Alternative y will defeat x (i.e., have a higher Borda count) if

$$1 * n_{1a} + 1 * n_{2a} + 2 * n_{2b} > 2 * n_{1a} + 2 * n_{1b}$$

which is easily simplified:

$$n_{1a} + 2 * n_{1b} - n_{2a} - 2 * n_{2b} < 0.$$

Keeping in mind that $0 \leq n_{1b} \leq 60$ and $n_{1b} = 60 - n_{1a}$, as well as $0 \leq n_{2a} \leq 40$ and $n_{2a} = 40 - n_{2b}$, the Borda count for y will be superior if

$$n_{1a} + n_{2b} \geq 80.$$

In the above example, where we set $n_{1a} = 50$ and $n_{2b} = 35$, this inequality was satisfied and so y was the winner.

Recall that the premise of this example is that the new alternative, z, is similar to (a clone of) y, and z is inserted into the preferences in the spot immediately preceding or following y. In order for the introduction of z to cause a change in the Borda count that benefits y, it is vital that y be more desirable than z in the eyes of many (at least 80) of the voters. One might criticize this example by pointing out that y and z are not truly similar because a vast majority of the voters see y as more desirable than z, and if they were really similar, the split would be more even. This argument, while interesting, does not rehabilitate the Borda count. Rather, it simply refocuses our attention on the problem of irrelevant alternatives. If the overall ranking of x against y depends on the question of whether y is more desirable than z, then one ought to be cautious about the result from a Borda count. If x loses, its supporters must surely feel abused because the position of their favorite against y is not changed in the transition from Table 3.7 to Table 3.8. On the other hand, if y is the loser and x is the winner, then the supporters of y are unhappy and feel unjustly treated because their loss was caused by the high placement of z in the rankings of some voters. It just doesn't seem right. It seems this kind of illogic is fundamental about procedures based on the Borda method of voting. And, in case you were wondering, that means the BCS football ranking system will likely produce exasperating results.

Digression on the Use of Cardinal Preferences. The element of the Borda Procedure that is most often thought to be the culprit is the simple method in which preferences are registered. The preferences of the voters are registered as integer-valued rankings. If a voter thinks that x is much better (or just a little better) than y, and y is much (or just a little bit better) than z, the ranking will be submitted as 1-2-3 and the magnitude of preference is ignored. A proponent of cardinal preferences would rather have us figure out a way to precisely measure these differences so that the voter could submit a vote like 1-1.8-2.0 to signify the fact that the first one is the best and the other two are far worse. In the advanced literature on social choice theory, there are in fact some highly prestigious authors, such as Nobel Prize winners John Nash (*Nash 1950*) and John Harsanyi (*Harsanyi 1955*) who have advocated decision-making based on these fine-grained evaluations. John Nash, the game theorist whose life story was the basis of the movie A Beautiful Mind, was probably the first to contend that these cardinal scores should be collected and multiplied together to create a social ranking. In his honor, Riker calls this the Nash method (see *Riker 1982*).

The use of cardinal scores has intuitive appeal. Harsanyi offers a rigorous mathematical argument in favor of this system, contending that it optimizes the welfare of the community in a utilitarian sense. That may be, but, as Riker demonstrates (*Riker 1982*, pp. 110-111), the problem of irrelevant alternatives remains. In fact, the problem may be more severe in the Nash method than in the Borda method. In the Borda method, the fact that voters are restricted to casting votes in integer values limits the extent to which the placement of z might influence the comparison of x and y. In a cardinal election, the votes can range across a continuum, seemingly opening up a much larger range of outcomes that are influenced by irrelevant alternatives.

Exercises 3.4

1. Consider a two candidate race for the presidency of a social club. There are 135 members of the club and 81 prefer *Fred* to *Barney* while the remaining members are of the opposite opinion. Write down the Borda vote vectors that would be cast and find the winner of the Borda count. Is the same candidate going to win a majority rule election?
2. This example has 65 voters and 3 alternatives.

Exercise 3.4.2	Voters		
Groups	1	2	3
Number of Members	22	12	31
Ranking			
First	x	y	z
Second	y	x	y
Third	z	z	x

   (a) Which candidate would win the plurality election?
   (b) What would be the results of a majority/runoff election?
   (c) What would the Borda vector for each group of voters be?
   (d) Create a Borda count table. What would be the Borda count for each candidate?
   (e) Is there a Condorcet winner in this example?

3. This is a set of preferences for a society of 1000 voters with 4 alternatives.

Exercise 3.4.3	Voters					
Groups	1	2	3	4	5	6
Number of Members	200	150	150	300	100	100
Ranking						
First	z	y	x	w	y	x
Second	x	z	y	z	x	y
Third	y	x	z	x	w	z
Fourth	w	w	w	y	z	w

   (a) Is there a Condorcet winner? If so, which candidate?
   (b) Create a Borda count table. Which candidate would win according to the Borda count?
   (c) Delete candidate w and re-calculate the Borda count.
   (d) Does the Condorcet winner change after deleting w?

4. Calculate the Borda count for the following table. Then add an alternative that finishes in last place in the Borda count, but also causes the ranking of the top-placed alternatives to change.

	Borda Votes							Borda Count
Voter:	1	2	3	4	5	6	7	
Alternatives								
x	2	2	0	0	1	1	2	?
y	1	1	2	2	0	0	1	?
z	0	0	1	1	2	2	0	?

5. Condorcet's original example that was used to criticize the Borda count involved an election with candidates named Jacques, Pierre, and Paul.

	Borda Votes					
Voter:	1	2	3	4	5	6
Number of Voters	30	1	10	29	10	1
Alternatives						
x	2	2	1	1	0	0
y	1	0	0	2	2	1
z	0	1	2	0	1	2

(a) Find the Condorcet winner.
(b) Find the Borda winner.
(c) How many ballots would you have to change in order to make the outcome of the two procedures match up? (Hint: You can solve this by trial and error, but you should not have to. Recall that the Borda count equals the aggregated pairwise vote.)

3.5 Sequential Pairwise Comparisons

The critics of the Bowl Championship Series often contend that college football should adopt a tournament format to select a national champion. A tournament is used in many other college sports as well as professional baseball, basketball, and football. Would a tournament solve the controversy over "who's number one?" There are good reasons to be skeptical. A tournament might increase advertising revenue, but we doubt it would put an end to questions about whether the best team was actually ranked number one at the end.

A Single Elimination Tournament. The teams are eliminated in a series of games, the final of which would be the national championship game. A hypothetical bracket for such a tournament is presented in Figure 3.1. The teams that are defeated at each stage are eliminated, and the winners advance to play each other.

The people who want a Division IA college football tournament usually claim that this is the best way to select the "true champion." In the hypothetical example in Figure 3.1, the University of Sasnak wins. Whether or not Sasnak is greeted as the "true champion" may be open to question, however. Suppose, for example, that Sasnak could be defeated by Sasnak State, Aksarben, Amohalko State, or Iruossim.

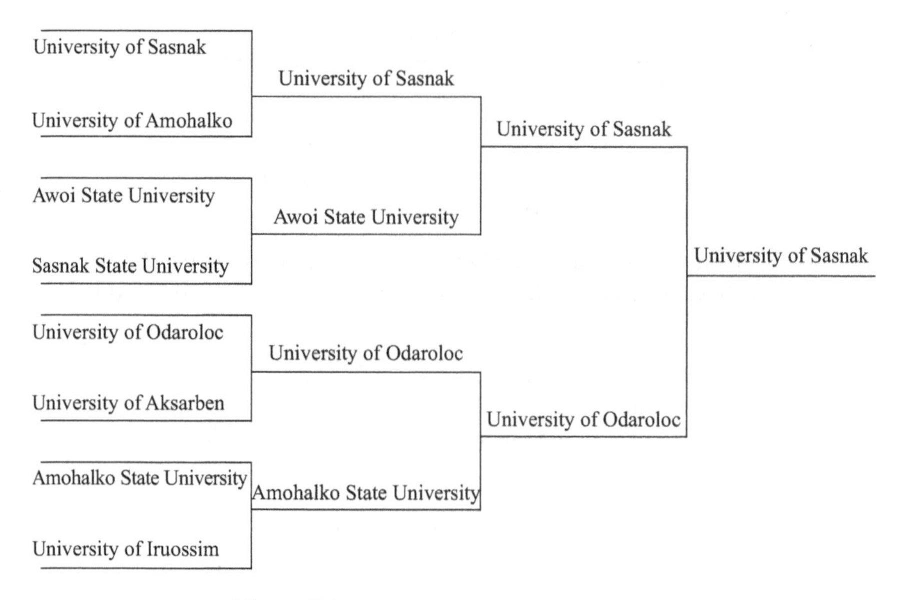

Figure 3.1: Hypothetical Football Tournament.

Those other teams are not given their "fair chance" to prove their superiority because the tournament structure governs their opportunities. If even one team can defeat the eventual winner of the tournament, it seems we are going to be forever debating what would have happened if the contests had been re-organized. The tournament does not so much solve the problem of deciding "which team is best" as it does create a new set of arguments for people in bar rooms and sports talk shows.

The tournament concept, it turns out, is widely used, not just for sporting events, but for political decision-making as well. For example, in American elections, the candidates of the top two political parties face each other in the general election (the "championship game"). Those candidates are the winners of earlier contests. Like the winners of sporting tournaments, these winners are no less open to after-the-fact challenges from contenders who did not get their fair chance.

There is one case in which a tournament structure produces an unequivocal, certain winner. If there is a Condorcet winner, one candidate (or team) that defeats each of the others in a one-on-one competition, then that winner will be the only survivor of the tournament.

Dominated Winner Paradox. Like the Borda count, the majority rule tournament format can produce some truly bizarre outcomes. The tournament never rejects a Condorcet winner, if there is one, but interesting things can still happen. Perhaps the most well-known is the "dominated winner paradox." It is possible that the winner of a tournament can be *unanimously defeated* by another alternative. We mean to say not just that the tournament picks the second-best or third-best, but rather, that the tournament winner is unanimously considered inferior. The preferences in

		Voter Preferences		
Rankings	Voters:	1	2	3
First		w	y	x
Second		x	w	z
Third		z	x	y
Fourth		y	z	w
Sequence of Votes		$w \succ_M x$	$y \succ_M w$	$z \succ_M y$

Table 3.9: Dominated Winner Paradox.

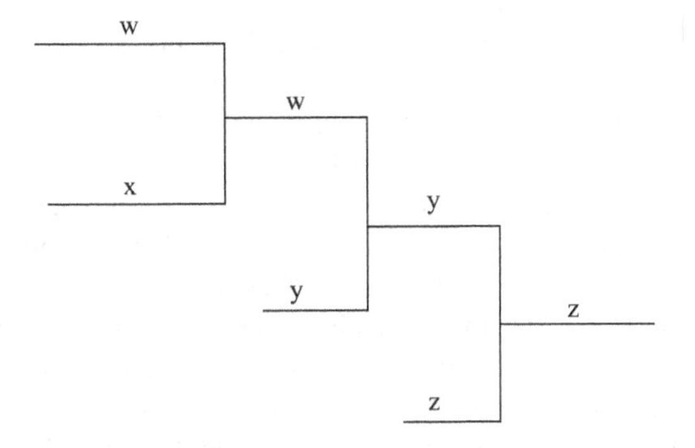

Figure 3.2: Tournament Structure for the Dominated Winner Paradox.

Table 3.9 are used to illustrate this paradox. There are three voters and four alternatives, w, x, y, and z. In the tournament structure that is drawn in Figure 3.2, there is a sequence of three votes. First, alternative x is defeated by w. Then, in the second stage, the winner w is defeated by y, and then the final vote pits y against z. Note that the first outcome to be eliminated, x, is preferred to w by every single voter.[2]

The Intransitivity of Majority Rule. In the first section of this chapter, we drew the reader's attention to the concept of transitivity. Now that concept comes to the forefront again. There are majority rule examples in which $x \succ_M y$, and $y \succ_M z$, but (and here's the nontransitive part) $z \succ_M x$. Social comparisons sometimes lack the fundamental properties of reason and logic summarized by the principle of transitivity that is present in the tastes of voters.

Consider an example with three candidates (two Democrats, one Republican) and three groups of voters. The Democrats, named *Thing 1* and *Thing 2*, are pitted in a tightly contested electoral campaign, and the winner will face the Republican candidate, named *Cat in the Hat*. Let's suppose there are three equally sized groups of voters. The transitive orderings of the voters are presented in Table 3.10. Readers

[2] This is called the "dominated winner paradox" because, in the field of cooperative game theory, one alternative is said to dominate another if every participant prefers it to the other.

	Voter Groups		
Rankings	1	2	3
First	*Thing 1*	*Cat in the Hat*	*Thing 2*
Second	*Thing 2*	*Thing 1*	*Cat in the Hat*
Third	*Cat in the Hat*	*Thing 2*	*Thing 1*

Table 3.10: Preferences for three candidates.

will note that this kind of table is slightly different than the one used for the Borda count, but the information that it contains is the same.

It is especially important to note that the preferences of the voters are transitive. Consider the preferences in Table 3.10. For group 1 we are told that *Thing 1* is preferred to *Thing 2*, and *Thing 2* is preferred to *Cat in the Hat*. Then transitivity of preference implies we are safe in concluding that *Thing 1* is preferred to *Cat in the Hat* by members of group 1.

The tournament begins with *Thing 1* and *Thing 2* facing each other and all three groups of voters are allowed to have their say. Since *Thing 1* is preferred to *Thing 2* by groups 1 and 2, *Thing 1* is the winner of that stage. The final stage pits *Thing 1* against *Cat in the Hat*, and with the backing of groups 2 and 3, *Cat in the Hat* wins.

If majority rule were transitive, then we could conclude that *Cat in the Hat* is socially preferred to *Thing 2*. It is a startling and truly paradoxical fact that majority rule does not support that prediction. Majority rule does not obey the property of transitivity.

Imagine what would happen if, in the time-honored tradition of politicians, *Thing 2* hires a large team of lawyers who appeal the case to the highest court. The court might order another election pitting *Thing 2* against *Cat in the Hat*. The *Cat in the Hat* should do everything he can to avoid another vote. Note that groups 1 and 3 prefer *Thing 2* to *Cat in the Hat*, so *Thing 2* would win that final electoral competition.

Even though the individual voter preferences are transitive, the majority rule ordering is not! This peculiar finding, sometimes called **The Voter's Paradox**, means that where we expect to find logic and reason: $x \succ_M y, y \succ_M z$, and $x \succ_M z$, instead we find nonsense and irrationality: $x \succ_M y, y \succ_M z$, and $z \succ_M x$. We could keep voting forever, with each winner being defeated in turn. This is called a **voting cycle**. A person who is intransitive lacks the fundamental properties of reason and logic and would be incapable of making decisions and managing personal affairs. Should a society that is intransitive be viewed in the same harsh light? For centuries, political scientists have been locked in debate over the issue.

Despite these problems, majority rule is very widely used in legislatures. There are some arguments that can be advanced in favor of the majority rule method and to justify its continued use. By far the most important justification for the use of majority rule is offered by May's theorem, which we have already discussed.

The second reason why majority rule is still widely used, even though cycles are possible, is that cycles are thought to be rare. If there are only three voters with three alternatives, and we ignore the possibility that voters may be indifferent, then we can make a list of all possible "societies." There are 216 possible three-voter societies, and a cycle can arise in 12, or approximately 5.6% (*Johnson 1998*, p. 23). Many computer simulation studies have been done to try to find out how likely a cycle is to arise when the numbers of voters and alternatives are increased. A review of several studies led *Riker* to contend that the probability of cycles rises dramatically (near 1.0) as the number of alternatives and voters increase, while *Jones et. al.* contended that if voter indifference is taken into account, then the probability of cycles is not quite so high.

If cycles are possible in the real world, why don't we observe them more often? Quite simply, we do not use voting procedures that are designed to discover them. Election procedures are generally written so that candidates are eliminated and are not given a second chance. Voting in legislatures is tightly controlled by party leaders who refuse to let the members propose frivolous alternatives in a "fishing expedition." In his discussion of the American Constitutional Convention of 1787, Riker tells the story of a cycle that was revealed when the delegates were attempting to formalize the system for presidential elections (*Riker 1986*, pp. 46-7). The cycle caused confusion and uncertainty, and eventually the matter was delegated to a subcommittee that was trusted to write up something good. That's how the Electoral College was created.

Exercises 3.5

1. The candidates in an electoral tournament are *Hamilton, Joe, Frank,* and *Reynolds*. In the first stage, *Reynolds* will be paired off against *Frank* and *Joe* will face off against *Hamilton*. The winners of those two contests will compete to decide the overall winner. Draw a figure that represents this tournament.
2. Here is a preference profile for the voters who are polled in the tournament described in the previous question.

	Voters		
Groups	1	2	3
Number of Members	22	12	31
Ranking			
First	*Hamilton*	*Joe*	*Frank*
Second	*Frank*	*Hamilton*	*Reynolds*
Third	*Joe*	*Reynolds*	*Joe*
Fourth	*Reynolds*	*Frank*	*Hamilton*

   (a) Calculate the winning candidate at each stage.
   (b) Check to find out if there is a Condorcet winner.
   (c) Is it possible for one of the losers in the first stage to defeat the eventual winner of this tournament?
   (d) Figure out if the winner of your tournament is the same as the result of the plurality, majority/runoff, and Borda count procedures.

3. Here is a preference profile:

	Voters		
Groups	1	2	3
Number of Members	33	33	33
Ranking			
First	x	y	z
Second	y	x	x
Third	z	z	y

   (a) Confirm there is a Condorcet winner.
   (b) Suppose there is a tournament in which x and y face off against each other, and then z challenges the winner. Does the Condorcet winner also win the tournament?
   (c) Can you think of any way in which to design the tournament so that the Condorcet winner does not win the final contest of the tournament?

4. Consider the following preferences.

	Voters		
Groups	1	2	3
Number of Members	18	16	21
Ranking			
First	x	y	z
Second	y	x	y
Third	z	z	x

   (a) Design a single elimination tournament and find the winner.
   (b) Is there any way to re-design your tournament so that a different alternative will win?

5. In the following table, we provide preferences for 7 out of 11 voters.

	Voters					
Groups	1	2	3	4	5	6
Number of Members	3	3	2	1	1	1
Ranking						
First	z	y	x			
Second	x	z	w			
Third	y	w	z			
Fourth	w	x	y			

   (a) Fill in the preferences for the last three columns in such a way that there is a pairwise voting cycle in which $x \succ_P y$, $y \succ_P z$, and $z \succ_P x$. You can insert any transitive rankings for groups 4, 5, and 6.
   (b) Choose your favorite letter from the set $\{x, y, z\}$. Design a single elimination tournament in which your selected alternative wins.

3.6 Condorcet Methods: The Round Robin Tournament

On the basis of preceding analysis, the reader should believe that the following claims are correct.

1. *If there is a Condorcet winner, then a single elimination tournament format will select that alternative.*
2. *If there is no Condorcet winner, the tournament winner is determined by the pairings of the alternatives.*

The presiding officer of a town council might exercise the power to set the agenda to advantage some community groups over others. There's a charming essay about an economist and a lawyer who studied social choice theory and then used it to hornswaggle the voters in a club that purchased airplanes for recreational use (*Riker 1986*).

To address this problem, Condorcet's approach was to search for a method of voting that will give a meaningful result when all of the pairwise comparisons are considered. Condorcet suggested that we collect enough information from voters so that we can hold (what is now called) a "round robin" tournament. In a round robin, each alternative faces each of the others in a head-to-head competition.

The **Condorcet criterion** states that if there is a Condorcet winner–one alternative can defeat each of the others head-on–then it should win. If there is no Condorcet winner, then the problem is to find a way to summarize the pairwise information and choose or shape the results into a ranking. Duncan Black, a pioneer of modern social choice research, suggested, "The reasons may not seem so overwhelmingly convincing, but we are moving in a region where all considerations are tenuous and fine-spun; and the claims of the Condorcet criterion to rightness seem to us much stronger than those of any other" (*Black 1958*). In the time since Condorcet, many different schemes have been proposed to summarize the outcomes of pairwise comparisons. Black suggested the use of a Borda count. We consider just a few of the many interesting proposals.

Searching for an Unbeatable Set of Alternatives. Here's an obvious starting point: eliminate undesirable alternatives from consideration. Suppose the alternatives under consideration are $\{u, v, w, x, y, z\}$. In Figure 3.3, the relationships among the alternatives are represented by arrows. In this type of graph, an arrow from x to y means that x defeats y is a pairwise competition. In panel (a) of Figure 3.3, note that u, y, and x, the ones that are inside the dotted line, have a special quality. Each of them can defeat each of the others that are outside the dotted line in a head-to-head race. It seems clear that the winner must not be drawn from $\{v, w, z\}$. Each of these can be defeated by each of the proposals in $\{u, x, y\}$. The **Smith criterion** is based on the idea that, if we can spot some "rejects" in this way, then we ought to make sure they don't win.

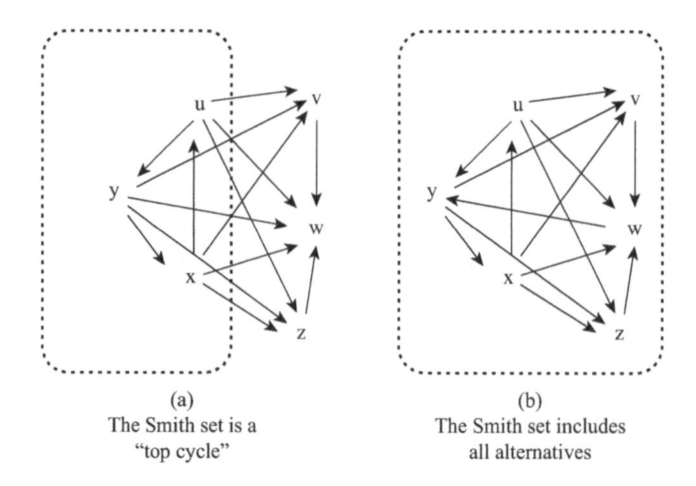

(a)
The Smith set is a
"top cycle"

(b)
The Smith set includes
all alternatives

Figure 3.3: The Smith Set.

Formally, the **Smith set** is defined as the smaller of two sets:

1. The set of all alternatives, X.
2. A subset $A \subset X$ such that each member of A can defeat every member of X that is not in A, which we call $B = X{-}A$. Formally, the Smith criterion states if $x \in A$ and $y \in B = X{-}A$, $x \succ_P y$.

Note we are using the plurality rule notation here, $x \succ_P y$ (see 3.3). Indifferent voters are ignored. As we have seen in Figure 3.3a, the Smith set may include a voting cycle, and that's why some authors refer to it as the "top cycle set." The top cycle includes only the "best of the best." The rest of the alternatives should be rejected.

One major shortcoming of the Smith criterion is illustrated in Figure 3.3b. We have turned around a few arrows, so there is no subset of alternatives in which each can defeat the rest. In such a case, the Smith set includes all alternatives. As a result, the Smith criterion sheds no light at all on the problem. The Smith criterion can be useful in ruling out alternatives in some examples, but not too many.

The Win-Loss Record. A second simple approach is to choose a winner on the basis of the win-loss records. This is often called the Copeland rule. We simply subtract the number of losses from the number of wins for each alternative and then rank the alternatives accordingly. If one reads the sports page during professional baseball or football seasons, the win-loss record is a very familiar concept. While simple in principle, this procedure has many shortcomings. Most importantly, if there is a voting cycle, then there are likely to be many ties in the win-loss column. Each alternative will have one win and one loss in the pairwise comparisons, so none can be distinguished from the others. Another problem with this approach is the sensitivity to the introduction of clones. One could pad the number of wins for an alternative by making copies of the alternatives that it can defeat. (For Three Stooges fans, if Curly is a loser, then nominate Moe and Larry as well so that the alternatives that defeat Curly also defeat Moe and Larry.)

Aggregated Pairwise Voting: The Borda Count Strikes Back! The aggregated pairwise vote is a method of evaluating a tournament. The alternatives accumulate votes in their head-to-head contests and the one with the most votes wins. It creates a transitive overall ranking, even if there is a pairwise majority rule cycle.

Suppose the voters are $N = \{1, 2, \ldots, n\}$ and that none of them are indifferent between any of the alternatives. Recall that the number of voters who prefer x to y can be represented as $|\{i \in N : x \succ_i y\}|$. The aggregated vote for alternative x is the sum of the support it receives against y and z in pairwise contests:

$$|\{i \in N : x \succ_i y\}| + |\{i \in N : x \succ_i z\}|$$

The aggregated pairwise vote has a number of interesting properties. Readers should easily convince themselves that the following claims are true:

1. A Condorcet winner (one who is undefeated by each of the others) is never ranked last by the aggregated pairwise vote, and
2. A Condorcet loser (one that is unable to defeat any of the others) is never ranked first by the aggregated pairwise vote.

The aggregated pairwise vote is never afflicted by the "dominated winner paradox" that was discussed above. While the Condorcet winner does not always come out on top in an aggregated pairwise vote, that candidate does not suffer the indignity of a last place finish either.

It turns out that these two properties of the aggregated pairwise vote have a powerful implication: only a restricted number of paradoxical voting outcomes is possible. With two alternatives, the aggregated pairwise system is really just plurality rule, so we might write either $x \succ_{AP} y$ or $x \succ_P y$. If the pairwise result is $x \succ_{AP} y$, then a supporter of Condorcet's view of the problem would expect that the social ranking should be $x \succ_{AP} y \succ_{AP} z$ or $z \succ_{AP} x \succ_{AP} y$. As long as the procedure applied to all three alternatives has x preferred to y, then everything makes sense. On the other hand, there is trouble if we observe social rankings like $y \succ_{AP} x \succ_{AP} z$ or $z \succ_{AP} y \succ_{AP} x$. Mathematician Donald Saari offers a way to catalogue these so-called anomalies. For a given set of voters (a preference profile), any voting procedure can be applied to rank the pairs, $\{x, y\}$, $\{y, z\}$, $\{x, z\}$, and also give an ordering of all three alternatives, $\{x, y, z\}$. Since the pairwise comparison is really just a plurality vote, let's use P as the subscript for pairwise comparisons, and we will use AP for the comparisons that take into account three or more alternatives. Saari uses the term *word* to refer to a combined set of pairwise and aggregated pairwise rankings, such as

$$word\ 1.\ x \succ_P y,\ y \succ_P z,\ x \succ_P z,\ x \succ_{AP} y \succ_{AP} z$$

or

$$word\ 2.\ x \succ_P y,\ y \succeq_P z,\ x \succ_P z,\ z \succ_{AP} y \succ_{AP} x.$$

There are 351 words possible with three alternatives (hint: 3·3·3·13 = 351). In word 1, there is no anomaly because each of the pairwise plurality decisions is the same as the ordering implied by the aggregated pairwise votes. On the other hand, word 2 is very anomalous. The pairwise comparisons are at odds with the overall ranking. In fact, recalling the first claim above, readers should notice that word 2 is impossible. The aggregated pairwise vote could not assign last place to the Condorcet winner, x.

Saari observes that only 135 out of the 351 words are actually possible with the aggregated pairwise vote applied to three alternatives (*Saari 1994*, p. 186). The driving force behind that result is the fact that the Condorcet winner can't finish last in the aggregated pairwise vote. The complete explanation of this result is given in the focus box on page 115.

In case you were wondering why the subtitle of this section is "the Borda count strikes back!" we are ready to give the answer: The aggregated pairwise voting procedure is equivalent to the Borda count (*Saari 1994*). The strengths of the aggregated pairwise vote are thus inherited by the Borda count. Saari advocates the use of the Borda count for a number of reasons, but the fact that it "is the 'natural' extension of the standard vote between two candidates" (*Saari 1994*, p. 178) to an election with more candidates is one of the most persuasive reasons.

How do we know that the two procedures are equivalent? Consider a voter i for whom $x \succ_i y \succ_i z$. The aggregation shows that i casts two votes for x (i prefers x to both y and z). Similarly, i votes for y against z, and the aggregation records one vote for y. This is, of course, exactly the same as casting a Borda ballot vector (2,1,0). The conclusion is that the Borda count has the same information that is collected by the aggregation of pairwise votes. The fundamental difference between plurality/majority voting and the Borda count is laid bare by this discovery. Recall the definition of plurality rule, which is the same as majority rule if none of the voters are indifferent.

$$\textit{Plurality rule}: x \succ_P y \textit{ if and only if } |\{i \in N : x \succ_i y\}| > |\{i \in N\ y \succ_i x\}| \tag{6}$$

Note that the first terms on each side of the following inequality are the same:

$$\begin{aligned}&\textit{Borda rule}: x \succ_{BC} y \textit{ if and only if}\\ &|i \in N\ : x \succ_i y\}| + |\{i \in N : x \succ_i z\}| > |\{i \in N\ : y \succ_i x\}| + |\{i \in N\ : x \succ_i y\}|\end{aligned} \tag{7}$$

The Borda comparison between x and y always includes some information about the irrelevant alternative z. The violation of the independence from irrelevant alternatives that we encounter in examples involving the Borda count appears to be inherent within it. At the same time, however, the Borda count avoids both intransitivities and the danger of allowing a Condorcet loser to win an election.

Feature Box: The Aggregated Pairwise Vote and the Borda Count. In *The Geometry of Voting*, mathematics professor Donald Saari argues that the Borda count is less prone to peculiarities than other voting methods that are based on ranked lists. The Borda count admits only 135 out of 351 possible words (combinations of pairwise and listwise outcomes). In contrast, other methods of tabulation allow anything (literally!) to happen. A central part of his argument is the fact that the Borda count is equivalent to aggregated pairwise voting, and in the latter we have observed that a Condorcet winner never places last.

To see why only 135 of the possible words can occur, begin by considering all of the Borda orderings with three alternatives. They are divided into three columns, representing social orderings in which the number of ties is zero, one, or two.

1. $x \succ_{BC} y \succ_{BC} z$	7. $x \succ_{BC} y \approx_{BC} z$	13. $x \approx_{BC} y \approx_{BC} z$
2. $x \succ_{BC} z \succ_{BC} y$	8. $x \approx_{BC} y \succ_{BC} z$	
3. $y \succ_{BC} x \succ_{BC} z$	9. $y \approx_{BC} z \succ_{BC} x$	
4. $y \succ_{BC} z \succ_{BC} x$	10. $y \succ_{BC} z \approx_{BC} x$	
5. $z \succ_{BC} x \succ_{BC} y$	11. $z \succ_{BC} x \approx_{BC} y$	
6. $z \succ_{BC} y \succ_{BC} x$	12. $z \approx_{BC} x \succ_{BC} y$	

We now have to figure out which pairwise outcomes are consistent with each of the Borda outcomes. The problem is attacked by the timeless method of "divide and conquer."

Consider outcomes 1-6, the ones in which there are no ties. Keeping in mind the fact that a Condorcet winner cannot finish in last place in the Borda count (and the Condorcet loser cannot finish first), we can figure out which pairwise comparisons are possible. We focus on this one case $x \succ_{BC} y \succ_{BC} z$ (and will generalize to the others).

Possible if $x \succ_{BC} y \succ_{BC} z$		Not possible if $x \succ_{BC} y \succ_{BC} z$
$x \succ_P y, y \succ_P z, x \succ_P z$	$x \approx_P y, y \succ_P z, x \succ_P z$	$x \approx_P y, y \succ_P z, x \approx_P z$
$x \succ_P y, y \succ_P z, z \succ_P x$	$x \approx_P y, y \approx_P z, x \succ_P z$	$x \approx_P y, y \succ_P z, z \succ_P x$
$x \succ_P y, y \succ_P z, z \approx_P x$	$x \approx_P x, z \succ_P y, x \succ_P z$	$x \approx_P y, y \approx_P z, z \succ_P x$
$x \succ_P y, z \succ_P y, z \approx_P x$	$y \succ_P x, y \succ_P z, x \succ_P z$	$x \approx_P y, y \approx_P z, z \approx_P x$
$x \succ_P y, y \approx_P z, z \succ_P x$	$y \succ_P x, y \approx_P z, x \succ_P z$	$x \approx_P y, z \succ_P y, z \succ_P x$
$x \succ_P y, y \approx_P z, z \succ_P x$	$y \succ_P x, z \succ_P y, x \succ_P z$	$x \approx_P y, z \succ_P y, z \approx_P x$
$x \succ_P y, y \approx_P z, z \approx_P x$	$y \succ_P x, z \succ_P y, z \succ_P x$	$y \succ_P x, y \succ_P z, z \succ_P x$
$x \succ_P y, z \succ_P y, x \succ_P z$	$y \succ_P x, z \succ_P y, z \approx_P x$	$y \succ_P x, z \succ_P y, z \succ_P x$
		$x \succ_P y, z \succ_P y, z \succ_P x$
		$y \succ_P x, y \succ_P z, x \approx_P z$

There are 17 sets of pairwise contests that are consistent with $x \succ_{BC} y \succ_{BC} z$. Since the setup of the problem is completely symmetric, the same must be true of Borda orderings 2-6. Since $6 \times 17 = 102$, we have found 102 of the 135 legal words.

Next consider the six Borda outcomes that have only one tie,which are items 7-12. There are 30 sets of pairwise outcomes that are consistent with these Borda results (five for each one). Considering Borda ordering 7, we find that the following are the legal words.

$$
\begin{array}{ll}
1. & x \succ_P y,\ x \succ_P z,\ y \approx_M z,\ x \succ_{BC} y \approx_{BC} z \\
2. & x \succ_P y,\ x \succ_P z,\ y \succ_P z,\ x \succ_{BC} y \approx_{BC} z \\
3. & x \succ_P y,\ z \succ_P x,\ y \succ_P z,\ x \succ_{BC} y \approx_{BC} z \\
4. & x \succ_P y,\ x \approx_P z,\ y \succ_P z,\ x \succ_{BC} y \approx_{BC} z \\
5. & x \approx_P y,\ x \succ_P z,\ y \succ_P z,\ x \succ_{BC} y \approx_{BC} z
\end{array} \tag{8}
$$

With five more legal words for each of the six Borda outcomes, we have thus added $5 \times 6 = 30$ valid words. The proof of this result is left as an exercise for the reader (but is very informative, and so a complete answer is included at the end of the book).

Finally, consider Borda ordering 13. If the Borda count yields total social indifference, $x \approx_{BC} y \approx_{BC} z$, only three sets of pairwise matchings are possible

$$
\begin{array}{ll}
1. & x \succ_P y,\ y \succ_P z,\ z \succ_P x \\
2. & y \succ_P x,\ z \succ_P y,\ x \succ_P z \\
3. & x \approx_P y,\ y \approx_P z,\ z \approx_P x
\end{array} \tag{9}
$$

The first two are the cyclical pairwise comparisons, and the final is generalized indifference. With those three legal words, out total is now 135.

The Schulze Method. If you search the Internet, this one is often found under the unseemly titles "Cloneproof Schwartz Sequential Dropping" or the "Beatpath Method." We have named it for its originator, Markus Schulze, who has been an active proponent of the method since 1997 and has shown that it has many desirable qualities (*Schulze 2003*). This method evolved in a sequence of email and Internet postings by a group of enthusiasts who sought to develop workable voting methods that can actually be put to use in "real life." Schulze notes that this method is used in organizations that have an aggregate membership of more than 1,700 and that it is the most widely used of all of the Condorcet round-robin methods. People who use the Linux operating system might be interested to know that the Schulze method is used in decision making in the Debian Linux project, one of the most widely disseminated Linux distributions.

Of all of the methods we have considered so far, this is the most difficult to understand and interpret at face value. In order to justify it, the proponents do not rely on intuition. Rather, they rely on the mathematical fact that this procedure meets many of the criteria against which voting procedures are judged. If there is a Condorcet winner, this method selects it as the winner. The Schulze method satisfies many of the other most important priorities, including undifferentiatedness (same as neutrality), anonymity, monotonicity, independence of clones (clone-proofness) and several others.

Schulze's method also satisfies a property known as Smith-IIA. Recall the Smith criterion, which held that the winner should be a member of the Smith set. Smith-IIA adds the additional stipulation that if a candidate is added in an election, it can only affect the election outcome if it is in the Smith set. If it is not in the Smith set, then it is an irrelevant alternative (so far as the choice among alternatives in the Smith set is concerned) and it should not matter.

The voters submit rankings of as many candidates as they wish, and the ones they do not rank are assumed to be interchangeable and less desirable than the ones they do rank. There are two mathematically equivalent descriptions of the Schulze method for picking the winner in a competitive election. These methods employ the idea of "transitive defeat." A "transitive defeat" (also known as a chain) allows x to assert itself against y through a chain of comparisons, like $x \succ_P w$ and $w \succ_P y$. Even if y defeats x in a head-to-head comparison, if a chain from x to y exists, then x can "transitively defeat" y. You might think of the chain as a sort of "self-defense" or "political countermeasure" for electoral candidates. The chains can be quite long, and to save space we might as well write $x \succ_P w \succ_P z \succ_P y$ to mean that x transitively defeats y with w and z in the middle of the chain.

As the old saying goes, a chain is only as strong as its weakest link. The strength of a chain from x to y is measured by the smallest margin of victory between any two alternatives in the chain. The symbol $P[x, y]$ represents the strength of the strongest chain from x to y. If there are two chains leading from x to y, we measure the strength of each one (i.e., find the smallest margin of victory), and $P[x, y]$ is the largest value. If there is no chain from x to y, then the strength is $P[x, y] = 0$.

The "beatpath method" is the one that Schulze uses as the definition of the social ranking process. The aim is to find an alternative with the strongest chains through which it can defeat all other alternatives. If the chain from x to y is stronger than the chain from y to x, $P[x, y] > P[y, x]$, then x disqualifies y from further consideration. If we collect up all the alternatives that are not disqualified, then we have a set of "potential winners." If we are lucky enough to have only one candidate left, then it is the beatpath winner. If we have several left, then a tie breaking procedure is needed. Quite frankly, the emphasis in this method is to narrow the alternatives down to this set of unbeaten candidates, and the tie breaking procedure is something of an afterthought.

To bring this down to earth, consider the application of this method to choosing a national football champion after gathering votes from sports writers and coaches. The voters rank three undefeated teams, Auburn (AU), the University of Southern California (USC), and the University of Oklahoma (OU). In Figure 3.4, the arrow from USC to OU indicates that USC is preferred to OU by 44 of the voters. Here, we are using the number of voters as the indicator of the strength of the defeat

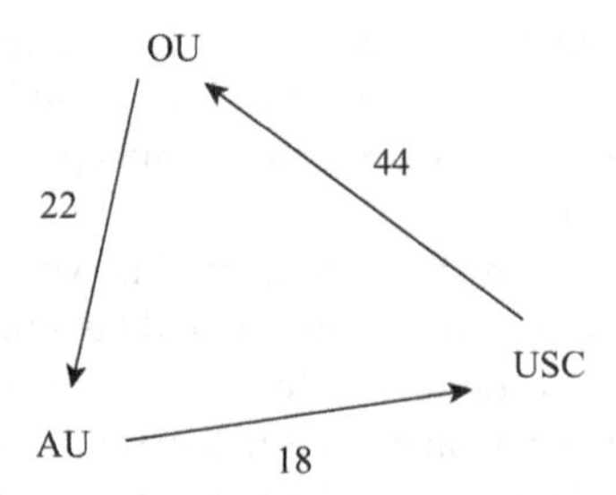

Figure 3.4: Beatpath for College Football.

(we could use the margin instead). Note there is a cycle, where OU is ranked higher than AU (strength = 22) and AU is preferred to USC (strength = 18). This example is particularly simple because there is only one path from each alternative to each of the others. The beatpaths are:

$$
\begin{array}{ll}
1 & P[OU, AU] = 22 \\
2 & P[OU, USC] = 18 \\
3 & P[AU, USC] = 18 \\
4 & P[AU, OU] = 18 \\
5 & P[USC, OU] = 44 \\
6 & P[USC, AU] = 22
\end{array}
$$

Because $P[USC, OU] > P[OU, USC]$ and $P[USC, AU] > P[AU, USC]$, USC is the beatpath winner.

The implications of the beatpath calculations are difficult to grasp for many people. Why does the procedure satisfy so many of the desirable qualities? How can we make sense of it? Shortly after Schulze proposed his method in 1997, it was shown that it is equivalent to a method in which we sequentially measure the strength of the candidates and then remove the weak ones, and then re-evaluate. The sequential evaluation process is organized around an enhancement of the Smith set which is called the Schwartz set. This approach is often called Cloneproof Schwartz Sequential Dropping (CSSD) because we will sequentially re-calculate the Schwartz set after penalizing the weak candidates.

In order to be a member of the **Schwartz set**, an alternative must either be undefeated in pairwise competition (it may defeat or tie each other alternative) or it must be able to transitively defeat every other alternative that can transitively defeat it. To be in the Schwartz set, a candidate must have "electoral self-defense" against each of the other candidates. More formally, x is in the Schwartz set if for any y such that $P[y, x] > 0$, then $P[x, y] > 0$. Even if x transitively defeats y, y can be in the Schwartz set if there is a chain of plurality decisions reaching from y to x. An alternative y is rejected from the Schwartz set if y cannot transitively defeat another alternative which can transitively defeat y. If there are no pairwise tie votes, the

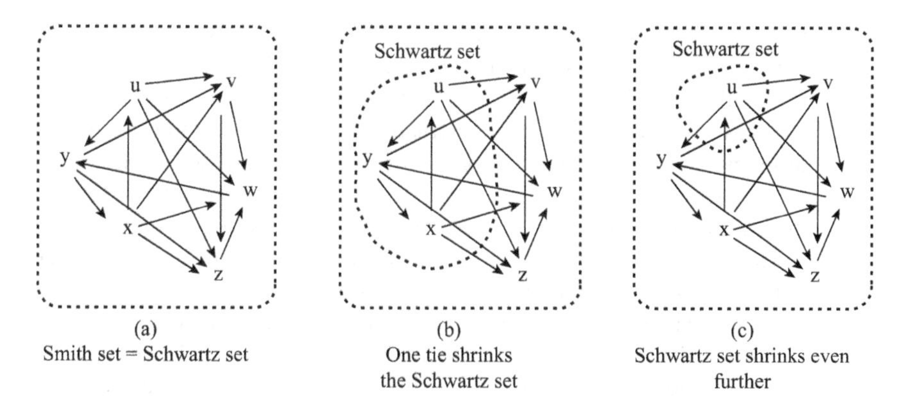

Figure 3.5: Schwartz sequential dropping.

Schwartz set equals the Smith set. If there are ties, then the Schwartz set can be much smaller than the Smith set. In Figure 3.5, ties are represented by dotted lines that connect alternatives. The ties make a difference because the transitive defeat concept is based on pairwise victories and ties are treated the same as losses. In other words, ties break chains.

The sequential approach can be thought of as a way to find the strongest chains. The Schwartz sequential dropping version of the Schulze procedure is the one that is described in the Constitution of the Debian software organization (http://www.debian.org/vote/ 2003/vote_0002). In Figure 3.5, we illustrate the sequential narrowing of the Schwartz set that results when ties (dotted lines) are inserted.

Schwartz sequential dropping proceeds iteratively, repeatedly narrowing down the alternatives. It is easy to make this sound more complicated than it really is, but we will try not to.

Step 1. Collect the ranking information from the voters. Voters are allowed to rank all of the alternatives or just some of them. They are allowed to register their indifference between alternatives. From the ballots, we can calculate the pairwise outcomes for all alternatives, either $x \succ_P y$ or $x \approx_P y$ or $y \succ_P x$.

Now the decision-making starts.

Step 2. Eliminate rejects. Begin by simplifying the problem by isolating "rejects," alternatives that are losers in the round-robin tournament. These losers are the ones that are left out of the Schwartz set. Recall that y is rejected if it is transitively defeated by some other proposal and there is no chain that leads from y to that other alternative. The rejects are eliminated from consideration, permanently.

Step 3. Evaluate the remaining contenders (members of the Schwartz set).

A. If there is only one remaining contender, then it is the winner.

B. If there is more than one contender, check the pairwise comparisons among all of the contenders. If all pairs are tied, it means, in some sense, that we have found all of the alternatives that are "equally good." If all are tied, proceed to the tie-breaking procedure in Step 4.

C. If all contenders are not tied pairwise, then we need to make some changes so we can reject more alternatives. Here is where the "sequential dropping" kicks in. Among all pairs of contenders, find the "weakest pairwise defeat." (To us, a name like "least impressive victory" would be more fitting.) The idea is to spot the least worthy pairwise victor and demote it. How to spot the least impressive winner? Look for the alternative that wins with the smallest number of votes.[3] When the weakest pairwise defeat is found, that victory can be erased by changing it to a tie in the data. That is, we change the result in the round robin data for the weakest defeat from $\succ_P$ to $\approx P$. (Caution: contrary to what some websites say, neither the weakest defeat nor the tied alternatives are dropped literally. The weakest defeat is converted to a tie and the Schwartz set is recalculated.) If there are several alternatives that are equally weak winners–many weakest defeats exist–then choose the one with the most votes for the loser and declare that one to be the weakest defeat. If there are several pairs tied in both the number of votes for the winner and the loser, treat them all in the same way: change the pairwise comparison to a tie. This generates new round robin data.

Go back to Step 2 and repeat the procedure. Use the new round robin data, of course. The rejects that were eliminated in previous stages are never re-admitted. Since we have inserted some ties into the results, the members of the Schwartz set will change: some contenders will become rejects and then they are dropped. The procedure will repeatedly loop between Steps 2 and 3 until the condition described in Step 3B is obtained.

Step 4. If we arrived at this stage, it means we have narrowed down the alternatives to the most appealing set of more-or-less equally good outcomes. Schulze calls these "potential winners." A winner is selected by a tie breaking rule.

The tie breaking rule that is suggested by Schulze is, more-or-less, going to rank the tied outcomes by listening to the opinions of randomly drawn voters. Suppose the tied candidates are $\{w, x, y, z\}$. Randomly draw a ballot and copy down the preferences of the voter. If that ballot states a ranking for all of the alternatives, then that is the end of the procedure. The group result will match that ballot. If the ballot is only partially filled in, for example, it states only $x \succ_i y$, then we take that information and mark down $x \succ_{Schulze} y$ (it is "set in stone," unalterable) and we draw another ballot. If that ballot has preference information about candidates that are

[3] Some versions of this procedure use the margin of victory, rather than the absolute number of votes received by the winner. Schulze recommends the total number of votes because he feels it gives voters a positive incentive to report their rankings as completely as possible. Otherwise, they might have an incentive to report just the rankings of the top one or two alternatives in order to "game" the election system.

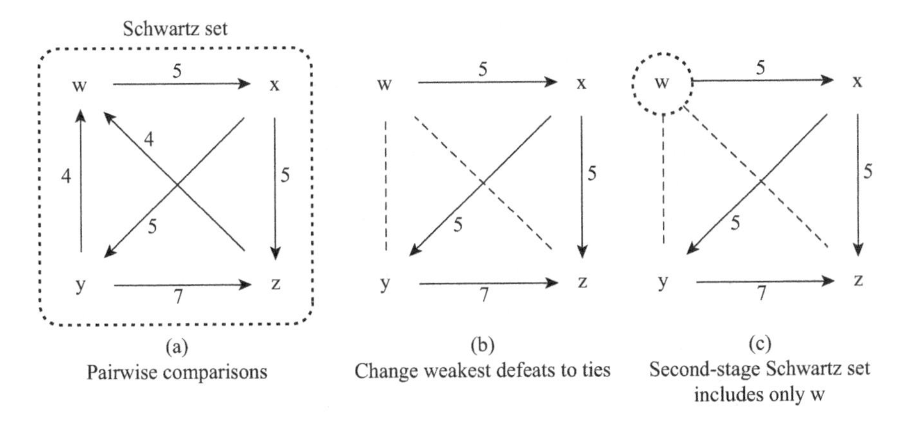

Figure 3.6: Schulze method applied to Table 3.3.

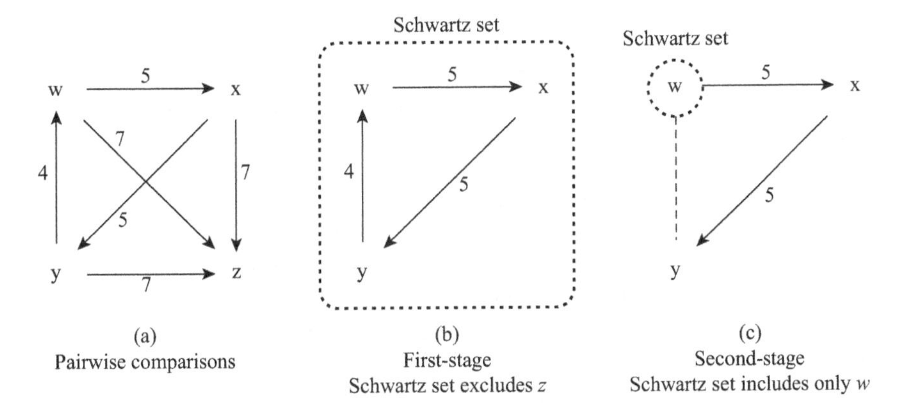

Figure 3.7: Schulze method applied to Table 3.4.

not already sorted by $\succ_{Schulze}$, such as w and y, then use that information $w \succ_{Schulze} y$. We draw ballots at random until every pairwise comparison is filled in, and we never change any of the Schulze comparisons we have already completed. If we happen to go through all ballots without ranking all of the alternatives, then the final ranking is assigned at random.

The Schulze method can be applied to any of the voting examples we have considered. Of course, if the preferences imply the existence of a Condorcet winner, then that alternative is chosen and the solution is simple. To find a more interesting case, we need to select an example in which there is a voting cycle. Recall the Borda paradox considered in Section 3.4.2. In Figures 3.6 and 3.7, we have used the preferences from Tables 3.3 and 3.4 with the Schwartz sequential dropping approach.

In Figure 3.6a, we note that the original preferences allow a majority rule cycle across all of the alternatives, so the first stage Schwartz set includes all of the alternatives. There are two comparisons tied for the weakest defeat, each with the support of 4 voters, so in 3.6b, those arrows are changed to dotted lines representing ties. After those ties are taken into account, the Schwartz set is recalculated, and it includes only w.

Recall that alternative z is exposed as a fraud in that example. The effect of dropping z to the bottom of the voter preference orderings appears in Figure 3.7. Note that the initial Schwartz set now includes three alternatives, $\{w, x, y\}$. Alternative z is excluded (it is a Condorcet loser). The weakest defeat is between w and y. After that is changed to a tie, then w is the only potential winner remaining. The same alternative, w, wins with the Schulze method in both cases.

The Schulze method has the apparent strength of solving the paradox in the Borda count: w wins in both cases. The Schulze method is thus more stable and, in some sense, more logical. The fact that the Borda method chose y in the original example (Table 3.3) and Schulze chooses w deserves mention. From a beatpath point of view, it appears that we did not correctly understand the Borda problem in Section 3.4.2. We thought the paradox was that y was turned from a winner into a loser when the fortunes of z changed. Now it appears that the original Borda winner, y, was "wrong." Borda chose the wrong winner because it allowed z to play the role of a spoiler, somehow distracting voters from their purpose, thus allowing y to win. When z is moved to the bottom in Table 3.4, its role (its influence as an irrelevant alternative) is eliminated, and the "correct alternative" w wins.

The Schulze method leads to a full, transitive ordering of all of the candidates. The beatpath values $P[x, y]$ can be used to sort alternatives from top to bottom. As a result, the method could be used to determine the winner of the Division 1A College football championship on the basis of the rankings submitted by coaches, writers, or computers.

The reader might be wondering, "If this method is so great, why isn't everyone using it?" Well, if you ask its proponents, we expect they will say, "Because everyone is not very smart!" Ordinary people – not academics – are not particularly interested in election procedures, much less complicated ones. As a result, the most widely used argument against the method seems to be that it is too complicated. To academics, this is not a very good argument, but it carries a lot of weight with real life policy makers. In response, the Schulze proponents will point out that, from a voter's point of view, it is no more complicated than a Borda count. All the voters need to do is submit rankings of some alternatives. The counting process is more complex, of course, but the effort is justified. The proponents can list a long list of favorable qualities that we have already mentioned, plus some more.

The beatpath approach is not completely free from the anomalies that infect the other voting mechanisms. Schulze draws our attention to the fact that his method violates the **Participation Criterion**. The Participation Criterion says that if an election has been decided, and then you add voters who prefer the winner to another alternative, then adding those new voters must not result in the victory of

Group:	1	2	3	4	5	6	7	8	9	10	11	12	13
Number of Voters:	4	2	4	2	2	2	4	12	8	10	6	4	3
Preferences													
Favorite	u	u	u	u	v	w	w	x	y	z	z	z	u
Second	v	v	y	y	z	x	x	y	w	u	u	y	y
Third	w	z	v	z	u	v	v	w	x	v	v	x	z
Fourth	x	x	z	v	w	y	z	u	v	w	x	v	w
Fifth	y	y	w	w	x	z	y	v	z	x	y	w	v
Sixth	z	w	x	x	y	u	u	z	u	y	w	u	x

Table 3.11: The Schulze method.

that other (less favored) alternative. More formally, the winner is x, hence socially $x \succ_{Schulze} y$. Add more voters such that $x \succ_i y$. Adding those voters should not change the social result to $y \succ_{Schulze} x$. The criterion does not require that x must still win, only that adding voters who do not prefer y should not make y the winner. Please note that the Participation Criterion is different from the monotonicity principle. Monotonicity states that *among existing voters*, a change in voter preferences that favors x against y must not hurt x's position in the final decision. Participation concerns the addition of new voters. Schulze is a monotonic procedure, but examples can be found to show that it violates the participation criterion.

Schulze's example of the violation of the Participation Criterion is presented in an example that has six candidates, $\{u, v, w, x, y, z\}$ (see Table 3.11). The winner is u when the electorate includes the 60 voters from groups 1-12. The violation of the Participation Principle is shown when group 13, which has three members, is added to the electorate. For all of the members of group 13, note that u is the most preferred alternative and x is the least desirable. One would expect that their favorite, the winner u, would remain in good position. Unfortunately, the Schulze rule applied to this 63 voter electorate declares x to be the winner. After the sequential dropping process, u is the only contender (the only member of the final Schwartz set), no use of the "tie breaker" is required.

Among the collection of peculiar election outcomes that we have been accumulating, this is surely one of the most bizarre. In order to investigate it more carefully, consider the graphs that illustrate the sequential dropping process in Figure 3.8. On the left, the pairwise data for the 12 group problem is illustrated. Note there are many ties in this example. The weakest defeat is the step $y \succ_P z$ which is supported by 32 voters. After that is changed to a tie, the Schwartz set shrinks to $\{u, v, w\}$, and from there it is easy to drop the weakest defeat and see that u wins. On the right side of the figure, note that when group 13 is added, there are no pairwise ties. When the weakest defeats are changed to ties, as shown in the top figure, then the pattern of ties and pairwise comparisons in the initial results on with 12 groups re-appear. The ties are the same, but the margins of victory are different. With 13 groups, the weakest defeat is now $v \succ_P w$. After a tie is drawn between v and w, the Schwartz set shrinks

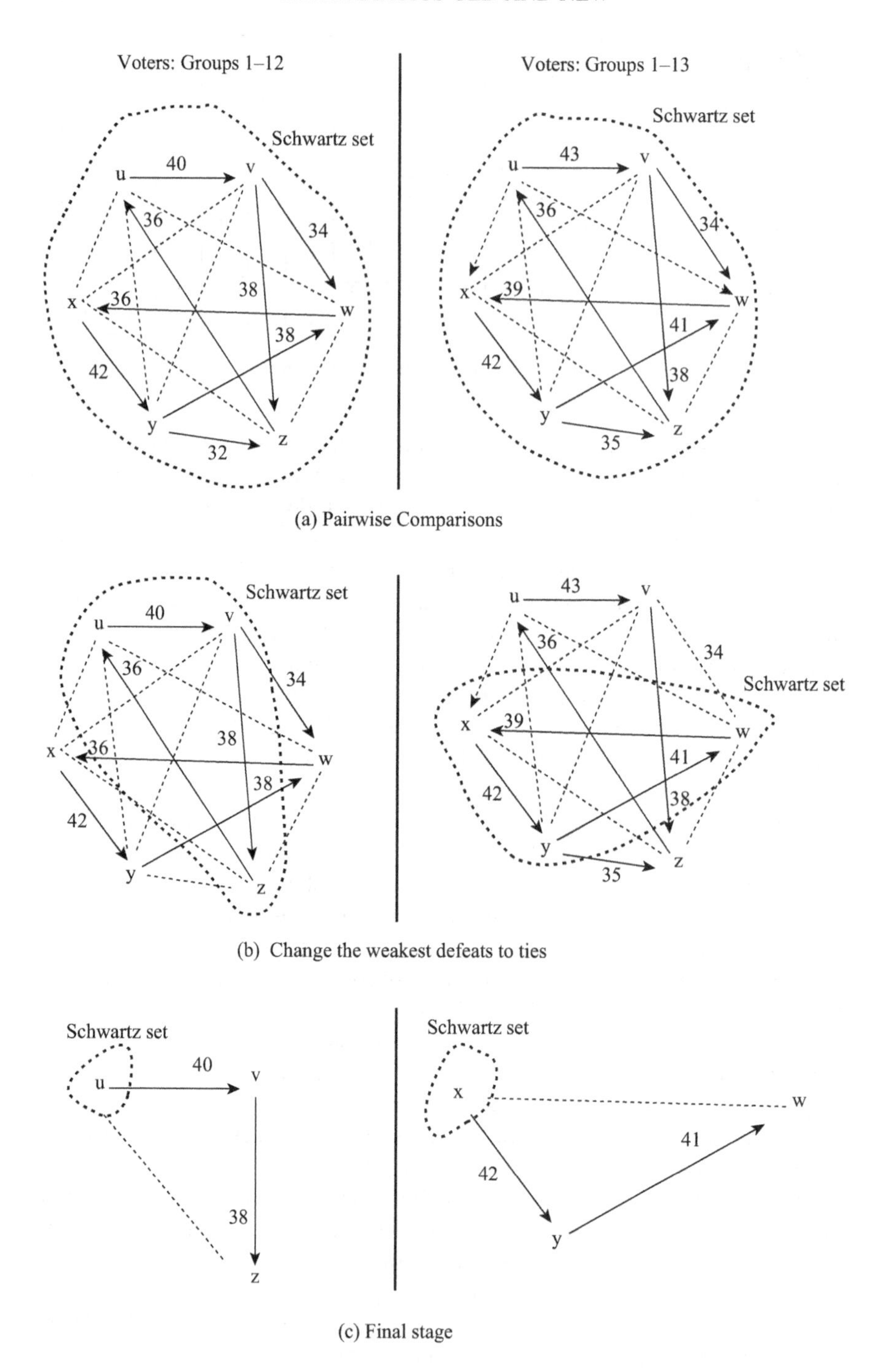

Figure 3.8: Schulze method with preferences in Table 3.11.

to $\{x, w, y\}$, and from there the result is easy to see: x wins. Why did u lose? When the path from v to w was removed, then u lost its chains of self-defense. The alternatives x, y, and z can all transitively defeat u, and there is no path leading from u back to them if $v \approx_p w$.

This is a variant of the problem of independence from irrelevant alternatives. What is driving the problem here is not the relative evaluation of u and x, but rather the careful placement of u, v, w, and z in the orderings of group 13 voters. The careful placement ends up changing the magnitude of pairwise victories, thus impacting the order in which the sequential dropping eliminates candidates. If you think of the sequential dropping process as a series of games in a tournament, then we have simply found another example of the way in which the ordering of the matches affects the outcome.

Exercises 3.6

1. The following table describes the tastes of three voters. Conduct a round robin tournament and use the win-loss records to try to decide which candidate should win the tournament.

	Voters		
	1	2	3
Ranking			
First	x	y	z
Second	y	z	x
Third	z	x	y

2. Consider the table in the previous question. Describe a couple of changes you might make in voter preferences (either by changing voters or adding new ones) if you want z to become the winner in the round robin.
3. For the following preference profile, conduct a round robin tournament and calculate the aggregated pairwise vote totals for the candidates. Confirm that the results are the same as the Borda count.

	Voters		
Groups	1	2	3
Number of Members	22	12	31
Ranking			
First	x	y	z
Second	y	x	y
Third	z	z	x

4. Consider the preferences used in the previous problem. Create a graph of the type shown in Figure 3.4. Use that graph to figure out which candidate would win with the Schulze procedure.
5. Consider the following preference profile. Create a beatpath graph of the type shown in Figure 3.4.

	Voters					
Groups	1	2	3	4	5	6
Number of Members	200	150	150	300	100	100
Ranking						
First	z	y	x	w	y	x
Second	x	z	y	z	x	y
Third	y	x	z	x	w	z
Fourth	w	w	w	y	z	w

6. Use the graph you created in the previous question to figure out which alternatives are in the Schwartz set. If the Schwartz set contains more than one element, proceed with the sequential dropping process (change the weakest defeats to ties and re-calculate the Schwartz set).
7. This is a challenging problem, one which we refer to in the feature box on page 3.44. Show that if the Borda count (or, equivalently, the aggregated pairwise vote) outcome is $x \succ_{AP} y \approx_{AP} z$, then the only pairwise outcomes that are consistent with that outcome are the ones listed in equation (8).

3.7 Single Vote Systems: Cousins of Plurality and Majority Rule

We began this chapter with the plurality problem. A plurality election with twenty or thirty candidates might award victory to a very unpopular candidate. The poster children for the problem might be the Democratic and Republican candidates in Minnesota's gubernatorial election who were edged out by former professional wrestler Jesse "The Body" Ventura. In the words of mathematician Keith Devlin, Jesse Ventura "won not because the majority of the voters chose him, but because plurality voting effectively thwarted the will of the people. Had the voters been able to vote in such a way that, if their preferred candidate were not going to win, their preference between the remaining two would be counted, the outcome would have been quite different" (*Devlin 2004*). The alternative schemes, the Borda rank order voting plan and the pairwise Condorcet systems, offer some hope. They have not won the hearts and minds of electoral reformers with the same power as the alternative to which we now turn our attention.

This system requires the voters to come to the polls only once. The voters rank the alternatives and then the election officials can proceed "as if" there really were a runoff election. When this procedure is used to select a single winner in an election, it is generally referred to as an Instant Runoff Vote (IRV) or an Alternative Vote system. When the method is used to select a group of candidates, such as the top five nominees for the Grammy Awards, or in a proportional legislative election, the procedure is called the Single Transferable Vote (STV). The procedures are the same for choosing a single winner, but the details get complicated for selecting the additional winners. There are two prominent implementations of the STV concept, the Hare system (in honor of Thomas Hare, an English proponent of this system) and the Hare-Clark system (add an honorific for Andrew Inglis Clark, a Tasmanian Attorney-General who pushed a variant of the STV into usage). The transferable vote is in use in national elections in Australia, Malta and Ireland as well as in local elections in Cambridge, Massachusetts, San Francisco, California, and cities in Northern Ireland and New Zealand. The method is used in faculty elections at Cornell University.

We will focus our attention on the selection of a single winner from a field of candidates, so we will detail only that part of the procedure.

The winner must obtain a target number of votes, dubbed the quota. If there is only one spot open–only one candidate can win–the usual quota is the majority of the votes:

$$quota = majority \geq \frac{(Number\ of\ voters + 1)}{2}$$

On election day, the voters go to the polls and rank the candidates from first to last in order of desirability. Then election managers name the winner by applying the following algorithm.

Step 1. A voter's support on the first ballot is given to the voter's top-ranked candidate. If one of the candidates has a majority, then that candidate is the winner. If no candidate has a majority, proceed to Step 2.

Step 2. The candidate with the smallest number of votes is eliminated from consideration. The ballots of the voters who had supported that eliminated candidate are reconsidered. Their votes are transferred to their second-ranked candidates. Then the votes are re-tabulated. If a candidate has a majority, then it is declared the winner. Otherwise, repeat Step 2. (The candidate that was eliminated in this step is excluded throughout the remainder of the process.)

If the election is choosing five members of a commission or a legislative delegation, the procedure is slightly more involved. Not only are the votes cast for the losers transferred, but also the "excess" votes (ones which exceed the quota) received by the winners can be transferred.[4] The calculations in that re-allocation are a bit involved; that is where the variants of STV diverge. The main idea in the procedure is that candidates are removed one-by-one, and the votes cast for the removed candidates are either counted for them (as part of their quota) or transferred and counted for other candidates that the voter also supports.

The proponents of this method claim that it does not throw away information in the same way as a plurality election. It saves voters from the terrible choice that is forced on them by a plurality vote. If the voters who favor either the Democrat or the Republican against Jesse Ventura have a chance to declare that priority, then

4

$$quota = \frac{Number\ of\ votes\ cast}{Number\ of\ winners\ to\ be\ chosen + 1} + 1$$

Note that if there is one winner to be chosen, the winner must have a bare majority.

Ventura might not end up winning in the Instant Runoff. The "Ventura problem," coupled with the on going use of the procedure in some places, gives the STV advocates some persuasive ammunition. Krist Novoselic, most well known as a member of the band Nirvana, has written a book that combines his band memoir with an endorsement of transferable votes, *Of Grunge and Government: Let's Fix This Broken Democracy* (2004). He also hosts a website (http://www.fixour.us). He eagerly watched the introduction of STV in San Francisco for the 2004 Board of Supervisors elections. Steven Hill, of the Center for Voting and Democracy, a proponent of the transferable vote systems, enthusiastically proclaimed, "More voters picked their supervisor, and fast results mean that San Francisco avoided a December runoff election, saving millions in taxes (the cost of the second election)" (*Novoselic 2004*).

There is no doubt that the STV might work to correct bad outcomes in some particular examples. Perhaps Jesse Ventura might not have been governor of Minnesota. There are example problems that we can consider in which the STV does not seem to perform too badly. If the STV procedure is applied to the preferences of groups 1-12 in Table 3.11 (the data for the Schulze example that violated the participation criterion), the STV stages eliminate v, then w, then y, and then u. The two remaining candidates, x and z, are tied with 30 votes. After adding group 13 into the calculations, the elimination process rejects the alternatives in the same order (v, w, y, and then u). The winner is z, which earns 33 votes against the 30 for x. The new voters prefer z to x, and it seems quite reasonable that adding them to the electorate would tip the scales in favor of z.

On the other side of the ledger, however, the STV has several properties that should cause concern. The fact that the STV system does not always select a Condorcet winner (when one exists) is a serious shortcoming. Academic research has often concentrated on another issue: STV is not monotonic. The principle of monotonicity has first come to our attention in the consideration of majority rule and May's Theorem. Now it comes to the forefront again. The basic idea of the Monotonicity Criterion is that an election system should be responsive, and not in a perverse way. If a voter changes his mind and decides to rank an alternative more highly, then that alternative's electoral fortunes should be helped, or at least not hurt. This is the property commonly called weak monotonicity.

Consider the preference rankings in Table 3.12. There are preferences for five groups, $\{1, 2, 3, 4, 5\}$ and four alternatives, $\{w, x, y, z\}$. The STV algorithm proceeds as follows.

1. The first place votes are, in order, $w = 9$, $x = 6$, $y = 6$, $z = 5$.
2. Eliminate z, so the candidates are w, x, and y. Re-allocate the votes of group 1 to their second-ranked candidate, x. The votes are retabulated: $x = 11$, $w = 9$, $y = 6$.
3. Eliminate y, so the final two candidates are w and x. Re-allocate the votes of groups 2 and 3 to x. The final vote totals are $x = 17$, $w = 9$. So x is declared the winner.

	Voters					Alternative Preferences
Groups	1	2	3	4	5	3′
Number of Members	5	4	2	6	9	2
Ranking						
First	z	y	y	x	w	x
Second	x	z	x	y	z	y
Third	y	x	z	z	x	z
Fourth	w	w	w	w	y	w

Table 3.12: Single transferable vote violates monotonicity.

We make only one small change in order to demonstrate the fact that STV is non-monotonic. We will replace group 3 with the votes in group 3′, so that the columns being used in the vote are {1, 2, 3′, 4, 5}. We hope the reader will agree that this seems to be an innocuous change. The voters in 3 think x is second best but the voters in 3′ think that x is the most desirable of all; x should still win, shouldn't it?

Observe how the STV result changes.

1. The first place votes are, in order, $w = 9$, z = 8, $y = 4$, $z = 5$.
2. Eliminate y, so the candidates are w, x and z. Re-allocate the votes of group 2 to their second-ranked candidate, z. The votes are thus retabulated: $z = 9$, $w = 9$, $x = 8$.
3. Eliminate x, so the final two candidates are z and w. Re-allocate the votes of groups 3′ and 4 from x to z. The final vote totals are $z = 17$, $w = 9$. The winner is z.

This is a clear violation of weak monotonicity (and, we might add, plain old common sense). The winner of the original STV was x, but after we raise x's popularity by replacing 3 with 30, the winner is z. Alternative x was eliminated in stage 3. This would be interpreted to mean that x finished in third place. And the only change between the first and second example was that the voters in group 3 became group 3′, meaning they raised x in their preferences. After noting this fact, William Riker concluded, "It is hard to believe that there is any good justification for the single transferable vote" (*Riker 1982*, p. 51).

Exercises 3.7

1. Consider the preferences used in problem 2 of Section 3.4. Conduct an STV election.
   (a) In what order will the candidates be eliminated? How will the votes be re-allocated?
   (b) Who will be the eventual winner?
   (c) If you conduct an ordinary runoff election with the top two vote getters, does the result agree with that of the instant runoff?

2. Repeat the previous exercise with the preferences used in problem 3 of Section 3.4.
3. Repeat the previous exercise with the preferences used in problem 2 of Section 3.5.
4. The previous problems should lead you to expect that the instant runoff gives the same results as the ordinary majority/runoff election procedure. With four candidates, however, the same is not true. Conduct elections with these voter preferences to see an example.

	Voters					
Groups	1	2	3	4	5	6
Number of Members	100	250	150	300	100	100
Ranking						
First	z	y	x	w	y	x
Second	x	z	y	z	x	y
Third	y	x	z	x	w	z
Fourth	w	w	w	y	z	w

5. Create a voter preference profile in which three candidates are considered and there is a Condorcet winner. Then calculate the instant runoff election winner. Experiment with your example until you find a profile in which the Condorcet winner does not win the instant runoff.
6. Suppose you are a member of the book of the month club. The club must choose between the classics *The Grapes of Wrath* and *Superman: Battle to the Finish with Luther!* You would love to have *Superman*, but you know for sure that at least 70% of the club members are snobs who would vote for *Grapes*. Think of two methods you might try in order to bring about the eventual selection of *Superman*. Note, you are not allowed to make a persuasive speech to change their minds or do anything that would change the pairwise comparison of the two books, but you can introduce new alternatives and different voting methods.

3.8 Conclusion

After reviewing the many voting systems and studying the many principles by which they are judged, the reader has a right to ask the author, "How do you rank these systems? Which is the best?"

We find ourselves wishing that there was some voting system that would meet all of the requirements that seem fair and logical. Wishing does not make it so, sadly enough. In fact, there is a theorem, known as Arrow's Possibility Theorem, which states, basically, that it is mathematically impossible to find a voting system that satisfies what appear to be the most basic, bare-minimum requirements. Arrow, an economist, was awarded the Nobel Prize, partly on the basis of this research (*Arrow 1963*).

Arrow's theorem can be stated in several ways. We think this is the most useful version. Suppose there is an election-based ranking system that meets these requirements:

Nondictatorial: The social ranking does not mirror the tastes of one voter without regard to the tastes of the other voters.

Pareto principle: If all voters favor one alternative over another, then the social ranking should respect that and rank the alternatives accordingly.

These seem like such simple, fundamental elements that they can be assumed as necessary ingredients in a social ranking procedure. Arrow's Possibility Theorem says the following. If an election system is nondictatorial and obeys the Pareto principle, then it is possible to find an example set of voter preferences such that:

1. The social ranking is intransitive (a voting cycle), and/or
2. Independence of irrelevant alternatives (IIA) is violated.

It is very important to emphasize that the Possibility Theorem does not mean that a ranking procedure is always intransitive or in violation of IIA. Rather, it is always possible to find an example society, a set of voter preferences, such that the result is somehow illogical. If the result is intransitive, then we have the paradox of "rational man, irrational society" (*Barry and Hardin, 1982*). If we adopt a system that blocks intransitivity, such as the Borda count, then we know for sure there are examples in which independence from irrelevant alternatives is violated.

Arrow's theorem is sometimes overstated. For example, some students think it says that democracy is always illogical. It does not mean anything of the kind. Rather, it means that it is not possible to find a democratic procedure that works in a sensible way all of the time.

If it is not possible to meet such basic, simple criteria as nondictatorship, of course it is not possible to meet even more elaborate properties like the participation criterion. As a result, the choice of electoral system has to be based not on which system meets all of the specified criteria, but rather on a value judgment about what sorts of peculiarities are the most harmful and which sorts of societal preference profiles are most likely to arise and cause peculiarities.

If the reader wants to know, "Which ranking system is best?" we can sympathize. We want to know the same thing. And so do most political scientists, economists, mathematicians, physicists, and psychologists.

Personally speaking, we have found pleasant results by using procedures that eliminate the obviously bad candidates (by some principle like the Smith set or the Schwartz set) and then choosing among the remainder by a positional method like the Borda count or the Schulze method. We have been very unhappy with decisions reached by pure plurality or a runoff election. The STV might be preferable to a plurality or majority rule election, but the fact that it is so badly nonmonotonic causes us to shy away from it. That is to say, plurality rule is truly bad, as both

Borda and Condorcet had contended, and we would rather have STV than plurality, but we'd rather have one of the more elaborate two-stage procedures instead of either one.

Suggested Readings. Readers who look into this further will no doubt find ten or twenty seriously proposed election procedures, all of which have advantages worth considering. Perhaps one would begin with the textbook by William Riker (*Riker 1982*) and then do some soaking and poking on the Internet. We have found many fascinating websites and would encourage readers to adventure, but keep in mind that these sites are not peer reviewed by experts and we have noticed a nontrivial number of mistakes in interpretation and mathematics. Perhaps the most rigorous analysis is to be found on the election-methods email list, which is hosted on http://electorama.com.

Chapter 4

Game Theory

The mathematical theory of games was first developed as a model for situations of conflict. It gained widespread recognition in the early 1940s when it was applied to the theoretical study of economics by the mathematician John von Neumann (1903–1957) and the economist Oskar Morgenstern (1902–1976) in their book *Theory of Games and Economic Behavior.* Since then its scope has been broadened to include cooperative interactions as well and it has been applied to the theoretical aspects of many of the social sciences. While the jury is still out on the question of whether this theory furnishes any valuable information regarding practical situations, it has stimulated much basic research in disciplines such as economics, political science, and psychology.

4.1 Introduction to Games

Situations of conflict, or any other kind of interactions, will be called *games* and they have, by definition, participants who are called *players*. We shall limit our attention to scenarios where there are only two players and they will be called Ruth and Charlie. The existence of a conflict is usually due both to the desire of each player to improve his circumstances, frequently by means of some acquisition, and the unfortunate limited nature of all resources. It is assumed that each player is striving to *gain* as much as possible, and that each player's gain is his opponent's loss. Finally, each player is assumed to have several *options* or *strategies* that he can exercise (one at a time) as his attempt to claim a portion of the resources. Because of the introductory nature of this text, most of the foregoing discussion is restricted to situations wherein the players make their moves simultaneously and independently of each other. It will be argued in Chapter 10 that this does not truly limit the scope of the theory and that the mathematical theory of games does have something to say about games, such as poker, in which the players move alternately and do possess a fair amount of information about their opponent's actions.

The foregoing discussion is admittedly vague and this chapter's remainder is devoted to the informal exposition of several examples of increasing complexity. The requisite formal definitions are postponed to the next chapter.

Penny-Matching. *Ruth and Charlie each hold a penny and they display them simultaneously. If the pennies match in the sense that both show heads or both show tails, then Ruth collects both coins. Otherwise, Charlie gets them.*

It is clear that this game is fair in the sense that neither player has an advantage over the other. Moreover, if the game is only played once, then neither player possesses a shrewd system, or strategy, that will improve his position, nor is there a foolish decision that will worsen it.

The situation changes if the game is played many times. While the repeated game remains symmetrical, and neither player has a strategy that will guarantee his coming out ahead in the long run, it is possible to play this game foolishly. Such would be the case were Ruth to consistently display the *head* on her penny. In that case Charlie would be sure to catch on and display the *tail* each time, thus coming out ahead. It is a matter of common sense that neither player should make any predictable decisions, nor should a player favor either of his options. In other words, when this game is repeated many times each player should play each of his options with equal frequency 1/2 and make his decisions unpredictable. One way a player can accomplish this is by flipping his coin rather than consciously deciding which side to show. With an eye to the analysis of more complex games, this game is summarized as

		Charlie	
		Heads	Tails
Ruth	Heads	1	–1
	Tails	–1	1

The entry 1 denotes a gain of one penny for Ruth, and the entry -1 denotes a loss of one penny for Ruth. Since Ruth's gain is Charlie's loss, and vice versa, this array completely describes the various outcomes of a single play of the game.

All the subsequent examples and most of the discussion will be phrased in terms of the same two players Ruth and Charlie. Payoffs will be described from Ruth's point of view. Thus, a payoff of a penny will always mean a penny gained by Ruth and lost by Charlie.

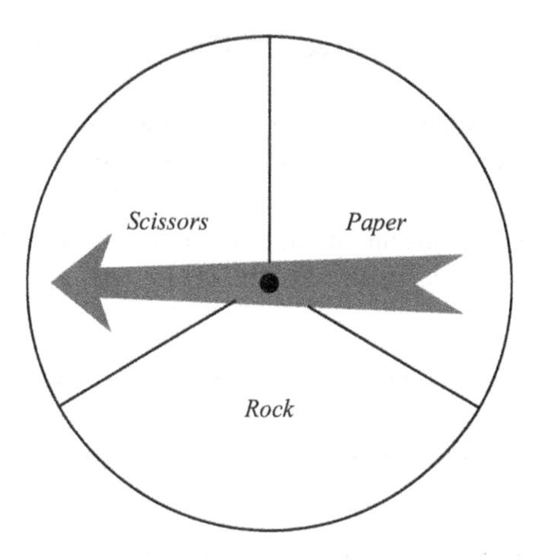

Figure 4.1

Rock-Scissors-Paper. *Ruth and Charlie face each other and simultaneously display their hands in one of the following three shapes: a fist denoting a rock, the forefinger and middle finger extended and spread to suggest scissors, or a downward facing palm denoting a sheet of paper. The rock wins over the scissors since it can shatter them, the scissors win over the paper since they can cut it, and the paper wins over the rock since it can be wrapped around the latter. The winner collects a penny from the opponent and no money changes hands in the case of a tie.*

This game has features that are very similar to those of Penny-matching. There are neither shrewd nor foolish decisions for a single play. If the game is repeated many times then players who favor one of the options place themselves at a disadvantage. The best strategy for each player is to play each of the options with the same frequency of 1/3 in a manner that will yield the opponent as little information as possible about any particular decision. For example, he could base each of his decisions on the result of a spin of the spinner of Figure 4.1.

The outcomes of the Rock-scissors-paper game are tabulated as

		Charlie		
		Rock	Scissors	Paper
	Rock	0	1	–1
Ruth	Scissors	–1	0	1
	Paper	1	–1	0

As before, a positive entry denotes a gain for Ruth whereas a negative entry is a gain for Charlie.

The biologists B. Sinervo and C. M. Lively have recently reported on a lizard species whose males are divided into three classes according to their mating behavior. The interrelationship of these three alternative behaviors very much resembles the Rock-scissors-paper game and merits a digression here. Each male of the side-blotched lizards (*Uta stansburiana)* exhibits one of three (genetically transmitted) mating behaviors:

(a) *highly aggressive* with a large territory that includes several females;
(b) *aggressive* with a smaller territory that holds one female;
(c) nonagressive *sneaker* with no territory who copulates with the others' females.

The highly aggressive male is stronger than the monogamous one who in turn is stronger than the sneaker. However, because the highly aggressive males must split their time between their various consorts, they are vulnerable to the sneakers. The observed consequence of this is that the male populations cycle from a high frequency of aggressives to a high frequency of highly aggressives, then on to a high frequency of sneakers and back to a high frequency of aggressives.

The next game is played in Italy using three fingers. For pedagogical reasons it is its two finger simplification that will be examined here.

Two-Finger Morra. *At each play Ruth and Charlie simultaneously extend either one or two fingers and call out a number. The player whose call equals the total number of the extended fingers wins that many pennies from the opponent. In the event that neither players' call matches the total, no money changes hands.*

It would be clearly foolish for a player to call a number that cannot possibly match the total number of displayed fingers. Thus, a player who extends only one finger would call either 2 or 3, whereas a player who extends two fingers would call only 3 or 4. Consequently, each player has in reality only four options and the game's possible outcomes are summarized as

		Charlie			
		(1,2)	(1,3)	(2,3)	(2,4)
Ruth	(1,2)	0	2	–3	0
	(1,3)	–2	0	0	3
	(2,3)	3	0	0	–4
	(2,4)	0	–3	4	0

where the option *(i, j)* denotes the extension of *i* fingers and a call of *j*.

This game, like its predecessors, is symmetrical. Neither player has a built-in advantage. It is tempting, therefore, to conclude that when this game is played many times, each player should again randomize his decisions and play them each with a frequency of 1/4. This, however, turns out to be a poor strategy. If Ruth did randomize this way then Charlie could ensure his coming out ahead in the long run by consistently employing option (1, 3). Note that in that case the expected outcome per play can be computed by the rules of probability (see addendum at the end of this chapter) as

$$\frac{1}{4}\cdot 2 + \frac{1}{4}\cdot 0 + \frac{1}{4}\cdot 0 + \frac{1}{4}\cdot(-3) = -\frac{1}{4}.$$

In other words, if Ruth plays each option with a frequency of 1/4 and Charlie consistently employs option (1, 3), Charlie can expect to win, on the average, 1/4 of a penny per play. Thus, this game differs drastically from the two previous ones. Whereas the symmetry of the game makes it clear that each player should be able to pursue a long-term strategy that entitles them to expect to come out more or less even in the long run, it is not at all clear this time what this strategy is. We shall return to this game and issue in Exercise 4.5.5.

The above considerations point out a difficulty that will have to be dealt with when the general theory of games is proposed. It was noted that if Ruth randomizes her behavior by using each of her four options with a frequency of 1/4, then Charlie can guarantee a long run advantage by consistently employing option (1, 3). However, should Charlie choose to do so, Ruth is bound to notice his bias, and she will in all likelihood respond by consistently opting for (1, 2), yielding her a win of 2 on each play. Charlie, then, will respond by playing (2, 3) consistently for a win of 3 on each play. Ruth, then, will switch to (2, 4) for a repeated gain of 4. Charlie, then, will switch back to (1, 3), thus beginning the whole cycle again. A reasonable theory of games should provide a stable strategy that avoids such "logical loops," and we will see that such is indeed made.

Bombing Sorties. *Ruth and Charlie are generals of opposing armies. Every day Ruth sends out a bombing sortie that consists of a heavily armed bomber plane and a lighter support plane. The sortie's mission is to drop a single bomb on Charlie's forces. However, a fighter plane of Charlie's is waiting for them in ambush and it will dive down and attack one of the planes in the sortie once. The bomber has an 80% chance of surviving such an attack, and if it survives it is sure to drop the bomb right on the target. General Ruth also has the option of placing the bomb on the support plane. In that case, due to this plane's lighter armament and lack of proper equipment, the bomb will reach its target with a probability of only 50% or 90%, depending on whether or not it is attacked by Charlie's fighter*. This information is summarized in the table below where the entries denote the probability of the bomb's delivery.

		Attack	
		Bomber	Support
Bomb placement	Bomber	80%	100%
	Support	90%	50%

Ruth knows that if the bomb is placed consistently on the bomber she can reasonably expect at least 80% of the missions to succeed. In all likelihood, Charlie's observers at the bombing site would notice this bias and he would direct his fighter plane pilot to always attack the bomber, thus holding Ruth's expectation down to 80% and no more. However, Ruth, who is an experienced poker player, decides to bluff by placing the bomb on the support plane occasionally. Let us say for the sake of argument that she does so $1/4 = 0.25$ of the time. Charlie now faces a dilemma. His observers have advised him of Ruth's new strategy and he suspects that it would be advantageous for him to attack the support plane some of the time, but how often should he do so?

Suppose Charlie decides to counter Ruth's bluffing by attacking the support plane half the time. In this case the situation is summarized as

			Attack frequencies	
			0.50	0.50
			Bomber	Support
Bomb placement frequencies	0.75	Bomber	80%	100%
	0.25	Support	90%	50%

Since Ruth and Charlie make their daily decisions independently of each other, it follows that on any single sortie, the probabilities of each of the four possible outcomes occurring are as displayed in Table 4.1. Hence, under these circumstances, wherein the bomb is placed on the support plane 1/4 of the time and this plane is attacked by the fighter 1/2 of the time, the percentage of successful missions, as computed on the basis of Table 1, is

Compound event	Probability of event	Likelihood of success of sortie
The bomb is on the bomber and the fighter attacks the bomber	$0.75 \cdot 0.50 = 0.375$	80%
The bomb is on the bomber and the fighter attacks the support plane	$0.75 \cdot 0.50 = 0.375$	100%
The bomb is on the support plane and the fighter attacks the bomber	$0.25 \cdot 0.50 = 0.125$	90%
The bomb is on the support plane and the fighter attacks the support plane	$0.25 \cdot 0.50 = 0.125$	50%

Table 4.1: Bombing sorties.

$$0.375 \cdot 80\% + 0.375 \cdot 100\% + 0.125 \cdot 90\% + 0.125 \cdot 50\%$$
$$= 30\% + 37.5\% + 11.25\% + 6.25\% = 85\%.$$

Thus, this response of Charlie's to Ruth's bluffing has created a situation wherein the bomb can be expected to get through 85% of the time. Since this figure amounts to only 80% when the bomb is placed consistently in the bomber, Ruth's bluffing seems to have paid off.

However, Charlie can change his response pattern. He could, say, decide to diminish the frequency of attacks on the support plane to only 1/5 of the time, leading to the situation.

			Attack frequencies	
			0.80	0.20
			Bomber	Support
Bomb placement frequencies	0.75	Bomber	80%	100%
	0.25	Support	90%	50%

In this case the percentage of successful sorties, computed on the basis of Table 4.2, is

Compound event	Probability of event	Likelihood of success of sortie
The bomb is on the bomber and the fighter attacks the bomber	$0.75 \cdot 0.80 = 0.60$	80%
The bomb is on the bomber and the fighter attacks the support plane	$0.75 \cdot 0.20 = 0.15$	100%
The bomb is on the support plane and the fighter attacks the bomber	$0.25 \cdot 0.80 = 0.20$	90%
The bomb is on the support plane and the fighter attacks the support plane	$0.25 \cdot 0.20 = 0.05$	50%

Table 4.2: Bombing sorties.

$$0.60 \cdot 80\% + 0.15 \cdot 100\% + 0.20 \cdot 90\% + .05 \cdot 50\%$$
$$= 48\% + 15\% + 18\% + 2.5\% = 83.5\%.$$

Thus, by diminishing the frequency of attacks on the support plane Charlie would create a situation wherein only 83.5% of the sorties would be successful. From Charlie's point of view this is an improvement on the 85% computed above. Could greater improvements be obtained by a further diminution of the frequency of attacks on the support plane? Suppose these attacks are completely eliminated. Then the state of affairs is

			Attack frequencies	
			1	0
			Bomber	Support
Bomb placement frequencies	0.75	Bomber	80%	100%
	0.25	Support	90%	50%

and the associated Table 3 tells us that the percentage of successful sorties is

$$0.75 \cdot 80\% + 0 \cdot 100\% + 0.25 \cdot 90\% + 0 \cdot 50\%$$
$$= 60\% + 0\% + 22.5\% + 0\% = 82.5\%.$$

This would seem to indicate that when Ruth bluffs 1/4 of the time Charlie should nevertheless ignore the support plane and direct his attacks exclusively at the bomber. Moreover, these calculations indicate that this strategy enables Ruth to improve on her bomber's 80% delivery rate by another 2.5%.

Compound event	Probability of event	Likelihood of success of sortie
The bomb is on the bomber and the fighter attacks the bomber	$0.75 \cdot 1 = 0.75$	80%
The bomb is on the bomber and the fighter attacks the support plane	$0.75 \cdot 0 = 0$	100%
The bomb is on the support plane and the fighter attacks the bomber	$0.25 \cdot 1 = 0.25$	90%
The bomb is on the support plane and the fighter attacks the support plane	$0.25 \cdot 0 = 0$	50%

Table 4.3: Bombing sorties.

These various analyses raise several questions that will guide the development of the subsequent sections. Can Ruth improve on the above 82.5% with a different strategy? What is the best improvement Ruth can obtain? What is Charlie's best response to any specific strategy of Ruth's? Does Charlie have an overall best strategy that is independent of Ruth's decisions?

An Addendum on Probabilistic Matters. Unless otherwise stated, it will always be assumed here that each time a game is played Ruth and Charlie make their respective decisions independently of each other. This comes under the general heading of *Independent Random Events*; i.e., random events whose outcomes have no bearing on each other. For example, if a nickel and a dime are tossed the outcomes will in general be independent, unless the two coins are glued to each other. Similarly, suppose Ruth draws a card at random from a standard deck, replaces it, shuffles the deck, and then Charlie draws a card from that deck. The two draws are then independent. On the other hand, if Ruth does not replace her card in the deck, then Charlie's draw is very much affected by Ruth's draw, since he cannot possibly draw the same card as she did. In this case the two random draws are not independent. The probabilities of independent events are related by the formula

---

If the two events E and F are independent then

probability of (E and F) = (probability of E) ·(probability of F).

Thus, the probability of the aforementioned coins both coming up *heads* is $0.5 \cdot 0.5 = 0.25$. Similarly, if a nickel and a standard six-sided die are tossed simultaneously, then the probability of the coin coming up *tails* and the die showing a 3 is $\frac{1}{2} \cdot \frac{1}{6} = \frac{1}{12}$.

Some random events have numerical values attached to their outcomes. Thus, the faces of the standard die have dots marked on them, a person selected at random has a height, and a lottery ticket has a monetary value (that is unknown at the time it is purchased). Loosely speaking, the *expected value* is the average of those numerical values when they are weighted by the corresponding probabilities. More formally:

If the random variable X assumes the numerical values $x_1, x_2, ..., x_n$ with probabilities $p_1, p_2, ..., p_n$, respectively, then the expected value (weighted average) of X is

$$p_1x_1 + p_2x_2 + ... + p_nx_n.$$

For example, suppose that the records of an insurance company indicate that during a year they will pay out for accidents according to the following pattern:

$100,000 with probability 0.0002
$50,000 with probability 0.0015
$25,000 with probability 0.003
$5,000 with probability 0.01
$1,000 with probability 0.03
$0 with probability 0.9553

Then the expected payment per car during the next year is

$$\$100{,}000 \cdot 0.0002 + \$50{,}000 \cdot 0.0015 + \$25{,}000 \cdot 0.003 + \$5{,}000 \cdot 0.01 + \$1{,}000 \cdot 0.03 + \$0 \cdot 0.9553 = \$250.$$

Similarly, if 5000 lottery tickets are sold of which one will win $10,000, two will win $1,000, five will win $100 and the rest will all receive a consolation prize worth one dime, then the expected value of each ticket is

$$10{,}000\frac{1}{5000} + 1{,}000\frac{2}{5000} + 100\frac{5}{5000} + 0.10 \cdot \frac{4992}{5000} \approx 2.60.$$

Exercises 4.1

1. Ruth is playing Morra and has decided to play the options with the following frequencies:

(1, 2) - 40%
(1, 3) - 30%
(2, 3) - 20%
(2, 4) - 10%

a) Design a spinner that will facilitate this pattern.
b) What will be the expected outcome if Charlie consistently plays (1,2)?

c) What will be the expected outcome if Charlie consistently plays (1,3)?
d) What will be the expected outcome if Charlie consistently plays (2,3)?
e) What will be the expected outcome if Charlie consistently plays (2,4)?
f) Which of the above consistent moves is best for Charlie?

2. Ruth is playing Morra and has decided to play the options with the following frequencies:

(1, 2) - 0%
(1, 3) - 50%
(2, 3) - 50%
(2, 4) - 0%

a) Design a spinner that will facilitate this pattern.
b) What will be the expected outcome if Charlie consistently plays (1,2)?
c) What will be the expected outcome if Charlie consistently plays (1,3)?
d) What will be the expected outcome if Charlie consistently plays (2,3)?
e) What will be the expected outcome if Charlie consistently plays (2,4)?
f) Which of the above consistent moves is best for Charlie?

3. Ruth is playing Morra and has decided to play the options with the following frequencies:

(1, 2) - 10%
(1, 3) - 20%
(2, 3) - 30%
(2, 4) - 40%

a) Design a spinner that will facilitate this pattern.
b) What will be the expected outcome if Charlie consistently plays (1,2)?
c) What will be the expected outcome if Charlie consistently plays (1,3)?
d) What will be the expected outcome if Charlie consistently plays (2,3)?
e) What will be the expected outcome if Charlie consistently plays (2,4)?
f) Which of the above consistent moves is best for Charlie?

4. General Ruth has decided that she will put the bomb on the support plane 10% of the time.
   a) Design a spinner that will facilitate this pattern.
   b) What will be the expected mission success rate if General Charlie persists in attacking the bomber exclusively?
   c) What will be the expected mission success rate if General Charlie persists in attacking the support plane exclusively?
   d) What will be the expected mission success rate if General Charlie attacks each plane 50% of the time?
   e) What will be the expected mission success rate if General Charlie attacks the support plane 40% and the bomber 60% of the time?

5. General Ruth has decided that she will put the bomb on the support plane 30% of the time.
   a) Design a spinner that will facilitate this pattern.
   b) What will be the expected mission success rate if General Charlie persists in attacking the bomber exclusively?
   c) What will be the expected mission success rate if General Charlie persists in attacking the support plane exclusively?
   d) What will be the expected mission success rate if General Charlie attacks each plane 50% of the time?
   e) What will be the expected mission success rate if General Charlie attacks the support plane 40% and the bomber 60% of the time?

6. General Ruth has decided that she will put the bomb on the support plane 40% of the time.
   a) Design a spinner that will facilitate this pattern.
   b) What will be the expected mission success rate if General Charlie persists in attacking the bomber exclusively?
   c) What will be the expected mission success rate if General Charlie persists in attacking the support plane exclusively?
   d) What will be the expected mission success rate if General Charlie attacks each plane 50% of the time?
   e) What will be the expected mission success rate if General Charlie attacks the support plane 40% and the bomber 60% of the time?

4.2 Formal Zero-Sum Games

It is time to make some formal definitions. For purely pedagogical reasons we begin with games in which each of the players has only two options, leaving the more general case for the end of the section.

Taking our cue from the Penny-matching and Bombing-sorties games of the previous section, we define a 2×2 *zero-sum game* as a square array of $2 \times 2 = 4$ numbers. Thus, the mathematical representation of the Penny-matching game is the array

1	–1
–1	1

and the mathematical representation of the Bombing-sorties game is the array

80%	100%
90%	50%

Just as it is convenient to represent the addition of 8 oranges to 6 oranges by the abstract equation $8 + 6 = 14$, we shall ignore the actual details of the games in most of the subsequent discussion and simply deal with arrays of unitless numbers. Thus, the most general 2×2 zero-sum game has the form

a	b
c	d

where a, b, c, d are arbitrary numbers. This abstraction has the advantages of succinctness and clarity. We shall, however, make a point of discussing some real games every now and then, and many more such games will be found in the exercises.

Each 2×2 zero-sum game has two players, whom we shall continue to call Ruth and Charlie. The mathematical analog of deciding on one of the options is the selection of either a row or a column of this array. Specifically, Ruth decides on an option by selecting a row of the array, whereas Charlie makes his decision by specifying a column. Thus, in Bombing-sorties, Ruth's placement of the bomb in the bomber, and Charlie's attacking the support plane, are tantamount to Ruth's selecting the first row and Charlie's selecting the second column of the array

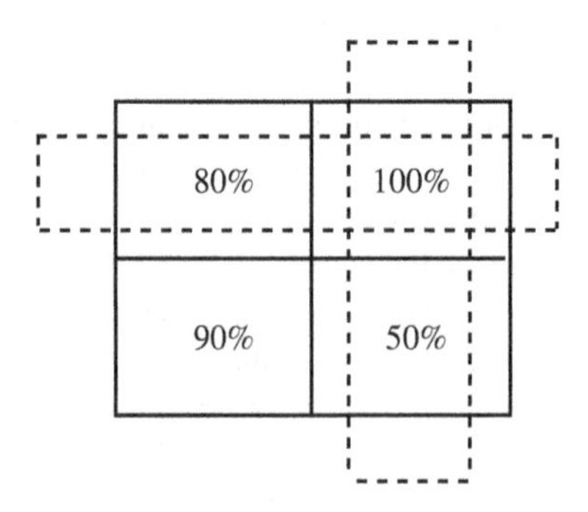

Each individual *play* of the game consists of such a pair of selections, made simultaneously and independently. The selected row and column constitute the *outcome* of the play and its *payoff* is the entry of the array that is contained in both of the selections. Thus, the payoff of the play illustrated above is 100%. On the other hand, had Ruth selected the second row and had Charlie stayed with the second column the payoff would have been 50%. This payoff of course represents Ruth's winnings (and Charlie's loss) from that play and she will in general wish to maximize its value, whereas Charlie will be guided by the desire to minimize this payoff.

Informally speaking, a players' strategy is a decision on the frequency with which each available option will be chosen. More formally, a *mixed strategy* is a pair of numbers $[a, b]$,

$$0 \leq a \leq 1, 0 \leq b \leq 1, a + b = 1,$$

where a denotes the frequency with which the first row (or column) is chosen, and b denotes the frequency with which the second row (or column) is chosen. Thus, Ruth's decision to place the bomb on the support plane 1/4 of the time is denoted by the pair [0.75, 025] and Charlie's strategy of attacking the two planes with equal frequencies is denoted by [0.5, 0.5]. When discussing general strategies it is convenient to denote Ruth's strategy by $[1–p, p]$ and Charlie's strategy by $[1–q, q]$.

This formalization of the intuitive concept of strategy as a list of probabilities was first offered by the mathematician Emile Borel (1871–1956) in series of papers that were written in the 1920s. Borel was also the first one to view these games as rectangular arrays. The decision to limit the notion of strategy in this manner could not have been an easy one and Borel's papers exhibit some ambivalence on this issue. It is very tempting to believe that one can gain some advantage over the opponent by varying one's probabilities at each play, as did Borel, but it is not at all clear to formulate such a variation. At the conclusion of one of his papers he wrote:

> *The function $\varphi(x, y)$ [the strategy] must then vary at each instant, and vary without following any law at all. One may well doubt if it is possible to indicate an effective and sure means of carrying out such counsel. It seems that, to follow it to the letter, a complete incoherence of mind would be needed, combined, of course, with the intelligence necessary to eliminate those methods we have called bad.*

It is our contention, and assumption, that, in practice, every player has a strategy in the sense of a pair of probabilities. Truly random behavior is impossible to attain within the context of a game. For how can any person repeatedly play a game such as Rock-scissors-paper in a truly random fashion, that is to say, without some pattern of frequencies emerging from his choices? If he makes each individual decision in his head, his past experience and personal preferences are sure to dictate a pattern. If, on the other hand, he uses some device such as a die or a spinner to implement the randomization, the device itself will shape the randomness into a frequency distribution. Each face of the die will come up approximately 1/6 of the time and the spinner's arrow will stop within each sector with a frequency that is proportional to that sector's central angle.

Another argument against the feasibility of a strategyless manner of playing is that each actual sequence of choices made by a player over his lifetime can be construed as a strategy. If a player of Penny-matching has been observed to have displayed *Heads* 83 times and *Tails* 39 times, it could be reasonably said that, since $83 + 39 = 122$, it follows that this player has the strategy

$$\left[\frac{83}{122}, \frac{39}{122}\right] \approx [\,0.68, 0.32\,].$$

In reality, of course, such statistics are rarely available. However, the point we are making here is that it is permissible to assume that <u>every</u> player is indeed employing some strategy that may or may not be known to the opponent.

More serious than the inability of people to behave in a truly random fashion is the theoretical impossibility of treating such a concept within a mathematical framework. Even in the mathematical theory of statistics, where random variables are as a commonplace as ants at a picnic, every such variable is assumed to have a probability distribution. Von Neumann and Morgenstern, who also identified strategies with probabilities as was done above, explicitly admitted that they were doing so simply because they had no theory that could handle anything else.

Thus, it will henceforth be assumed that

Every player employs a strategy.

It was seen in Bombing-sorties that the specification of each player's strategy resulted in a situation wherein an expected payoff could be computed. This *expected payoff* can be defined and computed for arbitrary 2×2 games in a similar manner. Thus, given the strategies $[1-p, p]$ and $[1-q, q]$ for the general 2×2 zero-sum game

	$1-q$	q
$1-p$	a	b
p	c	d

(1)

the likelihood of Ruth getting the payoff a is the probability of her choosing the first row and Charlie choosing the first column. Since these choices are made independently, we conclude that

the probability of Ruth getting payoff a is $(1-p)(1-q)$.

Similarly,

the probability of Ruth getting payoff b is $(1-p)q$,
the probability of Ruth getting payoff c is $p(1-q)$,
the probability of Ruth getting payoff d is pq.

As these four events are mutually exclusive and they exhaust all the possibilities it follows that Ruth's expected payoff is

$$(1-p)(1-q)a + (1-p)qb + p(1-q)c + pqd.$$

Diagrams such as that of (1) above, wherein the players' strategies are appended to the game's array, are called *auxiliary diagrams*. They turn the computation of the expected payoff into a routine task and so will be repeatedly used in the sequel as a visual aid.

Example 4.2.1. For the Bombing-sorties game Ruth's strategy [0.3, 0.7] and Charlie's strategy [0.6, 0.4] yield the auxiliary diagram

	0.6	0.4
0.3	80%	100%
0.7	90%	50%

The corresponding payoff is

$$0.3 \cdot 0.6 \cdot 80\% + 0.3 \cdot 0.4 \cdot 100\% + 0.7 \cdot 0.6 \cdot 90\% + 0.7 \cdot 0.4 \cdot 50\% = 14.4\% + 12\% + 37.8\% + 14\% = 78.2\%.$$

In other words, when the players employ the above specified strategies, Ruth can expect 78.2% of the missions to be successful.

Example 4.2.2. *Compute the expected payoff of the strategies [0.2, 0.8] of Ruth and [0.3, 0.7] of Charlie for the abstract 2×2 zero-sum game*

5	0
-1	2

The auxiliary diagram is

	0.3	0.7
0.2	5	0
0.8	-1	2

and so the expected payoff is

$$0.2 \cdot 0.3 \cdot 5 + 0.2 \cdot 0.7 \cdot 0 + 0.8 \cdot 0.3 \cdot (-1) + 0.8 \cdot 0.7 \cdot 2 =$$
$$0.3 + 0 - 0.24 + 1.12 = 1.18.$$

In other words, when the players use the specified strategies within the context of the given (repeated) game, Ruth can expect to win, on the average, 1.18 per play. Of course, this being a zero-sum game, Charlie should expect to lose the same amount per play.

We now turn to the formal definition of games in which one or both of the players have more than two options. If m and n are any two positive integers, then an $m \times n$ *zero-sum game* is a rectangular array of mn numbers having m rows and n columns. Thus,

0	1	-1
-1	0	1
1	-1	0

is a 3×3 zero-sum game (Rock-paper-scissors), and

0	2	-3	0
-2	0	0	3
3	0	0	-4
0	-3	4	0

is a 4×4 zero-sum game (Two-finger Morra). The array below is an example of an abstract 3×4 zero-sum game (an interesting concrete nonsquare game will be discussed in Chapter 7 in detail).

2	-1	-5	3
0	-2	3	-3
1	0	1	-2

(2)

It is again assumed that in each play Ruth selects a row and Charlie selects a column of the array. The selected row and column constitute the *outcome* of that play, and the entry of the array which is the intersection of Ruth's chosen row with Charlie's chosen column is the corresponding *payoff*. Thus, in the above abstract game, if Ruth selects the second row and Charlie selects the fourth column, the corresponding payoff is -3, a loss for Ruth.

Given an $m \times n$ zero-sum game, a *strategy* for Ruth is a list of numbers $[p_1, p_2, ..., p_m]$, such that

$$0 \leq p_i \leq 1 \text{ for each i} = 1, 2, ..., m, \text{ and } p_1 + p_2 + ... + p_m = 1,$$

where p_i denotes the frequency with which Ruth chooses the i-th row. Similarly, a strategy for Charlie is an ordered list of numbers $[q_1, q_2, ..., q_n]$, such that

$$0 \leq q_j \leq 1 \text{ for each j} = 1, 2, ..., n, \text{ and } q_1 + q_2 + ... + q_n = 1,$$

where q_j denotes the frequency with which Charlie chooses the j-th column. Thus, in the symmetrical Rock-scissors-paper game, the strategy [0.6, 0.3, 0.1] denotes the decision, by either player, to display *rock* 60% of the time, *scissors* 30% of the time, and *paper* only 10% of the time. In the above abstract 3 × 4 zero-sum game (or in any 3 × 4 zero-sum game for that matter) the strategy [0.2, 0.3, 0.5] denotes Ruth's choosing the first, second, or third rows 20%, 30%, 50% of the time, respectively. The strategy [0.4, 0, 0.1, 0.5] denotes Charlie's choosing the first, second, third, or fourth columns 40%, 0%, 10%, and 50% of the time. A *pure strategy* is one which calls for the exclusive use of a particular row or column. In Bombing-sorties each player has two pure strategies: [1, 0] which calls for the use of the first row or column only, and [0, 1] which calls for the exclusive use of the second row or column. In the abstract game of (2), Ruth has three pure strategies: [1, 0, 0], [0, 1, 0], and [0, 0, 1]. In the same game Charlie has four pure strategies: [1, 0, 0, 0], [0, 1, 0, 0], [0, 0, 1, 0], and [0, 0, 0, 1]. Strategies which are not known to be pure are called *mixed*. Thus, the strategies [0.3, 0.7] and $[1 - p, p]$ are mixed strategies, even though the latter may turn out to be pure when p is either 0 or 1.

Every such choice of specific strategies on the part of both players narrows the situation down to a point where an expected payoff can be computed. Suppose Ruth employs the strategy $[p_1, p_2, ..., p_m]$ and Charlie employs the strategy $[q_1, q_2, ..., q_n]$ in some abstract $m \times n$ zero-sum game. If $a_{i,j}$ denotes the payoff in the i-th row and the j-th column of this game (see below), then the likelihood of this payoff actually taking

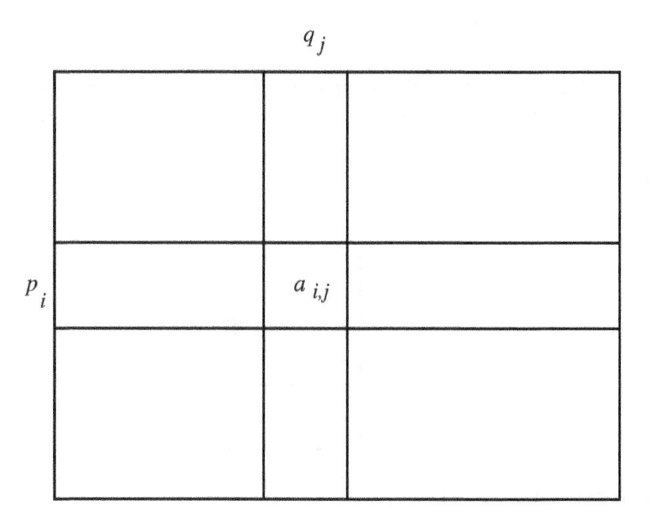

place is the probability of Ruth choosing the i-th row and Charlie choosing the j-th column, which, of course, equals $p_i\, q_j$. Thus, this specific outcome's contribution to the expected payoff is $p_i\, q_j\, a_{i,j}$. Consequently, the total *expected payoff*, being the sum of all these contributions, equals

---

Sum of all $p_i\, q_j\, a_{i,j}$ $\quad i = 1, 2, \ldots, m, j = 1, 2, \ldots, n$.

Example 4.2.3. *Compute the expected payoff when Ruth employs the strategy [0.2, 0.3, 0.5] and Charlie employs the strategy [0.1, 0.7, 0.2] in the Rock-scissors-paper game.*

The auxiliary diagram is

	0.1	0.7	0.2
0.2	0	1	–1
0.3	–1	0	1
0.5	1	–1	0

and the corresponding sum is

$$0.2 \cdot 0.1 \cdot 0 + 0.2 \cdot 0.7 \cdot 1 + 0.2 \cdot 0.2 \cdot (-1) +$$
$$0.3 \cdot 0.1 \cdot (-1) + 0.3 \cdot 0.7 \cdot 0 + 0.3 \cdot 0.2 \cdot 1 +$$
$$0.5 \cdot 0.1 \cdot 1 + 0.5 \cdot 0.7 \cdot (-1) + 0.5 \cdot 0.2 \cdot 0 =$$
$$0 + 0.14 - 0.04 - 0.03 + 0 + 0.06 + 0 - 0.35 + 0 = -0.17.$$

In other words, under these circumstances, Ruth should expect to lose 0.17 pennies per play.

Example 4.2.4. *Compute the expected payoff when the strategies [0.3, 0, 0.7] and [0.1, 0.2, 0.3, 0.4] are used by Ruth and Charlie respectively in the abstract game (1) above.*

Using the auxiliary diagram

	0.1	0.2	0.3	0.4
0.3	2	–1	–5	3
0	0	–2	3	–3
0.7	1	0	1	–2

we get an expected payoff of

$$0.3 \cdot\cdot .1 \cdot 2 + 0.3 \cdot 0.2 \cdot (-1) + 0.3 \cdot 0.3 \cdot (-5) + 0.3 \cdot 0.4 \cdot 3 + 0 \cdot 0.1 \cdot 0 + 0 \cdot 0.2 \cdot (-2)$$
$$+ 0 \cdot 0.3 \cdot 3 + 0 \cdot 0.4 \cdot (-3) + 0.7 \cdot 0.1 \cdot 1 + 0.7 \cdot 0.2 \cdot 0 + 0.7 \cdot 0.3 \cdot 1$$
$$+ 0.7 \cdot 0.4 \cdot (-2) = 0.06 - 0.06 - 0.45 + 0.36 + 0 + 0 + 0 + 0 + 0.07$$
$$+ 0 + 0.21 - 0.56 = -0.37.$$

Exercises 4.2

In each of Exercises 1–11 compute the expected payoff where ***R*** *denotes a strategy for Ruth and* ***C*** *denotes a strategy for Charlie for the given game* ***G***.

1. $\boldsymbol{R} = [0.2, 0.8]$, $\boldsymbol{C} = [0.7, 0.3]$, $\boldsymbol{G} =$

2	3
4	1

2. $\boldsymbol{R} = [0.6, 0.4]$, $\boldsymbol{C} = [0, 1]$, $\boldsymbol{G} =$

2	3
4	1

3. $\boldsymbol{R} = [0.2, 0.8]$, $\boldsymbol{C} = [0.7, 0.3]$, $\boldsymbol{G} =$

–1	3
4	–2

4. $\boldsymbol{R} = [0.6, 0.4]$, $\boldsymbol{C} = [0, 1]$, $\boldsymbol{G} =$

–1	3
4	–2

5. $\boldsymbol{R} = [0.6, 0.3, 0.1]$, $\boldsymbol{C} = [0.1, 0.4, 0.3, 0.2]$, $\boldsymbol{G} =$

1	0	–2	3
–3	4	2	–4
0	–1	0	1

6. $\boldsymbol{R} = [0.6, 0, .04]$, $\boldsymbol{C} = [0, 0.5, 0.3, 0.2]$, $\boldsymbol{G} =$

1	0	–2	3
–3	4	2	–4
0	–1	0	1

7. $\boldsymbol{R} = [0, 1, 0]$, $\boldsymbol{C} = [0.5, 0, 0, 0.5]$, $\boldsymbol{G} =$

1	0	–2	3
–3	4	2	–4
0	–1	0	1

8. $\boldsymbol{R} = [0.2, 0, 0.4, 0, 0.4]$, $\boldsymbol{C} = [0.1, 0.1, 0.8]$, $\boldsymbol{G} =$

1	–3	2
0	4	–4
–2	0	2
3	–3	–1
–3	5	1

9. $\boldsymbol{R} = [0, 0, 0, 0.4, 0.6]$, $\boldsymbol{C} = [0.1, 0, 0.9]$, $\boldsymbol{G} =$

1	–3	2
0	4	–4
–2	0	2
3	–3	–1
–3	5	1

10. $\boldsymbol{R} = [0, 0, 0, 1, 0]$, $\boldsymbol{C} = [0, 1, 0]$, $\boldsymbol{G} =$

1	–3	2
0	4	–4
–2	0	2
3	–3	–1
–3	5	1

11. $\boldsymbol{R} = [0, 1, 0, 0, 0]$, $\boldsymbol{C} = [0, 0, 1]$, $\boldsymbol{G} =$

1	–3	2
0	4	–4
–2	0	2
3	–3	–1
–3	5	1

Describe the following situations as mathematical zero-sum games.

12. **The River Tale** (J. D. Williams, 1986) Steve is approached by a stranger who suggests that they match coins. Steve says that it's too hot for violent exercise. The stranger says, "Well then, let's just lie here and speak the words *heads* or *tails* - and to make it interesting I'll give you $30 when I call *tails* and you call *heads,* and $10 when it's the other way round. And - just to make it fair - you give me $20 when we match."

13. **The Birthday** (J. D. Williams, 1986) Frank is hurrying home late, after a particularly grueling day, when it pops into his mind that today is Kitty's birthday! Or is it? Everything is closed except the florist. If it is not her birthday and he brings no gift, the situation will be neutral, i.e., payoff 0. If it is not and he comes in bursting with roses, and obviously confused, he may be subjected to the Martini test, but he will emerge in a position of strong one-upness - which is worth 1. If it is her birthday and he has, clearly, remembered it, that is worth somewhat more, say 1.5. If he has forgotten it he is down like a stone, say, -10.
14. **The Hi-Fi** (J. D. Williams, 1986) The firm of Gunning & Kappler manufactures an amplifier. Its performance depends critically on the characteristics of one small, inaccessible condenser. This normally costs Gunning and Kappler $1, but they are set back a total of $10, on the average, if the original condenser is defective. There are some alternatives open to them. It is possible for them to buy a superior quality condenser, at $6, which is fully guaranteed; the manufacturer will make good the condenser and the costs incurred in getting the amplifier to operate. There is available also a condenser covered by an insurance policy which states, in effect, "If it is our fault, we will bear the costs and you get your money back." This item costs $10. (This is a 3×2 game that Gunning & Kappler is playing against Nature whose options are to supply either a defective or a nondefective condenser.)
15. **The Huckster** (J. D. Williams, 1986) Merrill has a concession at Yankee Stadium for the sale of sunglasses and umbrellas. The business places quite a strain on him, the weather being what it is. He has observed that he can sell about 500 umbrellas when it rains and about 100 when it shines; and in the latter case he also can dispose of about 1000 sunglasses. Umbrellas cost him 50 cents and sell for $1 (this is 1954); glasses cost 20 cents and sell for 50 cents. He is willing to invest $250 in the project. Everything that isn't sold is a total loss (the children play with them).
16. **The Coal Problem** (J. D. Williams, 1986) On a sultry summer afternoon, Hans' wandering mind alights upon the winter coal problem. It takes about 15 tons to heat his house during a normal winter, but he has observed extremes when as little as 10 tons and as much as 20 were used. He also recalls that the price per ton seems to fluctuate with the weather, being $10, $15, and $20 a ton during mild, normal, and severe winters. He can buy now, however, at $10 a ton. He considers three pure strategies, namely, to buy 10, 15, or 20 tons now and the rest, if any, later. He will be moving to California in the spring and he cannot take excess coal with him.

4.3 Pure Strategies for Nonzero-Sum Games

In the zero-sum games discussed in the previous sections the payoffs are numerical, each player's gain is the other player's loss, and the players make their decisions simultaneously and independently of each other. There is, however, a considerable amount of interest in other types of games which may possess only some or even none of these properties: payoffs may be unquantifiable as is the case when loss of face is at stake, and in most wars both players sustain tremendous losses. While there are many different generalizations of zero-sum games, the remainder of this chapter is confined to a discussion of *two-person noncooperative nonzero-sum games*. The term *noncooperative* refers to the assumption that the players do not consult each other about ways and means for improving their payoffs. In the interest of brevity these games are referred to as simply *nonzero-sum games* or even as just *games* in the sequel.

Nonzero-sum games are also modeled by having Ruth select a row and Charlie select a column from a suitable rectangular array. However, since a player's gain is not necessarily his opponent's loss anymore, the payoff of each outcome can no longer be described by a single number. The result of Ruth choosing row i and Charlie choosing column j is the *payoff pair*

$$(a_i, b_j)$$

where a_i denotes Ruth's payoff and b_j denotes Charlie's payoff. Thus, the general 2×2 nonzero-sum game has the array

	Charlie	
Ruth	(a_1, b_1)	(a_2, b_2)
	(a_3, b_3)	(a_4, b_4)

(1)

It will be assumed that despite the presence of two payoffs in each pair, players only take their own payoffs into consideration when deciding on a move. This is commonly formalized as follows.

The Principle of Rationality: *Every player wishes to come out as well off as possible.*

In other words, cutting off one's nose in order to spite one's spouse is not rational behavior, and such impulses will be given no consideration in this text.

One of the most popular factors in the decision-making process is the issue of guarantees. Given a variety of options, what is the best outcome that can be guaranteed? In the context of games, the strategy that reflects this concern is called the *pure maximin* strategy. For Ruth this strategy dictates the choice of the row with the largest minimum first payoff; for Charlie this is the column with the largest minimum second payoff. The payoff guaranteed to a player by his maximin strategy is that player's *maximin value*. The payoff pair in the outcome determined by the two pure maximin solutions is called the *pure value pair* of the game. The pure maximin strategies and values as well as the pure value pairs of any nonzero-sum game are easily derived.

Example 4.3.1. *Find the pure maximin strategies and values and also the pure value pairs of the game*

(2, 3)	(3, 5)
(2, 2)	(1, 3)

To the right of each row record that row's smallest first entry and at the bottom of each column record that column's smallest entry. Circle the largest number on the right and

(2, 3)	(3, 5)	(2)
(2, 2)	(1, 3)	1
2	(3)	

the largest number at the bottom. The circled items indicate the maximin strategies for both players: row 1 for Ruth and column 2 for Charlie. The value pair is (3, 5) and happens to yield each player a payoff that is better than the guarantee sought.

Example 4.3.2. *Find the pure maximin strategies and values and also the pure value pairs of the game*

(5, 2)	(3, 0)	(8, 1)	(2, 3)
(6, 3)	(5, 4)	(7, 4)	(1, 1)
(7, 5)	(4, 6)	(6, 8)	(0, 2)

To the right of each row record that row's smallest first entry and at the bottom of each column record that column's largest entry. Circle the largest number on the right and the

(5, 2)	(3, 0)	(8, 1)	(2, 3)	⓶ 2
(6, 3)	(5, 4)	(7, 4)	(1, 2)	1
(7, 5)	(4, 6)	(6, 8)	(0, 2)	0
⓶ 2	0	1	⓶ 2	

largest numbers on the bottom. Ruth's maximin strategy is therefore Row 1. Charlie has two maximin strategies, Columns 1 and 4, yielding payoff pairs (5, 2) and (2, 3), respectively. Charlie may prefer to play Column 4 since it seems to promise a better payoff of 3, but this is no longer guaranteed. From the point of view of guarantees the two maximin strategies of Column 1 and Column 4 have the same value for Charlie.

We now go on to consider several games that are played very frequently and in many guises.

Prisoner's Dilemma. *Two people are arrested and held in connection with a certain robbery. The prosecution only has enough evidence to convict them of the robbery itself, but it is believed that the robbers were actually carrying guns at the time, making them liable to the more severe charge of armed robbery. The prisoners are held in separate cells and cannot communicate with each other. Each of them is offered the same deal:* If you testify that your partner was armed but he does not testify against you, your sentence will be suspended while he will spend 15 years in jail. *If both of you testify against each other, you will both get 10 years, and if neither of you does so, you both get 5 years in jail. Both prisoners are fully aware that both have been offered the same deal. They are given some time to think the deal over, but neither is aware of his ex-partner's decision.*

This game is clearly nonzero-sum and can be described as

	Refuse the deal	Accept the deal
Refuse the deal	(–5, –5)	(–15, 0)
Accept the deal	(0, –15)	(–10, –10)

where the payoff $-x$ denotes a jail term of x years. Each player's pure maximin strategy is easily derived:

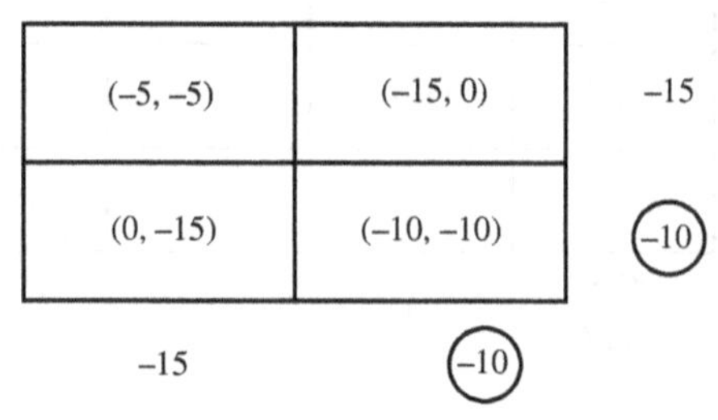

(–5, –5)	(–15, 0)	–15
(0, –15)	(–10, –10)	(–10)
–15	(–10)	

This maximin strategy recommends that both prisoners accept the deal offered by the prosecution, thereby guaranteeing that each will receive a 10-year jail term. Both experience and experimentation testify to the fact that this is frequently the decision made by people in such circumstances. This is unsettling because if both of the players had refused the deal, they would have each received a 5-year term, a distinctly better sentence for both.

Let us digress in order to formalize this vague notion of "unsettling." In any nonzero-sum game, a payoff pair (a, b) is said to be *better than* the payoff (a', b') if either of the following conditions hold:

$$a \geq a' \text{ and } b > b' \qquad \text{or} \qquad a > a' \text{ and } b \geq b'.$$

Thus, (3, 2) is better than each of the payoffs (1, 2), (3, 1) and (2, 2). An outcome of a game is *Pareto optimal* if the game possesses no outcome with a better payoff. In game (1) below, the outcomes with payoffs (2, 3) and (1, 5) are both Pareto optimal.

(2, 3)	(1, 5)
(2, 2)	(0, 3)

(1)

On the other hand, in the game of Example 1 above, the outcome with payoff (3, 5) is the only Pareto optimal outcome.

It is clear that when the players of a game are treated as a group, Pareto optimal outcomes are desirable. The unfortunate and tragic aspect of the Prisoners' dilemma game is that both the strategy recommended by game theory and the observed behavior of people result in the outcome with payoff (−10, −10), which is dominated by (−5, −5) and hence is not Pareto optimal.

Chicken. *Two adversaries are set on a collision course. If both persist, then a very unpleasant outcome, sometimes mutual annihilation, is guaranteed. If only one of the players swerves away (chickens) he loses the game. If both swerve, the result is a draw.*

Usually, the reward for winning this game is merely a sense of dominance, and swerving only entails loss of face, both equally unquantifiable. The possibility of mutual annihilation when both players persist is what makes this a nonzero-sum situation. In that case neither player gains anything; in fact, quite frequently both stand to lose their lives. We shall offer no solution and wish to discuss only some paradoxical aspects of the game. For this purpose, the game will be made concrete by transferring our attention to the Cuban missile crisis of October 1962, certainly one of the best known and most frightening Chicken games ever played.

In that month the United States discovered that the USSR was building a missile base in Cuba. As a small Soviet fleet was on its way to the island to bring in supplies, the Americans set up a naval blockade. It is quite likely that if these two fleets had met, the conflict would have escalated into a full fledged, and probably nuclear, war. In the event, the Soviets blinked, their ships withdrew, and the United States "won." It is our purpose here to discuss a curious wrinkle on this confrontation.

While it is impossible to realistically quantify the various outcomes of this game, it is possible to introduce some numbers into it by a preferential ranking of the various outcomes. There were four possible outcomes for the Cuban missile crisis:

(US swerves, USSR swerves), (US persists, USSR swerves)
(US swerves, USSR persists), (US persists, USSR persists).

The players would rank these outcomes preferentially as follows (1 denoting the least desirable and 4 the most desirable outcomes):

US		**USSR**
(US persists, USSR swerves)	4	(US swerves, USSR persists)
(US swerves, USSR swerves)	3	(US swerves, USSR swerves)
(US swerves, USSR persists)	2	(US persists, USSR swerves)
(US persists, USSR persists)	1	(US persists, USSR persists).

These preferences are summarized as the nonzero-sum game:

		USSR	
		Swerve	Persist
US	Swerve	(3, 3)	(2, 4)
	Persist	(4, 2)	(1, 1)

Here the "payoff" (i, j) associated with any outcome denotes the fact that the US assigns this outcome the ranking i in its preference list, whereas the USSR assigns it the ranking j. The maximin solution of this game recommends that both players swerve and has payoff (3, 3).

Suppose now that as the crisis was unfolding President Kennedy was notified that his staff was infiltrated by Soviet spies and that whatever decision he would make would immediately be made known to Premier Khrushchov. One would think this to be a cause for much consternation for the Americans. We will argue that the opposite is the case and that in fact:

In the game of Chicken, a player's foreknowledge of his opponent's decision works to that player's disadvantage, provided that the opponent is aware of the player's foreknowledge.

This conclusion will be drawn on the basis of the Principle of Rationality. Taking the infiltration into account, President Kennedy would have reasoned as follows:

If I decide to swerve, Khrushchov will know it and, choosing between (3, 3) and (2, 4), he will decide to persist, an outcome with payoff (2, 4);

if I decide to persist, Khrushchov will know it and, choosing between (4, 2) and (1, 1), he will decide to swerve, an outcome with payoff (4, 2).

Thus, if Kennedy swerves the outcome will be (2, 4), whereas if he persists the outcome will be (4, 2). He would therefore choose to persist, resulting in the payoff pair (4, 2) wherein the Soviets swerve.

This is, of course, exactly what transpired. On October 25, a Russian convoy headed on a collision course with the American blockade dispersed. It could be argued that while the USSR had in all likelihood not been privy to the American decision-making process, its leaders were wiser and more cognizant of the possible disastrous consequences of political games of Chicken. After all, only 17 years had passed since the conclusion of World War II in which 20 million Russians are estimated to have perished and major portions of their land were devastated whereas not a single one of that war's battles was fought on American soil.

The essential feature of the infiltration that was hypothesized in the Cuban Missile Crisis is that the USSR did not select its option until after the US had done so. In general this prerogative may or may not be beneficial to the player who possesses it.

Example 4.3.3. Consider the game

(5, 2)	(3, 0)	(8, 1)	(2, 3)
(6, 3)	(5, 4)	(7, 4)	(1, 1)
(7, 5)	(4, 6)	(6, 8)	(0, 2)

Suppose first that Charlie is informed of Ruth's decision when he makes his own, and that Ruth is aware of this fact. Then Ruth will reason as follows:

If I select row 1, Charlie will opt for the payoff pair (2, 3).
If I select row 2, Charlie will opt for one of the payoff pairs (5, 4) and (7, 4).
If I select row 3, Charlie will opt for the payoff pair (6, 8).

The maximin strategy then will direct Ruth to select row 3 and Charlie to choose column 3 resulting in an outcome with payoff pair (6, 8).

On the other hand, if Ruth is the one who is informed of Charlie's decision when she makes her own, and if Charlie is aware of this fact, then he will reason as follows:

If I select column 1, Ruth will opt for payoff pair (7, 5).
If I select column 2, Ruth will opt for payoff pair (5, 4).
If I select column 3, Ruth will opt for payoff pair (8, 1).
If I select column 4, Ruth will opt for payoff pair (2, 3).

The maximin strategy then will direct Charlie to choose column 1 and Ruth to choose row 3 for a payoff pair of (7, 5).

If neither player is informed of their opponent's decisions, then Ruth's maximin strategy calls for the selection of row 1 for a guaranteed payoff of at least 2. Charlie's maximin strategy calls for the selection of column 1 for a guaranteed payoff of at least 2. The actual outcome will then have payoff pair (5, 2). Thus, in this game, information about the opponent's decisions is advantageous to each player.

It was seen in the discussion of the Chicken game that the maximin analysis can fail to properly describe human behavior in some gamelike situations. In retrospect, in that particular case, the failure was due to the oversimplification that occurred when it was assumed that the players made their decision independently and simultaneously. In most instances of this game, however, the players' behaviors very much affect each other. They continuously observe each other and when one swerves the other will most certainly persist to the end. Rather than completely disqualify the maximin strategy for such games, a way was found to adapt it so as to continue to supply researchers with satisfactory solutions.

One of the features of the maximin strategy is that while it may not lead to the best of all possible payoffs, it does offer the players the consolation of having eliminated the possibility of their regretting their decisions. Each player knows that he has played so as to obtain the best possible guarantees, regardless of what the opponent did. This *No Regrets* policy can be applied to nonzero-sum games as well.

In a nonzero-sum game, an outcome is said to be a pure *Nash equilibrium point* if its payoff pair (a_i, b_j) is such that a_i is maximum for its column, and b_j is maximum for its row. Such a Nash equilibrium point has the desired *No Regrets* property since, given that Charlie selected column j, Ruth has no reason to regret having selected row i - no other selection of Ruth's would have yielded her a better payoff. Similarly, given that Ruth selected row i, Charlie has no reason to regret having selected column j - no other selection of his would have yielded him a better payoff. Note that in this definition of the Nash equilibrium each player is assumed to be concerned with regrets over her or his own actions only. Any feelings that they might entertain regarding the other players are ignored. The reason for this limitation is that the incorporation of such feelings into the considerations would result in too many complications.

Example 4.3.4. *Find the Nash equilibrium points of the game*

(5, 2)	(3, 0)	(8, 1)	(2, 3)
(6, 3)	(5, 4)	(7, 4)	(1, 1)
(7, 5)	(4, 6)	(6, 8)	(0, 2)

One method for finding these points is to examine all the payoff pairs (a_i, b_j) in succession and to star each a_i that is a maximum for its column and each b_j that is a maximum for its row. The Nash equilibrium points are those whose payoff pairs have

(5, 2)	(3, 0)	(8*, 1)	(2*, 3*)
(6, 3)	(5*, 4*)	(7, 4*)	(1, 1)
(7*, 5)	(4, 6)	(6, 8*)	(0, 2)

both entries starred. Thus, the Nash equilibrium points of the given game are those with payoffs (5, 4) and (2, 3). Note that one of these payoff pairs, namely (5, 4), is clearly preferable to the other. Nevertheless, this does not necessarily mean that either Ruth or Charlie will aim for this "better" payoff pair. After all, by selecting the second row (that of the pair (5, 4)), Ruth opens herself up to the possibility of obtaining a payoff of only 1 if Charlie selects column 4. Similarly, if Charlie were to select the second column (that of the pair (5, 4)), he would be opening himself up to the possibility of gain 0 if Ruth selects row 1.

The next example shows that pure Nash equilibrium points need not exist.

Example 4.3.5. *Find the Nash equilibrium points of the game*

(2, 1)	(1, 2)
(1, 2)	(2, 1)

When the row maximum of the first entry and the column maximum of the second entry of each payoff pair are starred we obtain the pattern

(2*, 1)	(1, 2*)
(1, 2*)	(2*, 1)

Since no payoff pair has both of its entries starred it follows that this game has no pure Nash equilibrium points.

The logic that underlies the Nash equilibrium is the expectation that as the players watch each other maneuver the situation will naturally gravitate toward a Nash equilibrium outcome. Such was the case in the Cuban Missile Crisis. As those October days passed the Russians became convinced of the Americans' determination to persist and so they swerved. The diagram below demonstrates that this actual outcome is indeed one of the two Nash equilibrium points of that confrontation.

		USSR	
		Swerve	Persist
US	Swerve	(3, 3)	(2*, 4*)
	Persist	(4*, 2*)	(1, 1)

Prisoner's Dilemma has only one Nash equilibrium point with

	Refuse the deal	Accept the deal
Refuse the deal	(–5, –5)	(–15, 0*)
Accept the deal	(0*, –15)	(–10*, –10*)

payoff pair $(-10, -10)$ which is consistent both with its maximin strategy and many observed solutions. Here, of course, the players cannot watch each other reach their decisions. Nevertheless, the oft reached outcome wherein both prisoners accept the deal does turn out to be a pure Nash equilibrium point.

The Nash equilibrium points have become a popular tool for theoretical economists. The next example should give the reader some feel for how this concept is used by them.

The Job Applicants. *Firms 1 and 2 have one opening each for which they offer salaries 2a and 2b, respectively (the 2 is only used in order to prevent fractions from appearing later). Each of Ruth and Charlie can apply to only one of the positions and they must simultaneously decide whether to apply to firm 1 or to firm 2. If only one of them applies for a job, he gets it; if both apply for the same position, the firm hires one of them at random.*

This situation is modeled as the 2×2 nonzero-sum game

		Charlie applies to	
		Firm 1	Firm 2
Ruth applies to	Firm 1	(a, a)	$(2a, 2b)$
	Firm 2	$(2b, 2a)$	(b, b)

In this array the entries $(2a, 2b)$ and $(2b, 2a)$ are self-explanatory. The entry (a, a) is obtained by reasoning as follows. If both Ruth and Charlie apply to the same firm, then, because it was stipulated that the firm will select an applicant at random, each can expect a payoff of half the salary, i.e., a. A similar line of thought justifies the entry (b, b).

Not surprisingly, the analysis of the game depends on the relationship between a and b. Suppose first that

$$a \leq 2b \text{ and } b \leq 2a.$$

In other words, suppose first that the two salaries are not too far out of line with each other in that neither exceeds double the other. Then the two entries $(2a, 2b)$ and $(2b, 2a)$ are both pure Nash equilibria. Such, for example, might be the outcome if Ruth and Charlie became aware of each other's existence and came to some mutual agreement.

On the other hand, if the salaries are out of line with each other, say

$$b \geq 2a,$$

then it is the entry (b, b) that constitutes the unique pure Nash equilibrium point. This corresponds to both Ruth and Charlie applying for the better job. The case where $b = 2a$ is intermediary and has all of the above three outcomes as its Nash equilibria.

Thus, this highly simplified model predicts that if the disparity between the salaries is not too great then some reasonable distribution of the positions may happen. If one job is much better than the other, then both will apply for it resulting in a situation where one of them remains unemployed.

This model will be discussed again in greater depth in the next two sections.

We mention in closing this chapter that the 2×2 nonzero-sum games have been classified in Rapoport, Guyer, and Gordon (1976) into 78 different types. This classification depends on the distribution of Pareto optimal outcomes, Nash equilibrium points, and their relationship to the maximin strategies.

Exercises 4.3

For each of the nonzero-sum games in Exercises 1 – 10 find the following:

a) all the pure maximin strategies,
b) the pure maximin values,
c) all the pure value pairs,
d) all the Pareto optimal payoffs,
e) all the pure Nash equilibrium points,
f) the outcome of the game if Charlie is aware of Ruth's decision when he makes his, and Ruth knows of this,
g) the outcome of the game if Ruth is aware of Charlie's decision when she makes hers, and Charlie knows of this.

1.

(2, 3)	(1,4)
(0, 5)	(4, 1)

2.

(0, 5)	(2, 3)
(4, 1)	(1, 4)

3.

(1, 3)	(3, 1)
(2, 2)	(2, 5)

4.

(3, 1)	(2, 5)
(1, 1)	(2, 2)

5.

(5, 4)	(4, 1)
(1, 4)	(3, 2)
(1, 1)	(5, 5)

6.

(1, 2)	(3, 4)
(5, 6)	(7, 8)
(7, 6)	(5, 5)

7.

(9, 0)	(8, 1)	(7, 2)
(6, 3)	(5, 4)	(4, 5)
(3, 6)	(2, 7)	(1, 9)

8.

(9, 0)	(8, –1)	(7, –2)
(6, –3)	(5, –4)	(4, –5)
(3, –6)	(2, –7)	(1, –9)

9.

(1, 2)	(1, –2)	(1, 1)	(0, 3)
(2, 1)	(0, 0)	(–3, 0)	(1, –2)
(3, –1)	(–2, 1)	(2, 1)	(2, 2)
(2, 1)	(2, 2)	(1, –1)	(–1, 1)

10.

(–1, 2)	(–2, 1)	(2, 0)	(1, 2)
(1, 1)	(3, 0)	(0, –3)	(–2, 1)
(1, –2)	(3, 0)	(1, 2)	(0, 3)
(3, 0)	(2, 2)	(–1, 1)	(–1, 1)

Describe the following situation as nonzero sum games.

11. **Leader** (A. Coleman) Two motorists are waiting to enter a heavy stream of traffic from opposite ends of an intersection, and both are in a hurry to get to their destinations. When a gap in the traffic occurs, each must decide whether to concede the right of way to the other or to drive into the gap. (Use the preferential ranking of the game of Chicken to assign numerical values to the payoffs.)
12. **Battle of the sexes** (Luce and Raiffa) Ruth and Charlie, who are happily married, are planning an evening's entertainment. Ruth would like to go to the concert at the Arts Center, and Charlie would

rather stay home and watch the ball game on TV. Still, both would rather spend their time together rather than separately. (Use the preferential ranking of the game of Chicken to assign numerical values to the payoffs.)

13. **Two leagues** (Dixit & Nalebuff) Two football leagues, USFL and NFL, are deciding whether to schedule their games in the fall or in the spring. They estimate that 10 million viewers will watch football in the fall, but only 5 million will watch in the spring. If one league has a monopoly during a season, it gets the entire market. If both leagues schedule their games for the same season, the NFL gets 70% and the USFL gets 30% of the market.
14. **Oil cartel** (Dixit & Nalebuff) Ruth and Charlie are the rulers of two countries that have formed an oil cartel. In order to keep the price of oil up, they have agreed to limit their productions, respectively, to 4 million and 1 million barrels per day. For each, cheating means producing an extra 1 million barrels each day. Depending on their decisions, their total output would therefore be 5, or 6, or 7 million barrels, with corresponding profit margins of $16, $12, and $8 per barrel.
15. (Dixit & Nalebuff) Ruth is a typical homeowner, and Charlie an average burglar. Ruth is trying to decide whether to keep a gun in her home and Charlie faces the options of whether or not to bring a gun to his next break-in. (Use preferential ranking.)
16. **Altruist's dilemma** (Heckathorn) Ruth and Charlie, who are married, are considering their Christmas gift strategy. To simplify matters, suppose they can each either spend a lot or a reasonable amount on each other's presents. Use preferential ranking to display this as a nonzero-sum game.
17. **Assurance game** (Heckathorn) Ruth and Charlie are the employees of a firm that has been remarkably successful over the last two years. They know that their boss has more than enough money to give them both a raise. Suppose each has the options of either not doing anything at all or else presenting their boss with the ultimatum: "If you don't give me a raise I quit!" Use preferential ranking to present this situation as a nonzero-sum game.
18. **Privileged game** (Heckathorn) This is essentially the same as Ex. 17 with the additional wrinkle that Ruth and Charlie are aware that when the boss caves in to one employee's ultimatum he will automatically give the other employee a smaller raise.
19. Does every nonzero-sum game have to have at least one Pareto optimal outcome? Justify your answer.
20. Give an example of a nonzero-sum game in which all the payoffs are distinct and each outcome is Pareto optimal.

4.4 Mixed Strategies for Nonzero-Sum Games

Just like their zero-sum kin, nonzero-sum games may be repeated and their strategies may be mixed. As was done in the discussion of zero-sum games, we begin with 2×2 nonzero-sum games. A *mixed strategy* for the general 2×2 game is a designation of the frequencies with which each option is exercised and consists of a pair $[a, b]$ where a

	Charlie	
Ruth	(a_1, b_1)	(a_2, b_2)
	(a_3, b_3)	(a_4, b_4)

and b are nonnegative numbers such that

$$a + b = 1.$$

Thus, the pair $[\frac{1}{4}, \frac{3}{4}]$ denotes the mixed strategies of using the first option one of four times whereas the second one is used three out of four times. The pairs [1, 0] and [0, 1] are really pure strategies. Each calls for the consistent use of one option to the exclusion of the second one. Ruth's and Charlie's general mixed strategies will again be denoted by $[1 - p, p]$ and $[1 - q, q]$, respectively. It will prove convenient to refer to such a pair of strategies as a *mixed strategy pair*. When Ruth and Charlie employ these mixed strategies in a nonzero-sum game the expected payoff pair is computed in much the same manner as it was for the zero-sum games, except that each player focuses on his own payoffs. When the mixed strategies pair is $([1 - p, p], [1 - q, q])$, Ruth's expected payoff is denoted by $e_R(p, q)$ and Charlie's expected payoff is denoted by $e_C(p, q)$.

Example 4.4.1. *Compute the expected payoffs when Ruth and Charlie employ the mixed strategy pair ([0.3, 0.7], [0.6, 0.4]) in the game*

(1, 2)	(3, –2)
(–1, 0)	(0, 4)

The auxiliary diagram

	0.6	0.4
0.3	(1, 2)	(3, –2)
0.7	(–1, 0)	(0, 4)

yields

$$e_R(0.7, 0.4) = 0.3 \cdot 0.6 \cdot 1 + 0.3 \cdot 0.4 \cdot 3 + 0.7 \cdot 0.6 \cdot (-1) + 0.7 \cdot 0.4 \cdot 0$$
$$= 0.18 + 0.36 - 0.42 + 0 = 0.12$$

and

$$e_C(.7, .4) = 0.3 \cdot 0.6 \cdot 2 + 0.3 \cdot 0.4 \cdot (-2) + 0.7 \cdot 0.6 \cdot 0 + 0.7 \cdot 0.4 \cdot 4$$
$$= 0.36 - 0.24 + 0 + 1.12 = 1.24.$$

In 1950, John Nash (b. 1928) succeeded in extending the No Regrets aspect of pure equilibria to the mixed context. The significance of this achievement was underscored in 1994 when he was awarded the Nobel Prize for economics, in which discipline the concept of an equilibrium point has become an indispensable theoretical tool. The formulation of this new mixed version of the No Regrets principle is very similar to the pure one:

> *After the repeated game is done, each player would like to rest assured that no other behavior on his part would have resulted in a better expected payoff.*

A pair of mixed strategies is said to constitute a mixed Nash Equilibrium of a mixed game if they result in payoffs such that neither player has cause to regret his choice of strategy. Nash's startling discovery was that unlike their pure kin, mixed Nash equilibria always exist. For the sake of completeness Nash's theorem is stated in its full generality even though its terms cannot be fully defined here. This general statement will be followed by a working restricted version.

Nash's Theorem 4.4.2. *Every* n*-person finite nonzero-sum game has an equilibrium point.*

The ensuing discussion and examples are confined mostly to two-person 2×2 nonzero-sum games. Our limited version of Nash's Theorem can now be stated as follows.

Theorem 4.4.3. *In any* 2×2 *nonzero-sum game, there is a mixed strategy pair* $([1 - \bar{p}, \bar{p}], [1 - \bar{q}, \bar{q}])$ *such that*
$e_R(\bar{p}, \bar{q}) \geq e_R(p, \bar{q})$ *for all* p, $0 \leq p \leq 1$ (1)

and

$e_C(\bar{p}, \bar{q}) \geq e_C(\bar{p}, q)$ *for all* q, $0 \leq q \leq 1$ (2).

Inequality (1) of this theorem says that if both players adopt the stipulated strategies, then Ruth would have gained naught by adopting any other strategy. In other words, having employed the recommended strategy $[1 - \bar{p}, \bar{p}]$, Ruth has no cause for regrets as long as Charlie sticks to $[1 - \bar{q}, \bar{q}]$. Inequality (2) makes a similar assertion about her opponent: Charlie has no cause to regret his employment of $[1 - \bar{q}, \bar{q}]$ as long as Ruth sticks to $[1 - \bar{p}, \bar{p}]$. The pair of strategies $([1 - \bar{p}, \bar{p}], [1 - \bar{q}, \bar{q}])$ described by this theorem constitute a *mixed Nash equilibrium*. The numbers $e_R(\bar{p}, \bar{q})$ and $e_C(\bar{p}, \bar{q})$ are the game's *mixed value pair*. Together, the mixed Nash equilibrium and the mixed value pair are a *Nash equilibrium solution* of the game.

All known proofs of Nash's Theorem are nonconstructive. In other words, while they assert the existence of the mixed Nash equilibrium, they provide no method for actually deriving its constituent mixed strategies. The present chapter is confined to a description of a method for recognizing Nash equilibrium pairs as well as several applications of this concept to the real world. First it is necessary to consider a preliminary issue.

Granted that in general neither player knows the opponent's strategy, still, after a game has been played, while trying to gauge whether he has cause for regrets, a player will be aware of the strategy that was used by his opponent. For this reason it is necessary at this point to consider the question of *optimal counterstrategies*. In other words, how should a player react if he knows his opponent's strategy? The guiding principle under these circumstances is given in the following theorem.

Theorem 4.4.4. *If one player of a nonzero-sum game employs a fixed strategy, then the opponent has an optimal counterstrategy that is pure.*

Example 4.4.5. *Suppose Ruth employs the strategy [0.2, 0.8] in the game*

(3, 2)	(2, 1)
(0, 3)	(4, 4)

Find an optimal response for Charlie.

We know that Charlie has an optimal counterstrategy that is pure and so we compute the expected payoffs that correspond to the two pure strategies that are available to him. The auxiliary diagrams below contain Charlie's

	1	0
0.2	2	1
0.8	3	4

	0	1
0.2	2	1
0.8	3	4

payoffs only and they yield

$$0.2 \cdot 2 + 0.8 \cdot 3 = 2.8 \text{ for } [1, 0],$$
$$0.2 \cdot 1 + 0.8 \cdot 4 = 4.4 \text{ for } [0, 1].$$

Since Charlie is interested in maximizing his payoff, it follows that the pure strategy [0, 1], which yields him a payoff of 4.4, is an optimal response for him.

Example 4.4.6. *Suppose Charlie employs the strategy [0.7, 0.3] in the game*

(3, 2)	(2, 1)
(0, 3)	(4, 4)

Find an optimal response for Ruth.

Since Ruth has an optimal counterstrategy that is pure, we compute the expected payoffs that correspond to the two pure strategies that are available to her. The auxiliary diagrams below contain Ruth's

	0.7	0.3
1	3	2
0	0	4

	0.7	0.3
0	3	2
1	0	4

payoffs only and they yield

$$0.7 \cdot 3 + 0.3 \cdot 2 = 2.7 \quad \text{for } [1, 0],$$
$$0.7 \cdot 0 + 0.3 \cdot 4 = 1.2 \quad \text{for } [0, 1].$$

Since Ruth is interested in maximizing her payoff, it follows that the pure strategy [1, 0], which yields her a payoff of 2.7, is an optimal response for her.

We now turn to the issue of recognizing mixed Nash equilibria. Suppose your mathematical consultant has provided you, at cost, with a mixed Nash equilibrium pair for a game that is of interest to you. How can you be sure that the mathematician has not made a mistake and that you are indeed getting your money's worth? Theorem 4 above provides us with a straightforward method for checking on the proposed strategy pair. One need merely verify that none of the pure strategies at each player's disposal provide that player with a better payoff than that resulting from the proposed Nash equilibrium pair.

Example 4.4.7. *Verify that the mixed strategy pair ([0.5, 0.5], [0.4, 0.6]) constitutes a mixed Nash equilibrium for the game*

(3, 2)	(2, 1)
(0, 3)	(4, 4)

Ruth's and Charlie's payoffs for the proposed Nash equilibrium pair are, respectively,

	0.4	0.6
0.5	(3, 2)	(2, 1)
0.5	(0, 3)	(4, 4)

$$e_R(0.5, 0.6) = 0.5 \cdot 0.4 \cdot 3 + 0.5 \cdot 0.6 \cdot 2 + 0.5 \cdot 0.4 \cdot 0 + 0.5 \cdot 0.6 \cdot 4$$
$$= 0.6 + 0.6 + 1.2 = 2.4,$$

and

$$e_C(0.5, 0.6) = 0.5 \cdot 0.4 \cdot 2 + 0.5 \cdot 0.6 \cdot 1 + 0.5 \cdot 0.4 \cdot 3 + 0.5 \cdot 0.6 \cdot 4$$
$$= 0.4 + 0.3 + 0.6 + 1.2 = 2.5.$$

Is it possible for Ruth to improve her payoff by switching to an alternate strategy? If so, then there must be a pure alternate strategy that allows her to accomplish this. However, the auxiliary diagrams below both yield payoffs of

	0.4	0.6
1	(3, 2)	(2, 1)
0	(0, 3)	(4, 4)

	0.4	0.6
0	(3, 2)	(2, 1)
1	(0, 3)	(4, 4)

$$0.4 \cdot 3 + 0.6 \cdot 2 = 2.4 \quad \text{for } [1, 0],$$
$$0.4 \cdot 0 + 0.6 \cdot 4 = 2.4 \quad \text{for } [0, 1],$$

both of which equal the value of $e_R(0.5, 0.6)$ computed above. This demonstrates that Ruth stands to gain nothing by switching to any alternate pure (or mixed) strategy. Thus, the proposed Nash equilibrium pair can cause Ruth no regrets.

Does Charlie have any cause for regrets? The auxiliary diagrams below yield payoffs

	1	0
0.5	(3, 2)	(2, 1)
0.5	(0, 3)	(4, 4)

	0	1
0.5	(3, 2)	(2, 1)
0.5	(0, 3)	(4, 4)

$$0.5 \cdot 2 + 0.5 \cdot 3 = 2.5 \quad \text{for } [1, 0],$$
$$0.5 \cdot 1 + 0.5 \cdot 4 = 2.5 \quad \text{for } [0, 1],$$

both of which equal the value of $e_C(0.5, 0.6)$ computed above. Thus, Charlie has no cause for regrets either.

The conclusion is that the strategy pair ([0.5, 0.5], [0.4, 0.6]) constitutes a Nash equilibrium pair for the given game.

The same method can also be used to recognize strategy pairs that are not Nash equilibria.

Example 4.4.8. *Decide whether* $\left(\left[\frac{1}{3}, \frac{2}{3}\right], \left[\frac{1}{6}, \frac{5}{6}\right]\right)$ *constitutes a Nash equilibrium pair for the game*

(5, 1)	(0, 0)
(0, 0)	(1, 5)

The expected payoffs corresponding to the proposed strategy pair are

	1/6	5/6
1/3	(5, 1)	(0, 0)
2/3	(0, 0)	(1, 5)

$$e_R\left(\frac{2}{3}, \frac{5}{6}\right) = \frac{1}{3} \cdot \frac{1}{6} \cdot 5 + \frac{2}{3} \cdot \frac{5}{6} \cdot 1 = \frac{5 + 10}{18} = \frac{15}{18} = \frac{5}{6},$$
$$e_C\left(\frac{2}{3}, \frac{5}{6}\right) = \frac{1}{3} \cdot \frac{1}{6} \cdot 1 + \frac{2}{3} \cdot \frac{5}{6} \cdot 5 = \frac{1 + 50}{18} = \frac{51}{18} = \frac{17}{6}.$$

To see whether Ruth has any cause for regrets we compute the payoffs of the auxiliary diagrams below as

	1/6	5/6
1	(5, 1)	(0, 0)
0	(0, 0)	(1, 5)

	1/6	5/6
0	(5, 1)	(0, 0)
1	(0, 0)	(1, 5)

$$\frac{1}{6}\cdot 5 = \frac{5}{6} \quad \text{for} \quad [1\ ,\ 0],$$
$$\frac{5}{6}\cdot 1 = \frac{5}{6} \quad \text{for} \quad [0\ ,\ 1].$$

As these alternate payoffs equal the value of $e_R(\frac{2}{3}, \frac{5}{6})$, Ruth stands to gain nothing by relinquishing the given strategy $[\frac{1}{3}, \frac{2}{3}]$. Passing on to Charlie, the auxiliary diagrams for the alternate pure strategies yield payoffs

	1	0
1/3	(5, 1)	(0, 0)
2/3	(0, 0)	(1, 5)

	0	1
1/3	(5, 1)	(0, 0)
2/3	(0, 0)	(1, 5)

$$\frac{1}{3}\cdot 1 = \frac{1}{3} \quad \text{for}[1\ ,\ 0],$$
$$\frac{2}{3}\cdot 5 = \frac{10}{3} \quad \text{for}[0\ ,\ 1].$$

Since the second of these, $\frac{10}{3} = 4.\overline{3}$, exceeds the value $\frac{17}{6} = 2.1\overline{6}$ of $e_C(\frac{2}{3}, \frac{5}{6})$, it follows that Charlie could improve his payoff by abandoning the proposed strategy of $[\frac{1}{6}, \frac{5}{6}]$ in favor of the pure strategy [0, 1]. Hence, the given mixed strategy pair does not constitute a Nash equilibrium.

We offer the reader an alternate way of visualizing the defining properties of the Nash equilibrium. Suppose the values of $e_R(p, q)$ and $e_C(p, q)$, $p = 0, 0.2, 0.4, 0.6, 0.8, 1$, $q = 0, 0.2, 0.4, 0.6, 0.8, 1$ are tabulated for the game

(3, 1)	(1, 5)
(1, 2)	(4, 1)

in the form

p \ q	0	0.2	0.4	0.6	0.8	1
0	(3, 1)	(2.6, 1.8)	(2.2, 2.6)	(1.8, 3.4)	(1.4, 4.2)	(1, 5)
0.2	(2.6, 1.2)	(2.4, 1.8)	(2.2, 2.4)	(2, 3)	(1.8, 3.6)	(1.6, 4.2)
0.4	(2.2, 1.4)	(2.2, 1.8)	(2.2, 2.2)	(2.2, 2.6)	(2.2, 3)	(2.2, 3.4)
0.6	(1.8, 1.6)	(2, 1.8)	(2.2, 2)	(2.4, 2.2)	(2.6, 2.4)	(2.8, 2.6)
0.8	(1.4, 1.8)	(1.8, 1.8)	(2.2, 1.8)	(2.6, 1.8)	(3, 1.8)	(3.4, 1.8)
1	(1, 2)	(1.6, 1.8)	(2.2, 1.6)	(2.8, 1.4)	(3.4, 1.2)	(4, 1)

Observe that in the $q = 0.4$ column, the first entry of all of the payoff pairs has the constant value of 2.2, and that in the $p = 0.8$ row, the second entry has the constant value of 1.8. Consequently the mixed strategy pair ([0.2, 0.8], [0.6, 0.4]) is a Nash equilibrium since neither player stands to gain anything by changing his mixed strategy.

While such a table can be a useful pedagogical tool, its construction is rather laborious and the details are greatly dependent on the exact numerical values of the constituent mixed strategies of the equilibrium. No more space will therefore be devoted to such tables in this text.

The next two examples are meant to demonstrate the wide range of applicability of the notion of a mixed Nash equilibrium. They come from the disciplines of economics and biology, respectively.

The Job Applicants. In the previous chapter we modeled a situation wherein Ruth and Charlie were allowed to apply to one of two positions offering salaries $2a$ and $2b$, respectively, as the game

		Charlie applies to	
		Firm 1	Firm 2
Ruth applies to	Firm 1	(a, a)	$(2a, 2b)$
	Firm 2	$(2b, 2a)$	(b, b)

It was noted above that this game always has pure Nash equilibria whose exact values depend on the relative sizes of a and b. Specifically, if neither salary exceeds double the other, i.e., if

$$b \leq 2a \text{ and } a \leq 2b,$$

then the outcomes corresponding to the payoffs *(2a, 2b)* and *(2b, 2a)* constitute pure Nash equilibria. In that case, however, the game also possesses a mixed Nash equilibrium (see Exercise 18) each of whose strategies is

$$\left[\frac{2a-b}{a+b}, \frac{2b-a}{a+b}\right].$$

The components of this common strategy can be interpreted as this model's predictions for the probabilities of Ruth (or Charlie) applying to the corresponding firm, i.e., if $a \leq 2b$ and $b \leq 2a$, then

$\frac{2a-b}{a+b}$ is the predicted probability of Ruth applying to firm 1,

$\frac{2b-a}{a+b}$ is the predicted probability of Ruth applying to firm 2.

For example, if firms 1 and 2 offer salaries of \$20,000 and \$30,000, respectively, then $a = \$10{,}000$, $b = \$15{,}000$, and this model predicts that the probability of a player applying for firm 2's position is

$$\frac{2 \cdot 15{,}000 - 10{,}000}{10{,}000 + 15{,}000} = \frac{20{,}000}{25{,}000} = 0.8.$$

On the other hand, if either $a \geq 2b$ or $b \geq 2a$, or, in other words, if one of the salaries is more than double the other, then all the Nash equilibria are the pure ones that were already discussed in the previous section.

This analysis provides economists with a starting point for an investigation of the question of what effect wage differentials have on the pool of applicants. The details of this investigation fall outside the scope of this book.

An Evolutionary Game. In recent years biological evolution has offered some applications of Nash equilibria as well. In many species, mating is preceded by a duel between the males. Stags shove each other with their antlers and snakes have wrestling matches. The winner gets to mate. Similar intraspecies contests result from territorial disagreements. A surprising aspect of these contests is that nature seems to be pulling its punches. Some individuals run rather than fight. Stags do not gore each other's vulnerable sides and in some species snakes do not bite each other. This is, of course, quite sensible behavior for the species, and the biologists J. Maynard Smith and G.R. Price have constructed a game theoretic model for these contests in which this moderated ferocity is explained as a Nash equilibrium. An unusual feature of this game is that it pits the species in question against itself.

To define the game, we stipulate a species whose individuals engage in intraspecies duels. Each such confrontation constitutes a play of the game. The individual members of the species are classified as either *Hawks* or *Doves.* A Hawk always attacks in a confrontation and a Dove always runs away. The winner of a confrontation gets to mate, or gets the better territory, and so he is in a better position to propagate his genes. In the event of an actual physical struggle, the loser sustains an injury. The payoff to the species of a single play consists of the effect the duel has on the individual's ability to reproduce, an elusive quantity called his *fitness*. It will be assumed that the confrontation winner's fitness is augmented by amount $2a$, and that a fight loser's fitness is reduced by amount $2b$ (the 2 is introduced here into the payoffs just in order to simplify the subsequent calculations). The precise payoffs are computed as follows:

Hawk vs. Hawk: The two contestants continue fighting until one is injured. Inasmuch as each has a 50% chance of winning (and gaining $2a$) or losing (and losing $2b$), each of them is assigned the payoff

$$50\% \cdot 2a + 50\% \times (-2b) = a - b.$$

Hawk vs. Dove: Since the Dove runs away, no physical struggle takes place; Hawk gains $2a$ and Dove neither gains nor loses anything.

Dove vs. Dove: Again each contestant has a 50% chance of adding $2a$ to his fitness, but as there is no physical fight, there is no question of injury. Thus, each player is assigned the payoff

$$50\% \cdot 2a = a.$$

The resulting nonzero-sum game has the array

	Hawk	Dove
Hawk	$(a-b, a-b)$	$(2a, 0)$
Dove	$(0, 2a)$	(a, a)

(4)

In this game both Ruth and Charlie represent the same species, and each play consists of a confrontation between some individuals. The pure strategy [1, 0] calls for the species to evolve Hawks only. The pure strategy [0, 1] calls for the species to evolve Doves only. A mixed strategy consists of evolving a mixture of both Hawks and Doves within the species. Such a strategy would presumably be encoded into the species' genes. However, mutations do occur with high frequency, and it is reasonable to assume that whatever strategy is now in effect could not be improved upon by any mutations. After all, if a change (that is internal to the species) could improve on its overall fitness, then, given the amount of time most species have been in existence, such a change would have in all likelihood taken place a long time ago. Thus, it is reasonable to conclude that the current ratio of Hawks to Doves in this species is a stable quantity that maximizes its overall fitness. It will now be argued that the optimality of this ratio implies that it actually comes from a mixed Nash equilibrium of the nonzero-sum game (4).

Let us consider the mathematical analog of this optimality argument. Suppose the species is currently using the mixed strategy $[1-\overline{p},\ \overline{p}]$, i.e., its Hawks-to-Doves ratio is now $(1-\overline{p}) : \overline{p}$. Each individual's confrontation then augments his fitness by the expected payoff computed from the auxiliary diagram

	$1-\overline{p}$	$\overline{p}$
$1-\overline{p}$	$(a-b, a-b)$	$(2a, 0)$
$\overline{p}$	$(0, 2a)$	(a, a)

This value is

$$\begin{aligned} e_R(\overline{p}, \overline{p}) = e_C(\overline{p}, \overline{p}) &= (1-\overline{p})^2(a-b) + \overline{p}(1-\overline{p})\,2a + \overline{p}^2 a \\ &= \overline{p}^2(a-b-2a+a) + \overline{p}(-2(a-b)+2a) + 1(a-b) \\ &= -\overline{p}^2 b + 2\overline{p}b + a - b. \end{aligned}$$

The reason $e_R(\overline{p}, \overline{p})$ and $e_C(\overline{p}, \overline{p})$ are equal is that both Ruth and Charlie represent the <u>same</u> species. The actual common value of these two expressions is in fact immaterial at this point. It is only necessary to keep in mind that $e_R(\overline{p}, \overline{p}) = e_C(\overline{p}, \overline{p})$.

Suppose now that a comparatively small subpopulation of this species mutates and begins breeding Hawks and Doves with a mixed strategy of $[1-p, p]$ for some $p \neq p$. Since this subpopulation is presumed small, its individuals will mostly be confronting normal individuals, and hence the expected increment to the mutated individual's fitness from each battle is the common value of $e_R(\overline{p}, \overline{p})$ and $e_C(\overline{p}, \overline{p})$ that correspond to the auxiliary diagrams

		Normal population	
		$1-\overline{p}$	$\overline{p}$
Mutated population	$1-p$	$(a-b, a-b)$	$(2a, 0)$
	p	$(0, 2a)$	(a, a)

or

		Mutated population	
		$1-p$	p
Normal population	$1-\bar{p}$	$(a-b, a-b)$	$(2a, 0)$
	$\bar{p}$	$(0, 2a)$	(a, a)

The normal individual, however, still confronts mostly other normal individuals and hence his fitness is augmented by the same quantity as before, namely,

$$e_R(\bar{p}, \bar{p}) = e_C(\bar{p}, \bar{p}).$$

The above Panglossian argument that mutated strategies will not improve upon the existing strategy entails, amongst others, the inequalities

$$e_R(\bar{p}, \bar{p}) \geq e_R(p, \bar{p}) \quad \text{and} \quad e_C(\bar{p}, \bar{p}) \geq e_C(\bar{p}, p).$$

These inequalities, however, are tantamount to the No Regrets guideline since, if we interpret game (4) as a game between the distinct opponents Ruth and Charlie, these inequalities say that neither Ruth nor Charlie could have improved their expected payoffs by changing from the strategy $[1-\bar{p}, \bar{p}]$ to any other strategy $[1-p, p]$. Thus, the existing Hawks and Doves breeding strategy $[1-\bar{p}, \bar{p}]$ constitutes a mixed Nash equilibrium strategy of game (4). Let us examine the Nash equilibria of this game.

If $a \geq b$, then $a-b \geq 0$ and the outcome with payoff $(a-b, a-b)$ constitutes the only (pure or mixed) Nash equilibrium (Exercise 35). The associated pure strategy is $[1, 0]$. This can be interpreted as saying that if the advantage of winning outweighs the injuries that accompany a loss, then the species will breed only Hawks.

If $a > b$, then (Exercise 36) there are two pure Nash equilibria with respective payoffs $(2a, 0)$ and $(0, 2a)$ and another mixed Nash equilibrium with strategies

$$[1-\bar{p}, \bar{p}] = [1-\bar{q}, \bar{q}] = \left[\frac{a}{b}, 1-\frac{a}{b}\right].$$

The pure Nash equilibria can be ignored since each has $[1, 0]$ as the strategy for one player and $[0, 1]$ as the other player's strategy - an impossible situation given that both players stand for one and the same species. This leaves us with the mixed Nash equilibrium as the one that describes the species' Hawks-to-Doves ratio. Note that as a diminishes in comparison to b, this model predicts that the species will breed fewer and fewer Hawks. In particular, if $a = 1$ and $b = 3$, i.e., if the effect of an injury from a duel outweighs the benefits of winning by a factor of 3-to-1, then this model predicts that only

$$\frac{a}{b} = \frac{1}{3}$$

of the individuals of the species will be Hawks.

Given the inherently inaccurate nature of the payoffs, the final case, $a = b$, because of the exact equality it demands, is of course unlikely to occur in reality. Exercise 37 asserts that in this case Hawks alone will be bred.

An alternate interpretation of the Nash-equilibrium strategy $[1 - \overline{p}, \overline{p}]$ in this evolutionary game is that it constitutes a ferocity index. On a scale of 0 (Hawk) to 1 (Dove), the quantity s denotes that species' willingness to fight that evolution has encoded into its genes. This could explain why, when fighting each other, the aforementioned snakes wrestle rather than bite.

***m × n* Nonzero-sum games.** All the concepts developed here for 2 × 2 nonzero-sum games apply to larger games as well. An $m \times n$ nonzero-sum game is a rectangular array of m rows and n columns in which each entry is a pair of numbers. The entry in the i-th row and j-th column is denoted by $(a_{i,j}, b_{i,j})$. Given two mixed strategy pair $([p_1, p_2, ..., p_m], [q_1, q_2, ..., q_n])$, Ruth and Charlie's expected payoffs, denoted by e_R and e_C, respectively, are computed as

$$e_R = \text{sum of all } p_i \cdot q_j \cdot a_{i,j} \quad i = 1, 2, ..., m, j = 1, 2, ..., n,$$
$$e_C = \text{sum of all } p_i \cdot q_j \cdot b_{i,j} \quad i = 1, 2, ..., m, j = 1, 2, ..., n.$$

Example 4.4.9. *Compute the expected payoffs when Ruth and Charlie use the mixed strategy pair ([0.5, 0.2, 0.3], [0.1, 0.2, 0.3, 0.4]) in the nonzero-sum game*

(1, 2)	(0, –1)	(3, 1)	(–2, 0)
(1, –3)	(0, 0)	(2, 1)	(–1, 1)
(3, 2)	(1, 1)	(–1, 1)	(3, –1)

The auxiliary diagram

	0.1	0.2	0.3	0.4
0.5	(1, 2)	(0, –1)	(3, 1)	(–2, 0)
0.2	(1, –3)	(0, 0)	(2, 1)	(–1, 1)
0.3	(3, 2)	(1, 1)	(–1, 1)	(3, –1)

yields the expected payoffs

$$\begin{aligned} e_R &= 0.5 \cdot 0.1 \cdot 1 + 0.5 \cdot 0.2 \cdot 0 + 0.5 \cdot 0.3 \cdot 3 + 0.5 \cdot 0.4 \cdot (-2) \\ &\quad + 0.2 \cdot 0.1 \cdot 1 + 0.2 \cdot 0.2 \cdot 0 + 0.2 \cdot 0.3 \cdot 2 + 0.2 \cdot 0.4 \cdot (-1) \\ &\quad + 0.3 \cdot 0.1 \cdot 3 + 0.3 \cdot 0.2 \cdot 1 + 0.3 \cdot 0.3 \cdot (-1) + 0.3 \cdot 0.4 \cdot 3 \\ &= .05 + 0.45 - 0.4 + 0.02 + 0.12 - 0.08 + 0.09 + 0.06 - 0.09 \\ &\quad + 0.36 = 0.58, \end{aligned}$$

$$\begin{aligned} e_C &= 0.5 \cdot 0.1 \cdot 2 + 0.5 \cdot 0.2 \cdot (-1) + 0.5 \cdot 0.3 \cdot 1 + 0.5 \cdot 0.4 \cdot 0 \\ &\quad + 0.2 \cdot 0.1 \cdot (-3) + 0.2 \cdot 0.2 \cdot 0 + 0.2 \cdot 0.3 \cdot 1 + 0.2 \cdot 0.4 \cdot 1 \\ &\quad + 0.3 \cdot 0.1 \cdot 2 + 0.3 \cdot 0.2 \cdot 1 + 0.3 \cdot 0.3 \cdot 1 + 0.3 \cdot 0.4 \cdot (-1) \\ &= .01 - 0.1 + 0.15 - 0.06 + 0.06 + 0.08 + 0.06 + 0.06 + 0.09 - 0.12 = 0.32. \end{aligned}$$

Theorem 5 offers the same guidance in the search for optimality in this wider context as well.

Example 4.4.10. *If Ruth employs the strategy [0.5, 0.2, 0.3] in the game of Example 4.4.9, find an optimal counterstrategy for Charlie.*

The auxiliary diagrams

	1	0	0	0
0.5	(1, 2)	(0, –1)	(3, 1)	(–2, 0)
0.2	(1, –3)	(0, 0)	(2, 1)	(–1, 1)
0.3	(3, 2)	(1, 1)	(–1, 1)	(3, –1)

	0	1	0	0
0.5	(1, 2)	(0, –1)	(3, 1)	(–2, 0)
0.2	(1, –3)	(0, 0)	(2, 1)	(–1, 1)
0.3	(3, 2)	(1, 1)	(–1, 1)	(3, –1)

	0	0	1	0
0.5	(1, 2)	(0, –1)	(3, 1)	(–2, 0)
0.2	(1, –3)	(0, 0)	(2, 1)	(–1, 1)
0.3	(3, 2)	(1, 1)	(–1, 1)	(3, –1)

	0	0	0	1
0.5	(1, 2)	(0, –1)	(3, 1)	(–2, 0)
0.2	(1, –3)	(0, 0)	(2, 1)	(–1, 1)
0.3	(3, 2)	(1, 1)	(–1, 1)	(3, –1)

yield the following payoffs for Charlie:

$$e_C = 0.5 \cdot 2 + 0.2 \cdot (-3) + 0.3 \cdot 2 = 1 - 0.6 + 0.6 = 1 \text{ for } [1, 0, 0, 0],$$

$$e_C = 0.5 \cdot (-1) + 0.2 \cdot 0 + 0.3 \cdot 1 = -0.5 + 0.3 = -0.2 \text{ for } [0, 1, 0, 0],$$

$$e_C = 0.5 \cdot 1 + 0.2 \cdot 1 + 0.3 \cdot 1 = 0.5 + 0.2 + 0.3 = 1 \text{ for } [0, 0, 1, 0],$$

$$e_C = 0.5 \cdot 0 + 0.2 \cdot 1 + 0.3 \cdot (-1) = 0.2 - 0.3 = -0.1 \text{for } [0, 0, 0, 1].$$

Since 1 is the largest of these payoffs, it follows that both [1, 0, 0, 0] and [0, 0, 1, 0] are optimal counterstrategies for Charlie.

The procedure used to check on proposed Nash equilibrium pairs for 2×2 games can be applied in the more general context of $m \times n$ games as well.

Example 4.4.11. *Show that* ([0.4, 0.4, 0.2], [0.2, 0.6, 0, 0.2]) *is a Nash equilibrium for the nonzero-sum game*

(0, 1)	(1, 0)	(1, 0)	(0, 1)
(1, 0)	(0, 1)	(1, 0)	(2, −1)
(1, 0)	(1, 0)	(1, 0)	(−1, 2)

The auxiliary diagram

	0.2	0.6	0	0.2
0.4	(0, 1)	(1, 0)	(1, 0)	(0, 1)
0.4	(1, 0)	(0, 1)	(1, 0)	(2, −1)
0.2	(1, 0)	(1, 0)	(1, 0)	(−1, 2)

yields the expected payoffs

$$\begin{aligned} e_R &= 0.4{\cdot}0.6{\cdot}1 + 0\ .4{\cdot}0.2{\cdot}1 + 0.4{\cdot}\ 0.2{\cdot}2 + 0.2{\cdot}0.2{\cdot}1 + 0.2{\cdot}0.6{\cdot}1 + 0.2{\cdot}0.2{\cdot}(-1) \\ &= 0.24 + 0.08 + 0.16 + 0.04 + 0.12 - 0.04 = 0.6, \end{aligned}$$

$$\begin{aligned} e_C &= 0.4{\cdot}0.2{\cdot}1 + 0.4{\cdot}0.2{\cdot}1 + 0.4\ {\cdot}0.6{\cdot}1 + 0.4\ {\cdot}0.2{\cdot}(-1) + 0.2{\cdot}0.2{\cdot}2 \\ &= 0.08 + 008 + 0.24 - 0.08 + 0.08 = 0.4. \end{aligned}$$

On the other hand, if Charlie sticks to his [0.2, 0.6, 0, 0.2] and Ruth experiments with her pure strategies, she gets the payoffs

$$\begin{aligned} 0.6 \cdot 1 = 0.6 \quad &\text{for} \quad [1, 0, 0], \\ 0.2 \cdot 1 + 0.2 \cdot 2 = 0.6, \quad &\text{for} \quad [0, 1, 0], \\ 0.2 \cdot 1 + 0.6 \cdot 1 + 0.2 \cdot (-1) = 0.6 \quad &\text{for} \quad [0, 0, 1], \end{aligned}$$

all of which equal the value of e_R. Thus, Ruth has no reason to abandon her given strategy. If Ruth sticks to her strategy of [0.4, 0.4, 0.2] and Charlie experiments with his pure strategies, he gets the payoffs

$$\begin{aligned} &0.4 \cdot 1 = 0.4 \text{ for } [1, 0, 0, 0], \\ &0.4 \cdot 1 = 0.4 \text{ for } [0,1, 0, 0], \\ &0 \qquad\qquad\quad \text{for } [0, 0, 1, 0], \\ &0.4 \cdot 1 + 0.4 \cdot (-1) + 0.2 \cdot 2 = 0.4 \text{ for } [1, 0, 0, 0], \end{aligned}$$

none of which is better than the value of e_C (note that one is actually worse). Thus, Charlie has no reason to regret staying with the given strategy either, and so the given strategy pair is indeed a Nash equilibrium.

Exercises 4.4

Compute the expected payoff when the games of Exercises 1–6 are played with the specified mixed strategy pairs.

1.

(4, 4)	(0, 1)
(1, 0)	(1, 1)

([0.7, 0.3], [0.2, 0.8])

2.

(4, 1)	(0, 2)
(1, 1)	(1, 0)

([0.2, 0.8], [0.5, 0.5])

3.

(4, 0)	(0, 1)
(1, 2)	(1, 5)

([0.3, 0.7], [0.8, 0.2])

4.

(4, 3)	(0, 2)
(1, 0)	(1, 1)

([0.7, 0.3], [0, 1])

5.

(4, 2)	(0, 1)
(1, 2)	(1, 0)

([1, 0], [0, 1])

6.

(1, 4)	(1, 1)
(2, 0)	(0, 1)

([0.1, 0.9], [0.5, 0.5])

For each of the games in Exercises 7–12:

a) Determine Ruth's optimal counterstrategy if Charlie employs the given strategy ***C*** *and she knows it.*

b) Determine Charlie's optimal counterstrategy if Ruth employs the given strategy ***R*** *and he knows it.*

7.

(3, 2)	(2, 4)
(2, 3)	(4, –3)

$\boldsymbol{R} = [\frac{1}{2}, \frac{1}{2}]$, $\boldsymbol{C} = [\frac{2}{3}, \frac{1}{3}]$

8.

(3, 2)	(2, 4)
(2, 3)	(4, –3)

$\boldsymbol{R} = [\frac{3}{4}, \frac{1}{4}]$, $\boldsymbol{C} = [\frac{1}{2}, \frac{1}{2}]$

9.

(3, 2)	(2, 4)
(2, 3)	(4, –3)

$\boldsymbol{R} = [\frac{3}{4}, \frac{1}{4}]$, $\boldsymbol{C} = [\frac{2}{3}, \frac{1}{3}]$

10.

(2, –3)	(–1, 3)	(–2, 0)
(0, 1)	(1, –2)	(0, 0)
(–1,–2)	(2, 3)	(3, –1)

$\boldsymbol{R} = [0.2, 0.5, 0.3]$, $\boldsymbol{C} = [0, 0.3, 0.7]$

11.

(2, –3)	(–1, 3)	(–2, 0)	(4, –1)
(0, 1)	(1, –2)	(0, 0)	(–2, 3)
(–1,–2)	(2, 3)	(3, –1)	(2, 2)

$\boldsymbol{R} = [0.8, 0.1, 0.1]$, $\boldsymbol{C} = [0.4, 0.3, 0.2, 0.1]$

12.

(2, –3)	(–1, 3)	(–2, 0)
(0, 1)	(1, –2)	(0, 0)
(–1,–2)	(2, 3)	(3, –1)
(4, –1)	(–2, 3)	(2, 2)

$\boldsymbol{R} = [0.4, 0.4, 0.1, 0.1]$, $\boldsymbol{C} = [0.4, 0.3, 0.3]$

For each of the games in Exercises 13-18, decide whether the mixed strategy pair constitutes a Nash equilibrium.

13.

(3, 2)	(2, 4)
(2, 3)	(4, –3)

$([\frac{1}{2}, \frac{1}{2}], [\frac{2}{3}, \frac{1}{3}])$

14.

(3, 2)	(2, 4)
(2, 3)	(4, –3)

$([\frac{3}{4}, \frac{1}{4}], [\frac{1}{2}, \frac{1}{2}])$

15.

(3, 2)	(2, 4)
(2, 3)	(4, –3)

$([\frac{3}{4}, \frac{1}{4}], [\frac{2}{3}, \frac{1}{3}])$

16.

(2, –3)	(–1, 3)
(0, 1)	(1, –2)

$([\frac{1}{3}, \frac{2}{3}], [\frac{1}{3}, \frac{2}{3}])$

17.

(2, –3)	(–1, 3)
(0, 1)	(1, –2)

$([\frac{1}{3}, \frac{2}{3}], [\frac{1}{2}, \frac{1}{2}])$

18.

$([\frac{3}{4}, \frac{1}{4}], [\frac{1}{3}, \frac{2}{3}])$

(2, –3)	(–1, 3)
(0, 1)	(1, –2)

19. Decide whether the strategy pair $([\frac{1}{8}, 0, \frac{5}{8}, \frac{1}{4}], [\frac{1}{2}, \frac{1}{2}, 0])$ constitutes a Nash equilibrium for the nonzero-sum game

(1, 1)	(0, 2)	(–1, 3)
(0, 2)	(–1, 3)	(2, 0)
(1, 1)	(0, 2)	(1, 1)
(–1, 3)	(2, 0)	(0, 2)

20. Decide whether the strategy pair $([\frac{1}{4}, \frac{1}{4}, 0, \frac{1}{2}], [\frac{1}{2}, 0, \frac{1}{4}, \frac{1}{4}])$ constitutes a Nash equilibrium for the nonzero-sum game

(1, 0)	(0, 1)	(1, 0)	(–1, 2)
(1, 0)	(1, 0)	(–1, 2)	(1, 0)
(–1, 2)	(1, 0)	(1, 0)	(1, 0)
(0, 1)	(1, 0)	(1, 0)	(1, 0)

21. Decide whether the strategy pair $([0, \frac{2}{7}, \frac{2}{7}, \frac{3}{7}], [\frac{2}{7}, \frac{3}{7}, \frac{2}{7}, 0])$ constitutes a Nash equilibrium for the nonzero-sum game

(–1, 2)	(1, 0)	(0, 1)	(1, 0)
(1, 0)	(–1, 2)	(1, 0)	(0, 1)
(0, 1)	(1, 0)	(–1, 2)	(1, 0)
(1, 0)	(–1, 2)	(1, 0)	(–1, 2)

22. Decide whether the strategy pair $([\frac{2}{7}, 0, \frac{2}{7}, \frac{3}{7}], [0, \frac{2}{7}, \frac{3}{7}, \frac{2}{7}])$ constitutes a Nash equilibrium for the nonzero-sum game

(–1, 2)	(1, 0)	(0, 1)	(1, 0)
(1, 0)	(–1, 2)	(1, 0)	(0, 1)
(0, 1)	(1, 0)	(–1, 2)	(1, 0)
(1, 0)	(–1, 2)	(1, 0)	(–1, 2)

23. Decide whether the strategy pair $([\frac{2}{7}, 0, \frac{2}{7}, \frac{3}{7}], [\frac{2}{7}, \frac{3}{7}, \frac{2}{7}, 0])$ constitutes a Nash equilibrium for the nonzero-sum game

(–1, 2)	(1, 0)	(0, 1)	(1, 0)
(1, 0)	(–1, 2)	(1, 0)	(0, 1)
(0, 1)	(1, 0)	(–1, 2)	(1, 0)
(1, 0)	(–1, 2)	(1, 0)	(–1, 2)

24. Suppose a = \$12,000 and b = \$25,000 in the Job Applicants game. What is the predicted probability of Ruth applying to the position offered by
a) Firm 1? b) Firm 2?
25. Suppose a = \$24,000 and b = \$20,000 in the Job Applicants game. What is the predicted probability of Ruth applying to the position offered by
a) Firm 1? b) Firm 2?
26. Suppose a = \$10,000 and b = \$12,000 in the Job Applicants game. What is the predicted probability of Ruth applying to the position offered by
a) Firm 1? b) Firm 2?
27. Suppose a = \$50,000 and b = \$20,000 in the Job Applicants game. What is the predicted probability of Ruth applying to the position offered by
a) Firm 1? b) Firm 2?
28. Suppose a = \$12,000 and b = \$18,000 in the Job Applicants game. What is the predicted probability of Ruth applying to the position offered by
a) Firm 1? b) Firm 2?

29. Suppose a = \$11,000 and b = \$26,000 in the Job Applicants game. What is the predicted probability of Ruth applying to the position offered by
    a) Firm 1? b) Firm 2?
30. Suppose $a = 2$ and $b = 3$ in the Maynard Smith & Price Evolutionary Game. What is the predicted proportion of
    a) Hawks? b) Doves?
31. Suppose $a = 3$ and $b = 2$ in the Maynard Smith & Price Evolutionary Game. What is the predicted proportion of
    a) Hawks? b) Doves?
32. Suppose $a = 15$ and $b = 9$ in the Maynard Smith & Price Evolutionary Game. What is the predicted proportion of
    a) Hawks? b) Doves?
33. Suppose $a = 6$ and $b = 15$ in the Maynard Smith & Price Evolutionary Game. What is the predicted proportion of
    a) Hawks? b) Doves?
34. Show that if $a \geq b$ in the Evolutionary Game, then there is only one pure Nash equilibrium and identify it.

4.5 The Solution of Zero-Sum Games

This exposition of game theory began with a discussion of zero-sum games in which each player's gain is the other player's loss and vice versa. The "solution" of these games will now be derived within the wider context of nonzero-sum games and their mixed equilibria. Historically, the solution of zero-sum games preceded Nash's theory by two decades. We have decided to give this nonchronological account because it allows for a faster development of the subject matter.

As was noted in Section 4.2, payoffs of zero-sum games are usually denoted by a single number which simultaneously designates both Ruth's gain and Charlie's loss. Hence the array of the typical 2 × 2 zero-sum game has the abbreviated form

	Charlie	
Ruth	a_1	a_2
	a_3	a_4

(a)

as well as the full form

		Charlie	
Ruth		$(a_1, -a_1)$	$(a_2, -a_2)$
		$(a_3, -a_3)$	$(a_4, -a_4)$

(b)

We shall re-examine the notions of pure maximin strategies and values, pure Nash equilibria, and mixed Nash equilibria in the context of zero-sum games.

Pure Maximin Strategies and Values. Since the entries in the array of a zero-sum game denote Ruth's payoffs, her *pure maximin strategy* and *value* remain the same, regardless of the context. On the other hand, the search for guarantees that lead to Charlie's maximin strategy for nonzero-sum games needs to be rephrased in terms of Ruth's payoffs. The logic of guarantees dictates that Charlie should select the column that gives him the best possibly guarantee. Once Charlie has selected a column, Ruth cannot gain more than the maximum entry of that column. Consequently, this logic directs Charlie to select that column whose maximum entry is as small as possible. For that purpose we record at the bottom of each column of the zero-sum array that column's maximum element and then circle the smallest of these maxima. Since what we have here is the smallest of a set of maxima, this strategy is called the *pure minimax strategy* and the corresponding circled entry is the *pure minimax value.*

Example 4.5.1. *Determine the pure maximin and minimax strategies and values of the zero-sum game with the array*

4	–2	2	–3	5
5	3	0	2	–1
6	3	2	3	1

To find Ruth's pure maximin strategy record to the right of each row its minimum entry and then circle the largest of these (see below). Accordingly,

					Row min
4	–2	2	–3	5	–3
5	3	0	2	–1	–1
6	3	2	3	1	(1)
Column max: 6	3	(2)	3	5	

this means that Ruth's maximin strategy is to play the third row and this guarantees her a payoff of at least 1. To find Charlie's pure minimax strategy record at the bottom of each column that column's minimum entry and then circle the smallest of these. Accordingly, this means that Charlie's minimax strategy is to play the third column and his minimax value is 2. What these numbers mean is that by playing the third row Ruth can guarantee a payoff of at least 1 and by playing the third column Charlie can guarantee that her payoff will never exceed 2.

Charlie's pure maximin and minimax strategies are always identical. His pure maximin and minimax values, however, are the negatives of each other. The reader should also take note of the following fact:

In a zero-sum game every column maximum is greater than or equal to every row minimum. (c)

This observation will soon prove its usefulness. It is justified by an examination of the zero-sum game below wherein a is a typical payoff, x is a typical column maximum, and y is a typical row minimum. By definition,

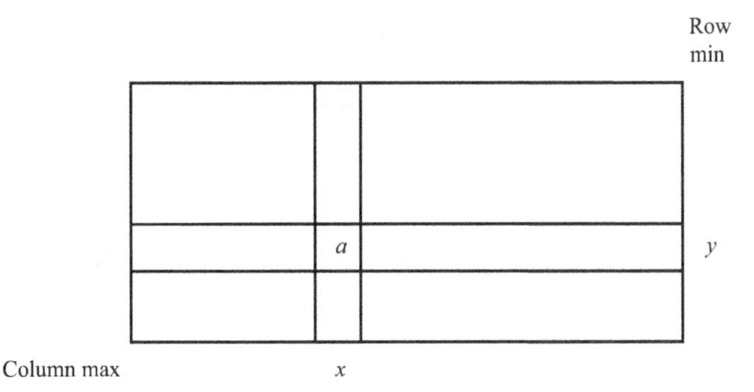

$$x \geq a \text{ and } a \geq y$$

and so it follows that

$$x \geq y$$

where x and y denote any column maximum and any row minimum, respectively.

Pure Nash Equilibria of Zero-Sum Games. Suppose, for the sake of argument, that $(a_1, -a_1)$ is the payoff of a pure Nash equilibrium of the general games (a) and (b) above. This means that

$$a_1 \geq a_3 \text{ and } -a_1 \geq -a_2$$

or, since multiplication by minus reverses inequalities,

$$a_1 \geq a_3 \text{ and } a_1 \leq a_2.$$

In other words, a_1 is the largest payoff in its column and the smallest in its row. These properties characterize the pure Nash equilibria of all zero-sum games, regardless of their dimensions. That is to say, that the payoff a of the abbreviated zero-sum array below corresponds to a pure Nash equilibrium of the full array provided it has the following two properties:

1. a is a largest entry of its column
2. a is a smallest entry of its row.

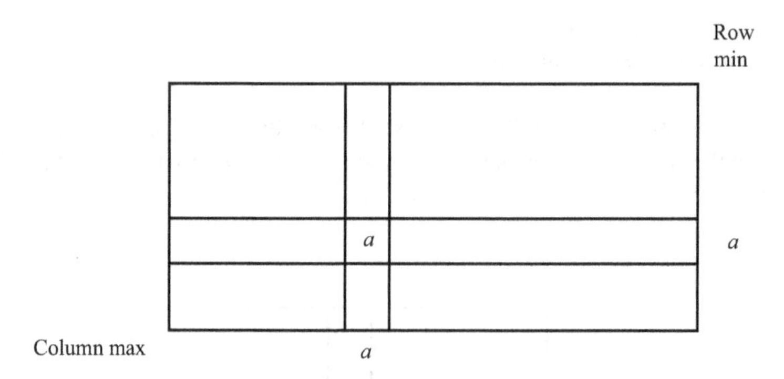

Such entries are called *saddle points* and the circled entries of the two zero-sum games below are both such saddle points. They are, of course, pure Nash

I

9	−2	−5
5	1	−9
3	2	5
−5	0	1

II

5	−2	1	−3	4
0	−2	5	0	−1
9	−1	0	2	1

equilibria of the full array, but in zero-sum games they are also simultaneously Ruth's pure maximin and Charlie's pure minimax values. The reason for this lies in observation (c) above. Since every column maximum is at least as great as every row minimum, and since a is both a column maximum and a row minimum, it follows that it must be both the least column maximum and the greatest row minimum, i.e., a is both the pure minimax and the pure maximin values (see examples below). This observation also provides a method for finding the saddle points of zero-sum games when such points exist. List the row minima and the column maxima as was done for the maximin and minimax values. If one of the row minima equals

I

9	−2	−5	−5
5	1	−9	−9
3	2	5	2
−5	0	1	−5
9	2	5	

II

5	−2	1	−3	4	−3
0	−2	5	0	−1	−2
9	−1	0	2	1	−1
9	−1	5	2	4	

one of the column maxima then the corresponding entry is a saddle point. If no such equality occurs then the game does not have a saddle point. Zero-sum games that do have saddle points are said to be *strictly determined.*

The Battle of the Bismarck Sea. In March 1943 a battle took place near New Guinea which can be described as a strictly determined zero-sum 2 × 2 game. The players, so to speak, were the Allies (Ruth) and the Japanese (Charlie).

The latter were consolidating their positions and found it necessary to transport troops from Rabaul to Lae (see map below) and the former were forewarned of these plans by intelligence reports which indicated that a Japanese troop and supply convoy was assembling at Rabaul. This convoy was expected to unload at Lae and could reach it by traveling either north or south of New Britain, each route requiring three days. Rain and poor visibility were expected for the northern route whereas clear weather and good visibility were forecast for south of the island.

The Allies also faced a choice. They could concentrate their reconnaissance on either the northern or the southern routes. It was estimated that if the Japanese chose the northern route then it would take the Allies either one or two days to locate the convoy, depending on whether the reconnaissance was concentrated on the northern or southern routes. On the other hand, if the Japanese chose the southern option, it would take either one day or no time at all to locate the convoy, again depending on the Allies' decision as to where to concentrate the search. For the remainder of the days the Japanese convoy would be exposed to Allied aerial bombing.

This battle can be summarized as the following 2×2 zero-sum game whose payoffs denote the number of bombing days left for the allies.

		Japanese choice of route	
		North	South
Allied concentration of reconnaissance	North	2 days	2 days
	South	1 day	3 days

This game has a saddle point whose pure maximin and minimax strategies are both [1, 0], i.e., North. In the event, this is exactly what transpired. The Japanese convoy employed the northern route and that was where the Allies expected them. General George C. Kenney, the Allied commander, offered the following description of the decision process: "We decided to gamble that the Nip would take the northern route." The Japanese general's rationale for not gambling and trying to fool the Allies by taking the unexpected southern route is unknown. In all likelihood, he was trying to avoid the worst possible scenario wherein his forces would take the southern route and be exposed to three days of punishing bombardment. In other words, he was probably using the maximin logic.

Mixed Nash Equilibria of Zero-sum Games. By the Nash Theorem 4.4.3, every game, zero-sum games included, has a mixed Nash equilibrium. Because the payoff pairs of the full arrays of zero-sum games add up to zero, the mixed value pairs of the equilibrium will also have a sum of zero. In other words, if Ruth's equilibrium payoff is e then Charlie's equilibrium payoff is $-e$. This expected payoff e has a meaning that is similar to that of the saddle point payoff of strictly determined games.

Suppose this mixed Nash equilibrium consists of the mixed strategies $= [1 - \bar{p}, \bar{p}]$ for Ruth and $C = [1 - \bar{q}, \bar{q}]$ for Charlie. Because of the No Regrets property of the Nash equilibrium, Charlie cannot improve his expected payoff by changing his mind. Consequently, by playing R Ruth guarantees herself a minimum expected payoff of e. Could there be another mixed strategy, say R', which would guarantee Ruth some larger minimum expectation $e' \geq e$? The answer is no and the reason is that by playing C Charlie can guarantee himself a payoff of at least $-e$; in other words, the strategy C enables Charlie to hold Ruth's expectation down to e. Hence, e is also the largest minimum expectation that Ruth can guarantee. Since this guarantee is provided by strategy R this strategy is called the mixed *maximin strategy*.

Similarly, again because of the No Regrets principle, Ruth cannot improve her expected payoff by changing her mind. In other words, by playing C Charlie can guarantee that Ruth's expectation will be held down to a maximum of e. Could there be another mixed strategy, say C', which would enable Charlie to hold Ruth's expectation down to some smaller value $e'' < e$? The answer is no and the reason is that by playing R Ruth can guarantee herself a payoff of at least e. Hence e is the smallest maximum expectation that Charlie can hold Ruth down to. Since this is provided to Charlie by strategy C, this strategy is called the *mixed minimax strategy*. The expectation e is the *value* of the game and this value together with the mixed maximin and minimax strategies constitute the *solution of the game*.

The following theorem is the central theorem of the theory of zero-sum games. While this theorem is presented here as a consequence of Nash's Equilibrium Theorem, it was first stated and proved by von Neumann as early as 1928, thus preceding Nash's theorem by over twenty years. It did not attain wide recognition until the publication of von Neumann and Morgenstern's seminal book *The Theory of Games and Economic Behavior* in 1944. It is now commonly known as *the Minimax Theorem*.

MINIMAX THEOREM *For every zero-sum game there is a number v which has the following properties:*

(a) *Ruth has a mixed strategy that guarantees her an expected payoff of at least v;*
(b) *Charlie has a mixed strategy that guarantees that (Ruth's) expected payoff will be at most v.*

The nature of the guarantees is such that if both players employ their recommended strategies, then the expected payoff will be exactly v. In other words, when

Ruth employs the maximin strategy and Charlie employs the minimax strategy, then Ruth can expect to win v units per play (on the average). For this reason the number v is called the *value* of the game. The value of a game together with its maximin and minimax strategies constitute the *solution* of the game.

In contrast with Nash equilibria of nonzero-sum games, zero-sum 2×2 games are easily solved. The solution of strictly determined games was already described above. For any nonstrictly determined game

a_1	a_2
a_3	a_4

(d)

Ruth's oddments are defined to be that one of the pairs

$$[a_4 - a_3, a_1 - a_2] \text{ or } [a_3 - a_4, a_2 - a_1]$$

which consists of positive numbers alone (the game being nonstrictly determined, one of these pairs will have this property). For example, for game

2	3
4	1

(e)

Ruth's oddments are

$$[4 - 1, 3 - 2] = [3, 1]$$

whereas for the game

5	−2
1	4

(f)

Ruth's oddments are

$$[4 - 1, 5 - (-2)] = [3, 7].$$

The maximin strategy is obtained from the oddments when they are each divided by their sum. In game (e), the maximin strategy is

$$\left[\frac{3}{3+1}, \frac{1}{3+1}\right] = \left[\frac{3}{4}, \frac{1}{4}\right] = [0.75, 0.25]$$

and the oddments [3, 7] of game (f) yield the maximin strategy

$$\left[\frac{3}{3+7}, \frac{7}{3+7}\right] = \left[\frac{3}{10}, \frac{1}{10}\right] = [0.3, 0.7].$$

Charlie's oddments in game (d) consist of that one of the pairs

$$[a_4 - a_2,\ a_1 - a_3] \text{ or } [a_2 - a_4,\ a_3 - a_1]$$

which consists of two positive numbers. These oddments are converted to the minimax strategy in the same manner as above. Thus, for game (e) Charlie's oddments are

$$[3 - 1,\ 4 - 2] = [2, 2]$$

and his minimax strategy is

$$\left[\frac{2}{2+2}, \frac{2}{2+2}\right] = \left[\frac{2}{4}, \frac{2}{4}\right] = [0.5, 0.5].$$

For game (f) Charlie's oddments are

$$[4 - (-2),\ 5 - 1] = [6, 4]$$

and his minimax strategy is

$$\left[\frac{6}{6+4}, \frac{4}{6+4}\right] = \left[\frac{6}{10}, \frac{4}{10}\right] = [0.6, 0.4].$$

The value of the game is computed by means of the usual auxiliary diagram. Thus, for game (f) we have

	0.6	0.4
0.3	5	–2
0.7	1	4

giving a value of

$$v = 0.3 \cdot 0.6 \cdot 5 + 0.3 \cdot 0.4 \cdot (-2) + 0.7 \cdot 0.6 \cdot 1 + 0.7 \cdot 0.4 \cdot 4$$

$$= 0.9 - 0.24 + 0.42 + 1.12 = 1.36$$

Caveat. It is important to stress that the oddments method only applies to non-strictly determined games. If the method is brought to bear on a strictly determined games, the derived answer, although it may look legitimate, will in general be wrong. Thus, the general procedure when solving a zero-sum 2×2 game calls for first deciding whether or not it is strictly determined. If it is, the decision process yields the solution. If the game is not strictly determined, then the oddments method should be applied.

The River Tale (J. D. Williams, 1986). *Steve is approached by a stranger who suggests that they match coins. Steve says that it's too hot for violent exercise. The stranger says, "Well then, let's just lie here and speak the words* heads *or* tails *- and to make it interesting I'll give you \$30 when I call* tails *and you call* heads, *and \$10 when it's the other way round. And - just to make it fair - you give me \$20 when we match."*

The payoffs may look fair, but they are not. This is the zero sum game

		Stranger	
		Heads	*Tails*
Steve	*Heads*	–20	30
	Tails	10	–20

Steve's oddments are

$$[10-(-20),\ 30-(-20)] = [30,\ 50]$$

and the stranger's are

$$[30-(-20),\ 10-(-20)] = [50,\ 30].$$

Hence their respective mixed maximin and minimax strategies are [3/8, 5/8] and [5/8, 3/8]. The auxiliary diagram is

	5/8	3/8
3/8	–20	30
5/8	10	–20

yielding the value

$$\nu = \frac{3}{8}\cdot\frac{5}{8}\cdot(-20) + \frac{3}{8}\cdot\frac{3}{8}\cdot 30 + \frac{5}{8}\cdot\frac{5}{8}\cdot 10 + \frac{5}{8}\cdot\frac{3}{8}\cdot(-20)$$

$$= \frac{-300 + 270 + 250 - 300}{8 \times 8} = -\frac{80}{64} = -1.25$$

In other words, should Steve chose to play this game, he will lose, on the average, \$1.25 each time he plays.

Bombing Sorties. This game, which was discussed in great detail in Section 4.1, can now be solved. If the immaterial % sign is dropped from the payoff then this game has array

80	100
90	50

Ruth's oddments are

$$[90-50,\ 100-80]=[40,\ 20]$$

yielding a maximin strategy of

$$\left[\frac{40}{60},\frac{20}{60}\right]=\left[\frac{2}{3},\frac{1}{3}\right].$$

Charlie's oddments are

$$[100-50,\ 90-80]=[50,\ 10]$$

yielding a minimax strategy of

$$\left[\frac{50}{60},\frac{10}{60}\right]=\left[\frac{5}{6},\frac{1}{6}\right].$$

The auxiliary diagram is

	5/6	1/6
2/3	80	100
1/3	90	50

and so the game's value is

$$\begin{aligned}\nu &= \frac{2}{3}\cdot\frac{5}{6}\cdot 80+\frac{2}{3}\cdot\frac{1}{6}\cdot 100+\frac{1}{3}\cdot\frac{5}{6}\cdot 90+\frac{1}{3}\cdot\frac{1}{6}\cdot 50\\ &= \frac{800+200+450+50}{18}=\frac{1500}{18}=83.3\end{aligned}$$

This solution is interpreted as follows. It is the game theoretic recommendation that Ruth place the bomb on the support plane one out of three times (on the average) and that Charlie attack the support plane one out of six times (again, on the average). If they follow these recommendations Ruth guarantees herself an expected bomb delivery rate of at least $83.\bar{3}\%$ whereas Charlie guarantees that Ruth's expectation will not exceed $83.\bar{3}$.

The next example describes a social zero-sum game whose array has more than two rows. While the solution of such games falls outside the scope of this text, the validity of such solutions can be verified by the same means used in the previous chapter for verifying Nash equilibria. All that is needed is to simply replace each payoff a of the zero-sum game by the payoff pair $(a, -a)$ and then proceed with the verification process.

The Jamaican Fishing Village. *The anthropologist W. C. A. Davenport observed the inhabitants of a certain Jamaican fishing village for two years in the early fifties. These fishermen possessed twenty-six fishing canoes manned each by a captain and two or three crewmen. The fishing took the form of setting pots (traps) and drawing from them. The fishing grounds were divided into inside and outside banks. The inside banks lay from 5 to 15 miles offshore, whereas the outside banks lay beyond. The crucial factor that distinguished between the two areas was the occasional presence of very strong currents in the latter, which rendered fishing impossible. Accordingly, each captain must decide each day on a fishing strategy for that day. He could: (1) set all his pots inside, (2) set all the pots outside, or (3) set some of the pots inside and some outside.*

Davenport modeled this situation as a 3 × 2 game whose players are the village and the environment. On any day, each of the twenty-six captains *decided on an option* for the village by adopting one of the three available fishing strategies for that day. The environment "decided" on its option by either sending a current or not. Based on his observations of the local market place and the costs accrued by the captains, Davenport estimated the payoffs of the various possibilities as follows:

		Environment	
		Current	No current
	Inside	17.3	11.5
Village	In-out	5.2	17.0
	Outside	–4.4	20.6

The monetary unit was the Pound and the reason for the one negative payoff was that the captain must pay his crew for each outing, regardless of the catch. Davenport went on to treat this as a zero-sum two person game, solved it as such, and compared the game's maximin strategy to the actual distribution of the captains' choices.

The solution of this games is (see Exercise 49):

Mixed maximin strategy:	[0.67, 0.33, 0]
Mixed minimax strategy:	[0.31, 0.69]
Value:	13.31

This is to be understood as saying that Game Theory's recommendation to the village is that 67% of its fishing should be done in the inside banks exclusively, 33% as an inside-outside combination, and none of the fishermen should dedicate themselves to fishing in the outside banks alone. This way the village can guarantee its fishermen an expected payoff of at least 13.31 per outing.

Davenport observed that 18 (69%) of the captains fished only in the inside banks, 8 (31%) adopted the inside-outside combination, and none restricted their fishing to the outside banks alone. A remarkable fit between theory and observation. On the other hand, during his two year stay Davenport observed that a current was present in the outside bank on 25% of the days. This could be interpreted as saying that the environment was employing the mixed strategy of [0.25, 0.75]. This is not as exciting a fit with the minimax strategy of [0.31, 0.69] as we had for the village's maximin strategy, but it is still close. While this is not to be taken as evidence for the anthropomorphic view of nature, it should be pointed out that the game's payoffs are dependent on many factors, including the frequencies of the currents and the village's fishing strategies. Is it possible that this dependence might cause the game's entries to gravitate toward values whose minimax strategy agrees with the actual frequency of the current?

Exercises 4.5

Solve the games in Exercises 1–40.

1.

1	3
2	1

2.

3	2
0	2

3.

2	1
3	1

4.

2	1
1	3

5.

2	1
4	6

6.

1	–3
–2	1

7.

1	3
–2	0

8.

0	–1
0	1

9.

2	4
0	–1

10.

1	1
2	−1

11.

1	1
−3	1

12.

−2	1
0	3

13.

5	−1
−4	3

14.

−1	3
6	−4

15.

8	2
2	−8

16.

3	1
1	2

17.

−3	1
−4	−2

18.

3	−3
−6	1

19.

0	1
0	−2

20.

−1	−2
1	3

21.

−1	−2
0	−3

22.

8	−2
2	8

23.

2	4
0	−1

24.

2	−2
2	2

25.

3	−1
0	3

26.

1	0
2	0

27.

1	−2
3	0

28.

2	1
2	4

29.

0	−1
−1	1

30.

−2	2
5	−5

31.

3	1	2
5	−1	3

32.

0	−2	7	2
3	−1	−1	2

33.

0	2
1	−1
2	3
−1	−2

34.

–5	5	0
2	–2	1
4	3	2

35.

1	–2	0	–3
–1	–2	0	1
5	–4	6	–3

36.

1	2	3
0	1	2
–1	0	1
–2	–1	0

37.

0	2	–4	0
–2	0	–2	4
4	2	0	6
0	0	–6	0

38.

–1	–2	0	1	0
3	4	2	2	4
2	0	1	–1	–3
1	–1	5	2	6

39.

2	1	1	3
6	–1	–7	8
–8	1	7	–6
5	0	–5	1
–1	–2	2	3

40.

5	2	–5	6	0
–2	0	2	0	–3
5	1	0	–1	–2
3	0	–1	2	–2
0	2	–1	0	–1

Describe the following situations as mathematical zero-sum games and solve them.

41. **The Birthday** (J. D. Williams, 1986) Frank is hurrying home late, after a particularly grueling day, when it pops into his mind that today is Kitty's birthday! Or is it? Everything is closed except the florist. If it is not her birthday and he brings no gift, the situation will be neutral, i.e., payoff 0. If it is not and he comes in bursting with roses, and obviously confused, he may be subjected to the Martini test, but he will emerge in a position of strong one-upness – which is worth 1. If it is her birthday and he has, clearly, remembered it, that is worth somewhat more, say 1.5. If he has forgotten it he is down like a stone, say, -10.
42. **The Hi-Fi** (J. D. Williams, 1986) The firm of Gunning & Kappler manufactures an amplifier. Its performance depends critically on the characteristics of one small, inaccessible condenser. This normally costs Gunning and Kappler $1, but they are set back a total of $10, on the average, if the original condenser is defective. There are some alternatives open to them. It is possible for them to buy a superior quality condenser, at

$6, which is fully guaranteed; the manufacturer will make good the condenser and the costs incurred in getting the amplifier to operate. There is available also a condenser covered by an insurance policy which states, in effect, "if it is our fault, we will bear the costs and you get your money back." This item costs $10. (This is a 3×2 game that Gunning & Kappler is playing against Nature whose options are to supply either a defective or a nondefective condenser.)

43. **The Huckster** (J. D. Williams, 1986) Merrill has a concession at Yankee Stadium for the sale of sunglasses and umbrellas. The business places quite a strain on him, the weather being what it is. He has observed that he can sell about 500 umbrellas when it rains and about 100 when it shines; and in the latter case he also can dispose of about 1000 sunglasses. Umbrellas cost him 50 cents and sell for $1 (this is 1954); glasses cost 20 cents and sell for 50 cents. He is willing to invest $250 in the project. Everything that isn't sold is a total loss (the children play with them).
44. **The Coal Problem** (J. D. Williams, 1986) On a sultry summer afternoon, Hans' wandering mind alights upon the winter coal problem. It takes about 15 tons to heat his house during a normal winter, but he has observed extremes when as little as 10 tons and as much as 20 were used. He also recalls that the price per ton seems to fluctuate with the weather, being $10, $15, and $20 a ton during mild, normal, and severe winters. He can buy now, however, at $10 a ton. He considers three pure strategies, namely, to buy 10, 15, or 20 tons now and the rest, if any, later. He will be moving to California in the spring and he cannot take excess coal with him.

45. The payoffs in the game below denote dollars. Ruth and Charlie are about to play this game 100 times.

3	–1
1	2

a) Does Ruth have a strategy that guarantees her winning something each time the game is played?

b) Does Ruth have a strategy that guarantees her winning a total of $80 or more?

c) Does Ruth have a strategy that guarantees her winning a total of $120 or more?

d) Ruth is playing this game in order to raise $120 which she needs very badly (she owes this money to a loan shark). What strategy should she use?

e) Does Charlie have a strategy that is guaranteed to hold Ruth's winnings to less than $90?

f) Does Charlie have a strategy that is guaranteed to hold Ruth's winnings to less than \$120?
g) It so happens that Charlie only has \$125. Should he lose more, he will have to ask his parents again for some money, something he is not willing to do anymore. If he still wants to play, what should his strategy be?

46. The payoffs in the game below denote dollars. Ruth and Charlie are about to play this game 100 times.

3	−2
−1	2

a) Does Ruth have a strategy that guarantees her winning something each time the game is played?
b) Does Ruth have a strategy that guarantees that her losses will not exceed \$120?
c) Does Ruth have a strategy that guarantees that her losses will not exceed \$80?
d) Ruth is playing this game in order to raise \$30 which she needs very badly (she owes this money to a loan shark). What strategy should she use?
e) Does Charlie have a strategy that is guaranteed to hold Ruth's winnings to less than \$90?
f) Does Charlie have a strategy that is guaranteed to hold Ruth's winnings to less than \$120?
g) It so happens that Charlie only has \$60. Should he lose more, he will have to ask his parents again for some money, something he is not willing to do anymore. If he still wants to play, what should his strategy be?

47. The payoffs in the game below denote dollars. Ruth and Charlie are about to play this game 100 times.

3	2
−1	2

a) Does Ruth have a strategy that guarantees her winning something each time the game is played?
b) Does Ruth have a strategy that guarantees her winning a total of \$200 or more ?
c) Does Ruth have a strategy that guarantees her winning a total of \$300 or more?

d) Ruth is playing this game in order to raise $30 which she needs very badly (she owes this money to a loan shark). What strategy should she use?
e) Does Charlie have a strategy that is guaranteed to hold Ruth's winnings to less than $90?
f) Does Charlie have a strategy that is guaranteed to hold Ruth's winnings to less than $250?
g) It so happens that Charlie only has $250. Should he lose more, he will have to ask his parents again for some money, something he is not willing to do anymore. If he still wants to play, what should his strategy be?

48. The payoffs in the game below denote dollars. Ruth and Charlie are about to play this game 100 times.

3	–3
–4	2

a) Does Ruth have a strategy that guarantees her winning something each time the game is played?
b) Does Ruth have a strategy that guarantees that her losses will not exceed $120?
c) Does Ruth have a strategy that guarantees that her losses will not exceed $300?
d) Charlie is playing this game in order to raise $30 which he needs very badly (he owes this money to a loan shark). What strategy should he use?
e) Does Charlie have a strategy that is guaranteed to hold Ruth's winnings to less than $180?
f) Does Charlie have a strategy that is guaranteed to hold Ruth's winnings to less than $250?
g) It so happens that Ruth only has $35. Should she lose more, she will have to ask her parents again for some money, something she is not willing to do anymore. If she still wants to play, what should her strategy be?

49. Verify the solution given for the Jamaican Fishing Village game. (In other words, verify that the stated strategies are indeed a Nash equilibrium of the full form of the game.)

4.6 Activities/Websites

1. Prisoner's Dilemma.
Distribute to each student a form containing the following:

> Name of student:
>
> You have the option of either marking the box below or not. If both you and your opponent mark the box you both get 10 points whereas if neither marks the box, each gets 20 points. Should one player mark the box and the other not, the one who marked it gets 30 points, whereas the other gets nothing.
>
> ☐

Have the students return these forms to you, shuffle them, pair them off at random, and play each student against his/her pair-mate. Return the forms to the students with the payoffs after the results are recorded. Compare the students' behavior to the pure Nash equilibrium of this game. Repeat the game several times.

In order to make this game more meaningful it is suggested that each player's payoff be converted to a certain amount of extra credit points.

The reason this is a variant of *Prisoner's Dilemma* is that the above payoffs were obtained by adding 15 to each of that game's payoffs and then doubling the sum.

2. The Job Applicants.
Distribute to each student a form containing the following:

> Name of student:
>
> You have the option of either marking the box below or not. If both you and your opponent mark the box you both get 20 points whereas if neither marks the box, each gets 10 points. Should one player mark the box and the other not, the one who marked it gets 30 points, whereas the other gets 20 points.
>
> ☐

Have the students return these forms to you, shuffle them, pair them off at random, and play each student against his/her pair-mate. Return the forms to the students with their payoffs after the results are recorded. Compare the students' collective behavior to the mixed Nash equilibrium of this game: $C = R = [0.2, 0.8]$. Repeat the game several times.

In order to make this game more meaningful it is suggested that each player's payoff be converted to a certain amount of extra credit points.

This game is essentially the same as that discussed on p. 4.63.

3. A Zero-sum Game (Instructor against class). Before meeting your class create your own box in your office and either mark it or not. Distribute to each student a form containing the following:

> Name of student:
>
> The student has the option of either marking the box below or not. If both the student and the instructor mark their boxes, or if neither of them marks their boxes, the student gets 0 points. If the student does not and the instructor does mark, the student gets 20 points. If the student marks and the instructor does not mark, the student gets 10 points.
>
> ☐

Have the students return these forms to you and play your decision against each. Return the form to them with their payoff. Repeat a total of 10 times or more. Assigning to the students the role of Ruth, compare their collective behavior with the maximin strategy. Try to hold the students' gains down as much as possible.

In order to make this game more meaningful it is suggested that each player's payoff be converted to a certain amount of extra credit points.

Chapter 5

Linear Programming

Outside of the physical sciences and the discipline of statistics, the most popular of mathematical applications is the topic of linear programming. This tool can be used wherever decisions are to be made that need to take into account the limited nature of all resources and the inevitability of competing requirements. It was developed by George B. Dantzig (1914–2005) during World War II in order to resolve the huge distribution problems faced by the armed forces in trying to keep the several fronts supplied with men, weapons, and supplies. The various examples and exercises of this chapter should convince the student of the wide range of applications of this tool.

5.1 Graphing Inequalities

We begin with the discussion of a problem that will motivate some definitions.

Example 5.1.1. *A small company produces chairs and full length sofas. The production of a chair requires 6 man-hours and the materials cost \$30. The production of a sofa requires 10 man-hours and the materials cost \$80. The company has 120 man-hours and a budget of \$870 available each week. If the profit per chair is \$45 and the profit per sofa is \$80, how many of each should be produced so as to maximize the weekly profit?*

Let x denote the number of chairs and y the number of sofas to be produced. In terms of these variables the profit is

$$45x + 80y.$$

The problem calls for finding those x and y that result in the largest possible profit, subject to the constraints of limited labor and limited budget. Specifically, x and y are constrained by the following inequalities:

Labor: $6x + 10y \leq 120$
Budget: $30x + 80y \leq 870.$

These elements are common to all *linear programming problems*. There is always a function that needs to be maximized (or minimized, as is the case when, say, the focus is on costs). This is called the *objective function*. There are also inequalities that delimit the variables on which the objective function depends. These are the *constraints*. Mathematical problems with such objective functions and constraints are called optimization problems and are in general very difficult. Linear programs are those optimization problems whose objective functions and constraints happen to be *linear* in the sense that none of the variables are multiplied by each other (or by themselves). The property of linearity makes this restricted class easier, and the consensus of the mathematical community is that the available tools solve linear programs in as efficient a manner as can be expected. A typical linear program may involve hundreds of variables and constraints, and one has been known to use as many as 800,000 variables. Not surprisingly, therefore, the solution requires both computers and efficient software packages. The fact that some of these packages sell for hundreds of thousands of dollars attests to their utility and demand.

In this text attention is limited to linear programs with two variables only, and their solution is geometrical. The inequalities will be graphed and the resulting figure will then be used to determine the best choice for the values of the variables. The processes of graphing the inequalities that constitute the constraints begins with the graphing of linear equations.

The graph of an equation of the form

$$Ax + By = C$$

in the Cartesian plane is a straight line. One method for sketching this graph is to find its intersections with the axes. The x-intercept is found by substituting 0 for y and solving the resulting equation for x; the y-intercept is found by substituting 0 for x and solving the resulting equation for y.

Example 5.1.2. *Sketch the graph of $2x + 3y = 6$.*

Setting $y = 0$ results in the equation

$$2x = 6 \text{ or } x = \frac{6}{2} = 3,$$

which yields the x-intercept (3, 0). Setting $x = 0$ results in the equation

$$3y = 6 \quad \text{or} \quad y = \frac{6}{3} = 2,$$

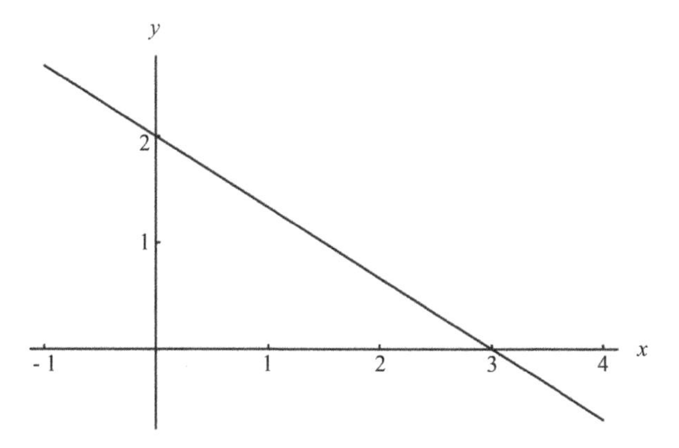

Figure 5.1

which yields the y-intercept (0, 2). The graph of the given equation is drawn in Figure 5.1.

There are some exceptional cases where this method fails. If $C = 0$ but neither A nor B are 0, then both the x- and y-intercepts are the origin, and so another point is needed to determine the graph. Such a point is easily obtained by substituting for x any non-zero value. Any value will do, but substituting $x = B$ will avoid annoying fractions.

Example 5.1.3. *Sketch the graph of $2x + 3y = 0$.*

Substituting either $x = 0$ or $y = 0$ yields the origin (0, 0) as an intercept. The recommended substitution $x = 3$ results in the equation

$$2 \cdot 3 + 3y = 0 \quad \text{or} \quad 3y = -6 \quad \text{or} \quad y = -2,$$

which yields the additional point (3, −2) on the graph. The graph of the given equation is therefore the straight line joining the points (0, 0) and (3, −2) (see Figure 5.2).

If $C = 0$ and exactly one of A and B is also 0, the resulting equation can be simplified to one of the two forms

$$x = a \quad \text{or} \quad y = b.$$

The first of these has the vertical line through $(a, 0)$ as its graph and the second has the horizontal line through $(0, b)$ as its graph.

Example 5.1.4. *Sketch the graphs of the equations $x = 2$ and $y = 3$.*

The graphs of these equations are displayed in Figure 5.3.

This chapter's introductory Example 5.1.1 illustrates the need for graphing inequalities. Such graphs are no longer straight lines. For example, by definition, the graph of $x \geq 2$ consists of all the points whose first coordinate is at least 2. This graph is the half-plane that lies to the right of the straight line $x = 2$ (see Figure 5.4). Similarly, since the graph of $x \leq 2$ consists of all the points whose first coordinate is at most 2, it is the half-plane that lies to the left of the straight line $x = 2$. In general the graphs of $x \geq a$ and $x \leq a$ are, respectively, the half-planes

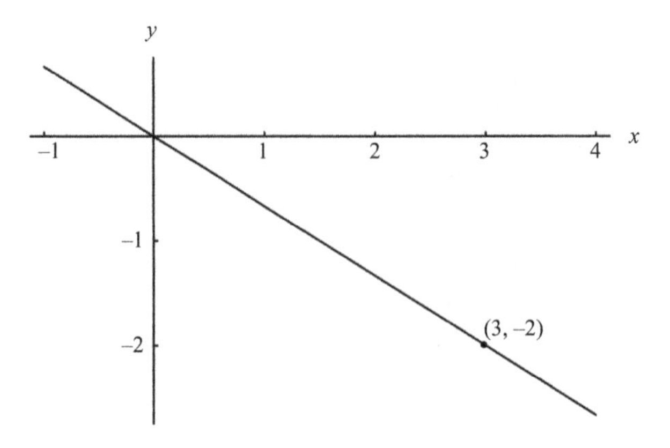

Figure 5.2

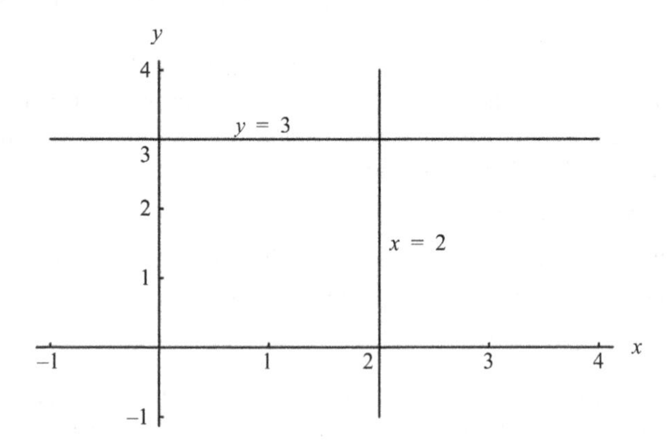

Figure 5.3

that lie to the right and the left of the vertical straight line $x = a$. The same reasoning, mutatis mutandis, leads to the conclusion that the graphs of $y \geq b$ and $y \leq b$ are, respectively, the half-planes that lie above and below the horizontal straight line $y = b$ (see Figure 5.5).

The graph of the inequality $x + y \geq 5$ becomes clear once it is noted (see Figure 5.6) that the sum of the coordinates of the points above and to the right of the graph of $x + y = 5$ add up to more than 5. The graph of $x + y \geq 5$ is therefore the half-plane above and to the right of the straight line of $x + y = 5$ (see Figure 5.7). Similarly, the graph of the inequality $x + y \leq 5$ is the half-plane below and to the left of the same straight line (see Figure 5.7).

The above examples generalize to the following rule:

The graphs of the inequalities $Ax + By \geq C$ and $Ax + By \leq C$ consist of the two half-planes that lie on either side of the straight line $Ax + By = C$.

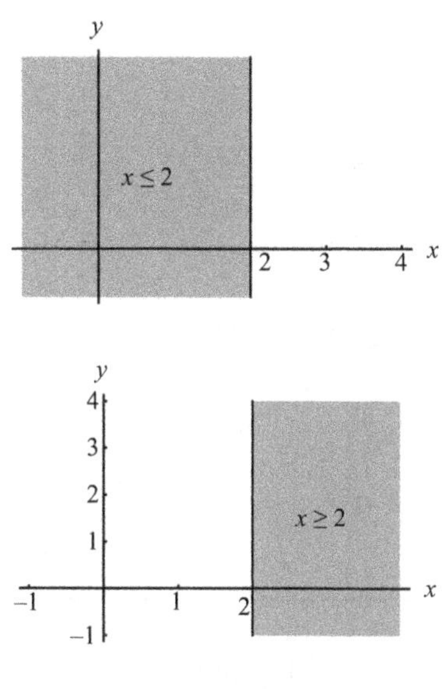

Figure 5.4

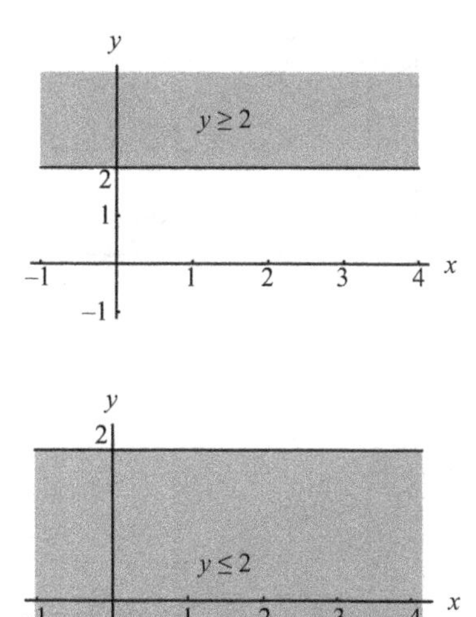

Figure 5.5

Just which of the two half-planes correspond to which inequality depends on the signs of A, B, and C. Rather than enunciate the complicated rule, it is recommended that the reader simply pick a particular point that lies off the line, and decide, by substituting its coordinates into the inequality, whether that point belongs to the graph. If the substitution results in a valid inequality, the graph

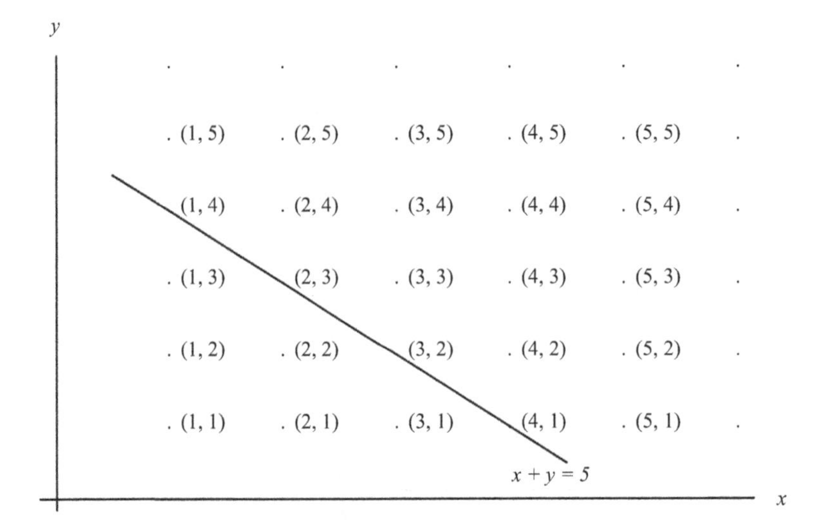

Figure 5.6

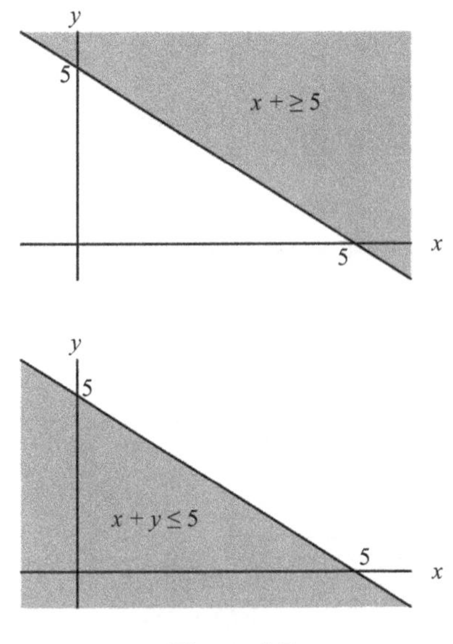

Figure 5.7

consists of that half-plane that contains the point; if it does not, the graph is the other half-plane.

Example 5.1.5. *Sketch the graph of* $2x - 3y \geq 6$.

First it is necessary to sketch the graph of the equation $2x - 3y = 6$. The x-intercept, obtained by setting $y = 0$, is $(3, 0)$. The y-intercept, obtained by setting

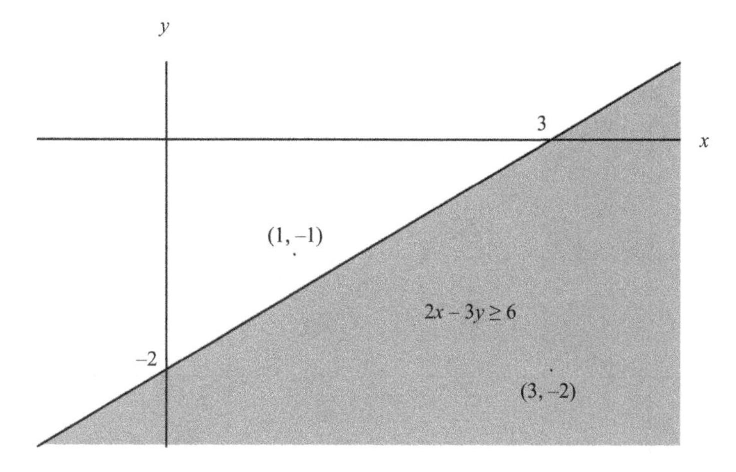

Figure 5.8

$x = 0$ is $(0, -2)$. To decide which side of the line constitutes the required graph we substitute the coordinates of the origin $(0, 0)$ into the given inequality and obtain

$$2 \cdot 0 - 3 \cdot 0 \geq 6 \quad \text{or} \quad 0 \geq 6$$

which is clearly false. This means that $(0, 0)$ is not in the graph of $2x - 3y \geq 6$ and hence the required graph consists of that side of the straight line which does not contain it (see Figure 5.8).

The decision to examine the origin was a matter of convenience. Had the point $(1, -1)$ be chosen instead, the substitution would have yielded

$$2 \cdot 1 - 3(-1) \geq 6 \quad \text{or} \quad 5 \geq 6$$

which is also false, leading to the same conclusion about the graph of the given inequality. Things would have turned out slightly differently, though, if the point $(3, -2)$ had been chosen. In that case substitution would have resulted in

$$2 \cdot 3 - 3(-2) \geq 6 \quad \text{or} \quad 12 \geq 6.$$

Since this is a valid inequality, this would have meant that the required half-plane must contain $(3, -2)$, which is consistent with the previous conclusion.

It is advisable to use the origin for this purpose whenever possible, that is, whenever the associated straight line does not pass through it. Should the latter be the case, it is recommended that a point on either of the axes be used.

We now turn to the graphing of systems of inequalities. The graph of such a system consists of all the points that satisfy each of the inequalities.

Example 5.1.6. *Sketch the graph of the system:*

$$x + y \geq 6$$

$$x - 2y \leq 0$$

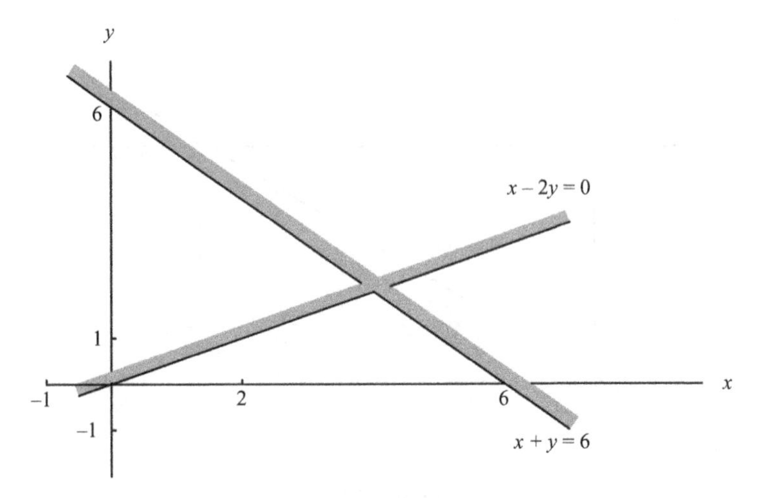

Figure 5.9

The straight line associated with the first inequality is $x + y = 6$, which has intercepts (6, 0) and (0, 6). Substitution of $x = 0$ and $y = 0$ into the inequality yields

$$0 + 0 \geq 6$$

which is false. Hence the graph of the first inequality is the partially shaded side of $x + y = 6$ indicated in Figure 5.9.

The straight line associated with the second inequality has equation $x - 2y = 0$. Its only intercept is (0, 0), but another point is obtained by substituting $y = 1$ so that

$$x - 2 \cdot 1 = 0 \quad \text{or} \quad x = 2.$$

This yields the additional point (2, 1) on the line. Since the point (0, 0) is on this line, (2, 0) is used instead to decide which of its sides constitutes the graph. The substitution of the coordinates then yields

$$2 - 2 \cdot 0 \leq 0 \quad \text{or} \quad 2 \leq 0$$

which is false. Hence the graph of $x - 2y \leq 0$ is that side of the line $x - 2y = 0$ which does not contain the point (2, 0) (see Figure 5.9).

The graphs of the two individual inequalities are sketched in Figure 5.9. The graph of the system consists of the intersection of these two graphs, namely, that one of the four regions formed by the intersection of the two straight lines both of whose sides are shaded (see Figure 5.10).

The graph of the constraints of Example 5.1.1, which was used to motivate this chapter's primary concepts, is now sketched. This graph will be used in Example 5.1.7 below to solve the optimization problem of Example 5.1.1.

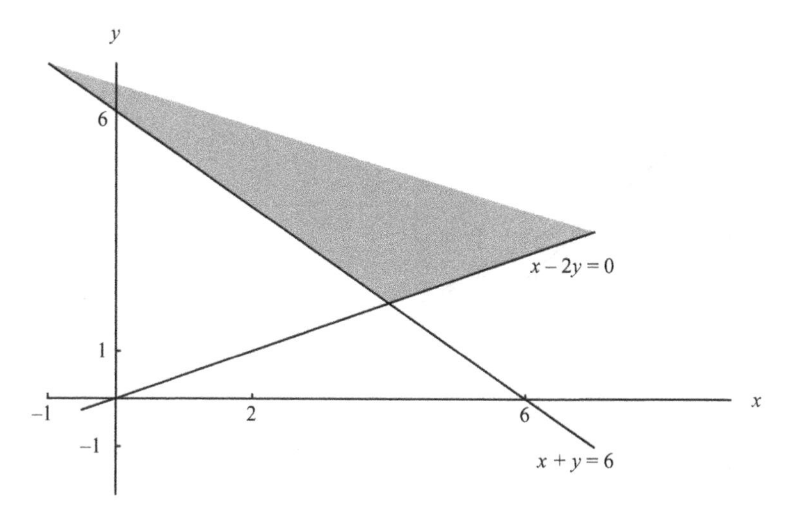

Figure 5.10

Example 5.1.7. *Sketch the graph of the system of inequalities:*

$$6x + 10y \leq 120$$

$$30x + 80y \leq 870$$

$$x \geq 0$$

$$y \geq 0$$

$6x + 10y \leq 120$:	The straight line $6x + 10y = 120$ has intercepts (20, 0) and (0, 12). The substitution of the coordinates of (0, 0) into the inequality yields $6 \cdot 0 + 12 \cdot 0 \leq 120$ which is valid. Hence the graph consists of the side that contains the origin.
$30x + 80y \leq 870$:	The straight line $30x + 80y = 870$ has intercepts (29, 0) and (0, 87/8). The substitution of the coordinates of (0, 0) into the inequality yields $30 \cdot 0 + 80 \cdot 0 \leq 870$, which is valid. Hence the graph consists of the side of $30x + 80y = 870$, which contains the origin.
$x \geq 0$:	This is the half-plane to the right of $x = 0$, which is the y-axis.
$y \geq 0$:	This is the half-plane above $y = 0$, which is the x-axis.

The four individual inequalities are graphed in Figure 5.11.

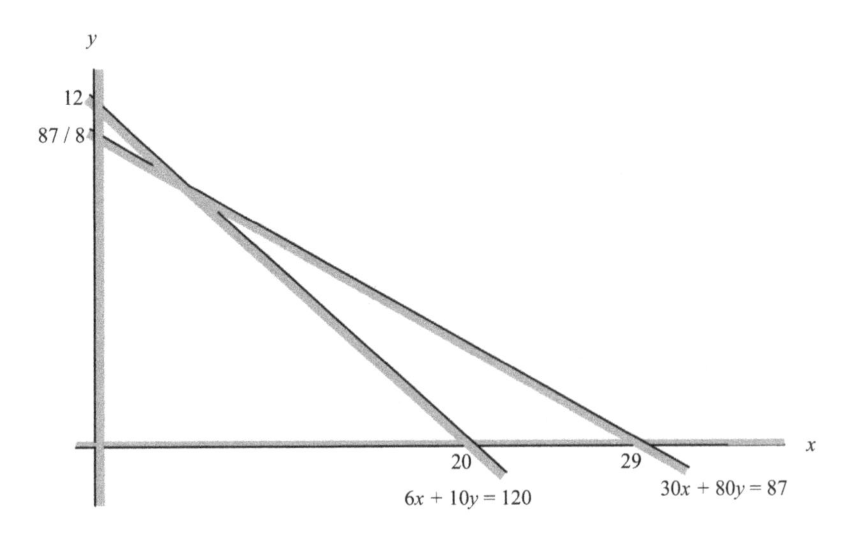

Figure 5.11

The graph of the system is depicted in Figure 5.12 where the vertex (5, 9) was derived as follows:

$$\begin{array}{ccccc} 6x + 10y = 120 & & 30x + 50y = 600 & & 48x + 80y = 960 \\ & \text{or} & & \text{or} & \\ 30x + 80y = 870 & & 30x + 80y = 870 & & 30x + 80y = 870 \\ \hline & & -30y = -270 & & 18x = 90 \\ & & y = (-270)/(-30) = 9 & & x = 90/18 = 5 \end{array}$$

Example 5.1.8. *Sketch the graph of the system of inequalities:*

$$x + y \leq 10$$

$$3x - 2y \geq 0$$

$$x + 5y \geq 20.$$

$x + y \leq 10$: The straight line $x + y = 10$ has the intercepts (10, 0) and (0, 10). The substitution of the coordinates of (0, 0) into the inequality yields the valid inequality $0 + 0 \leq 10$. Hence the graph of this inequality is the side that contains the origin.

$3x - 2y \leq 0$: The straight line $3x - 2y = 0$ has the single intercept (0, 0). The substitution of $x = 2$ yields the additional point (2, 3).

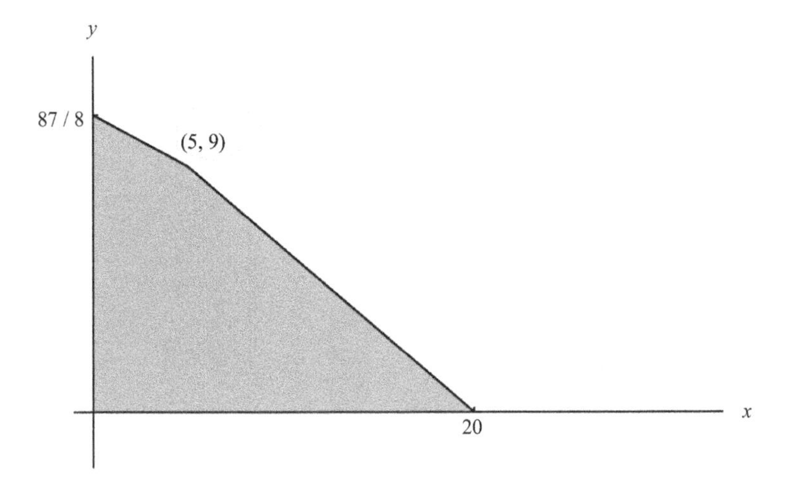

Figure 5.12

The substitution of the coordinates of (1, 0) into the inequality yields the valid inequality $3 \cdot 1 - 2 \cdot 0 \geq 0$. Hence the graph of this inequality is the side that contains the point (1, 0).

$x + 6y \geq 20$: The straight line $x + 6y = 20$ has the intercepts $(0, \frac{10}{3})$ and (20, 0). The substitution of the coordinates of (0, 0) into the inequality yields the false inequality $0 + 6 \cdot 0 \geq 20$. Hence the graph of this inequality is the side that does not contain the origin.

These three individual inequalities are graphed in Figure 5.13. The graph of the whole system consists of that region <u>all</u> of whose border lines are shaded. This is the triangle depicted in Figure 5.14. The vertices of the region also need to be computed for reasons that will become clear later. The coordinates of each such vertex are found by simultaneously solving the equations of the two lines that intersect at that vertex:

<u>(4, 6):</u>

$$\begin{array}{ccccc} x + y = 10 & & 2x + 2y = 20 & & -3x - 3y = -30 \\ & \text{or} & & \text{or} & \\ 3x - 2y = 0 & & 3x - 2y = 0 & & 3x - 2y = 0 \\ \hline & & 5x = 20 & & -5x = -30 \\ & & x = 20/5 = 4 & & x = (-30)/(-5) = 6 \end{array}$$

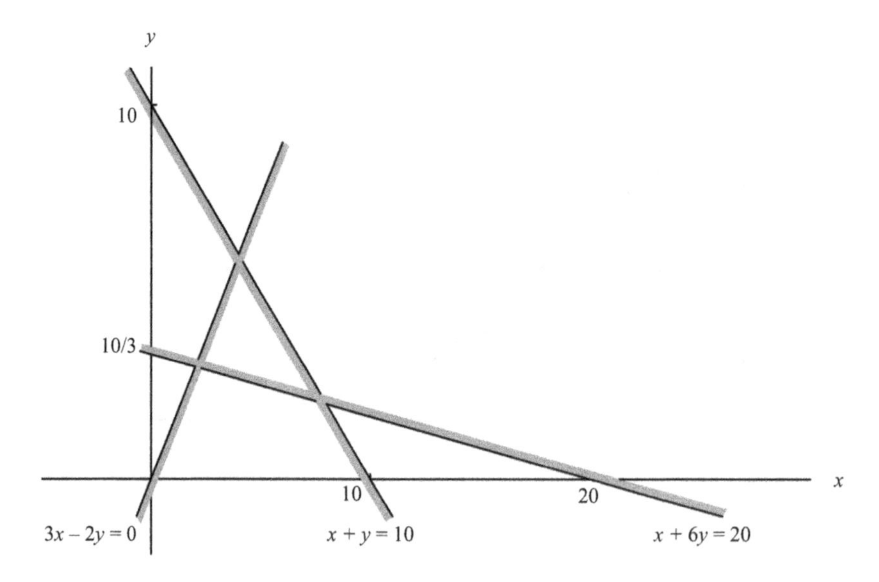

Figure 5.13

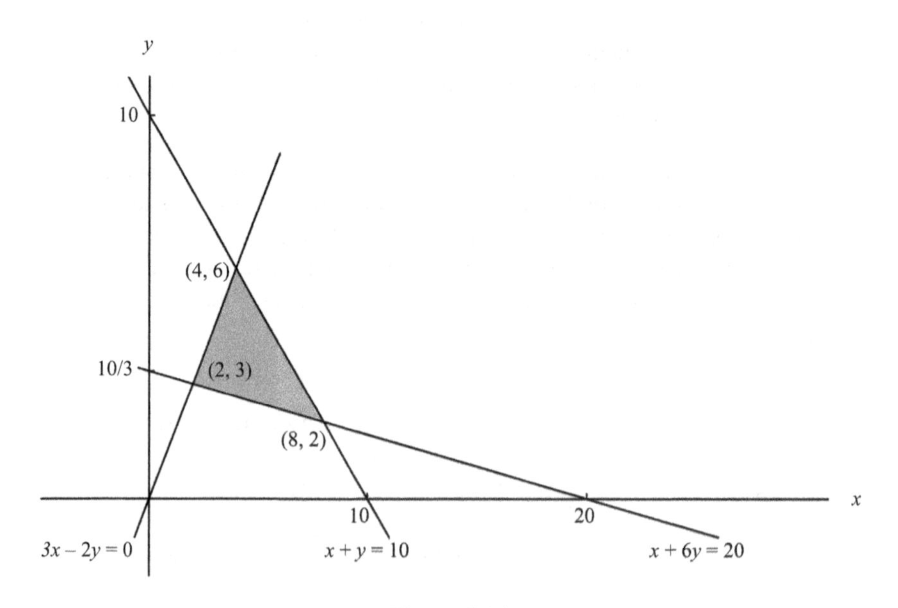

Figure 5.14

(2, 3):

$$3x - 2y = 0 \qquad 9x - 6y = 0 \qquad 3x - 2y = 0$$

or or

$$x + 6y = 20 \qquad x + 6y = 20 \qquad -3x - 18y = -60$$

$$10x = 20 \qquad -20y = -60$$

$$x = 20/10 = 2 \qquad y = (-60)/(-20) = 3$$

(8, 2):

$$x + y = 10 \qquad -x - y = -10 \qquad -6x - 6y = -60$$

or or

$$x + 6y = 20 \qquad x + 6y = 20 \qquad x + 6y = 20$$

$$5y = 10 \qquad -5x = -40$$

$$y = 10/5 = 2 \qquad x = (-40)/(-5) = 8.$$

Example 5.1.9. *Sketch the graph of the system of inequalities*:

$$x + y \geq 10$$

$$3x - 2y \geq 0$$

$$x + 6y \geq 20.$$

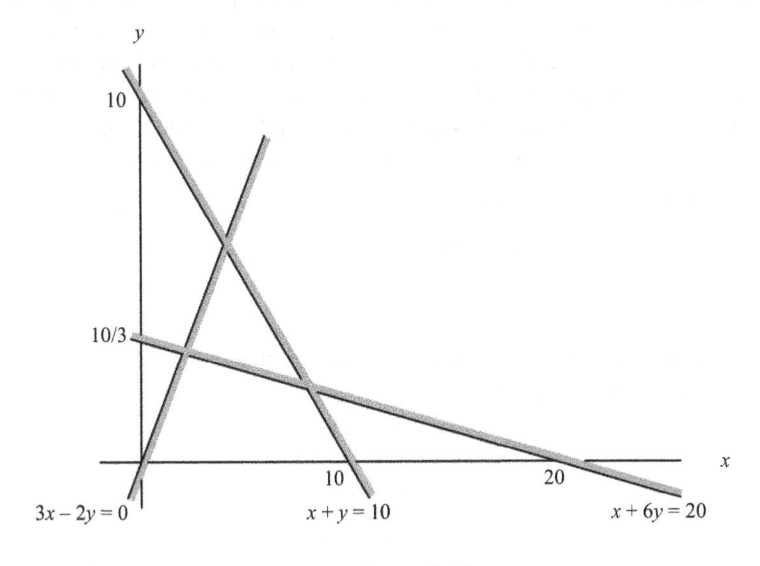

Figure 5.15

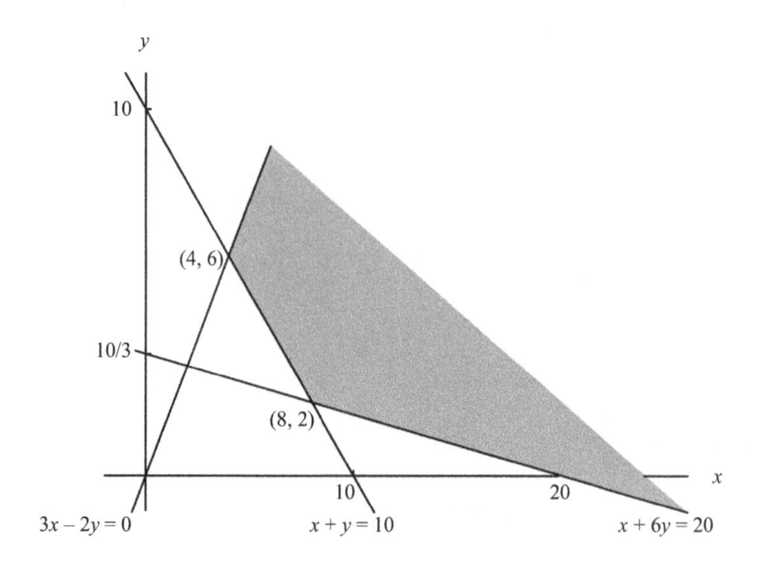

Figure 5.16

This example differs from the previous one only in that the first inequality has been reversed. This means that it is now necessary to use the other side of the straight line $x + y = 10$ whereas the rest of Figure 5.13 remains unchanged. We therefore obtain the preliminary sketch of Figure 5.15.

The new graph is depicted in Figure 5.16.

Section Appendix. *The solution of simultaneous equations.*

Given a system of simultaneous equations, one method for obtaining its solutions is to eliminate each of the unknowns in its turn. This is accomplished by modifying the equations so that the unknown has the same coefficient in both and then subtracting the two equations. This equalization can be obtained by multiplying each equation by that unknown's coefficient in the other equation. After the subtraction, a single equation with a single unknown is left, which is easily solved.

Example 5.1.10. *Solve the simultaneous system of equations*

$$3x + 4y = 25$$

$$5x - 3y = 3.$$

To eliminate the unknown x, multiply the first equation by 5, the second by 3, and subtract the resulting equations:

$$15x + 20y = 125$$

$$15x - 9y = 9$$

$$\overline{\qquad\qquad\qquad}$$

$$29y = 116$$

This yields

$$y = 116/29 = 4.$$

To eliminate the unknown y, multiply the first equation by -3, the second by 4, and subtract the resulting equations:

$$\begin{array}{r} -9x - 12y = -75 \\ 20x - 12y = 12 \\ \hline -29x = -87 \end{array}$$

This yields

$$x = (-87)/(-29) = 3.$$

The solution to the given system, therefore, is $x = 3$ and $y = 5$.

Exercises 5.1

In Exercises 1–10, graph the region of solutions of the given linear inequality.

1. $3x + y \leq 6$
2. $3x + 4y \geq 12$
3. $4x - 3y \leq 12$
4. $5x - 2y \geq 10$
5. $4x - 3y \geq 0$
6. $5x - 2y \leq 0$
7. $x \geq 3$
8. $x \leq -4$
9. $y \leq -2$
10. $y \geq 2$

In Exercises 11–20, graph the region of solutions of the given system of linear inequalities. Be sure to specify the corners of the region.

11. $2x + 3y \leq 17$
$3x - y \geq -2$
12. $5x - y \geq 6$
$2x - 3y \leq -5$
13. $x + 2y \geq 4$
$3x - 2y \geq -12$
$-2x + 5y \geq 10$
14. $x - y \leq 1$
$3x + 2y \leq -8$
$3x - y \geq 6$
15. $x + 2y \leq 4$
$3x - 2y \leq -12$
$-2x + 5y \geq 10$
16. $x - y \geq 1$
$3x + 2y \geq -8$
$3x - y \geq 6$
17. $2x + 5y \geq 60$
$5x + 3y \geq 60$
$x \geq 0, y \geq 0$
18. $x + 20y \geq 460$
$15x + y \leq 400$
$x \geq 0, y \geq 0$
19. $x + 2y \leq 16$
$3x + y \leq 30$
$x + y \leq 14$
$x \geq 0, y \geq 0$
20. $x + 3y \geq 180$
$3x + 2y \leq 190$
$2x + y \leq 120$
$x \geq 0, y \geq 0$

5.2 Optimization

Having described the graph of the constraints, attention is now shifted onto the objective function. The general rule for optimizing an objective function will be motivated by a simple example.

Example 5.2.1. *Find the largest and smallest values assumed by the objective function $x + y$ in the quadrilateral with vertices (4, 2), (1, 1), (1, 3), and (3, 4) (see Figure 5.17).*

Figure 5.18 contains a drawing of the quadrilateral in question and also shows how the value of $x + y$ varies inside as well as in the vicinity of the quadrilateral.

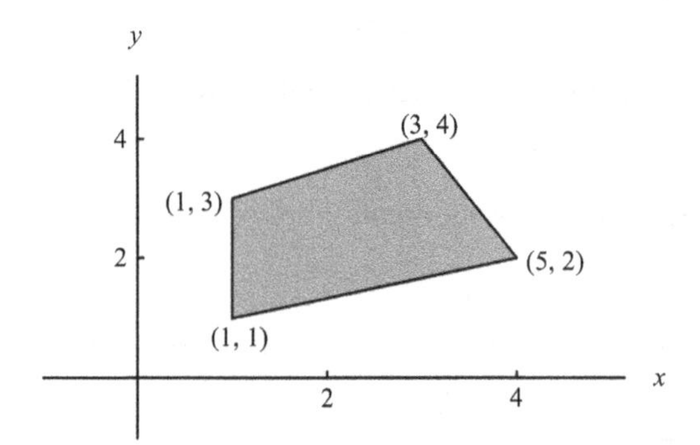

Figure 5.17

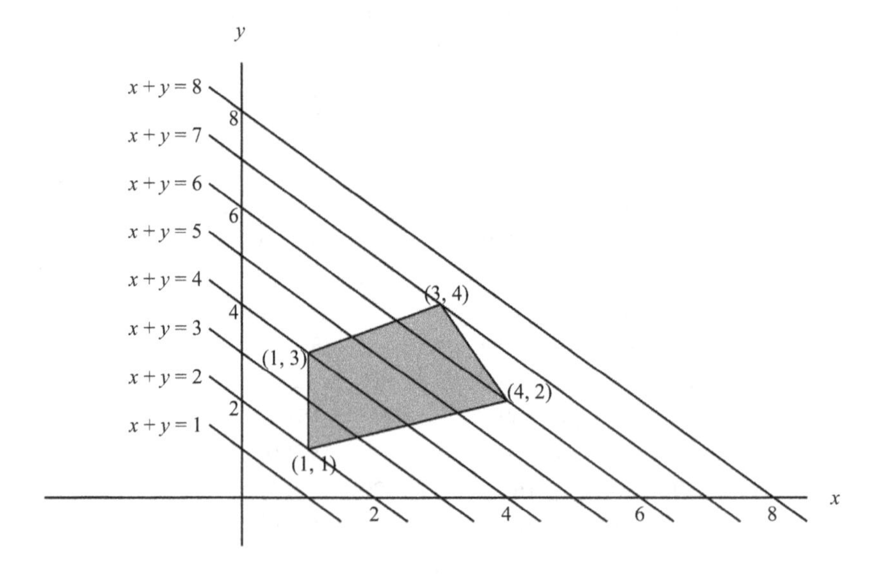

Figure 5.18

Note that, by definition, the value of $x + y$ along any of the lines $x + y = c$ is c, and that as c assumes larger values the graph of $x + y = c$ slides parallel to itself in the northeast direction. As long as this sliding line passes through the quadrilateral this means that there are points (x, y) in the quadrilateral such that $x + y = c$. For those values, such as 8, for which the line $x + y = c$ misses the quadrilateral, there are no points in the quadrilateral such that $x + y = c$. It is clear from the figure that the last value of c for which $x + y = c$ intersects the quadrilateral is $c = 7$. It follows that the largest value of $x + y$ in the quadrilateral is 7. Similarly, the first value for which the sliding line meets the quadrilateral is $c = 2$ and hence the minimum value of $x + y$ in the quadrilateral is 2.

This particular example is typical. Varying the value of C in the general straight line

$$Ax + By = C$$

is tantamount to sliding the line parallel to itself. It is also clear that as such a straight line slides across any polygon, in its last position, just as it is about to leave the polygon, it will contain a vertex of that polygon. Sometimes it may contain an entire side of the polygon, but even then it still contains one of the vertices. Consequently, we have the following *Vertex Principle:*

> *The maximum and minimum values of an objective function of the form $Ax + By$ in a polygon are assumed at some of that polygon's vertices.*

This principle reduces the problem of finding the optimal value of an objective function over an entire polygon to the apparently simpler task of computing its values at the polygon's vertices. We use the phrase "apparently simpler" because in actual problems, where the number of equations and variables reaches into the hundreds, the number of vertices is huge and may easily exceed the number of atoms in the universe. Checking the value of the objective function at all these vertices is impractical, even with computers, and the main task of any linear programming software package is to find an efficient way of locating the optimizing vertices. Of course, this text's examples and exercises will contain only a handful of vertices.

Example 5.2.2. *Find the maximum and minimum values of the objective function $2x - 3y$ over the pentagon with vertices (1, 3), (3, 4), (5, 3), (3, 1), and (2, 1).*

To find the answer we tabulate the values of $2x - 3y$ at the given vertices:

Vertex	$2x - 3y$
(1, 3)	$2 \cdot 1 - 3 \cdot 3 = 2 - 9 = -7$
(3, 4)	$2 \cdot 3 - 3 \cdot 4 = 6 - 12 = -6$
(5, 3)	$2 \cdot 5 - 3 \cdot 3 = 10 - 9 = 1$
(3, 1)	$2 \cdot 3 - 3 \cdot 1 = 6 - 3 = 3$
(2, 1)	$2 \cdot 2 - 3 \cdot 1 = 4 - 3 = 1$

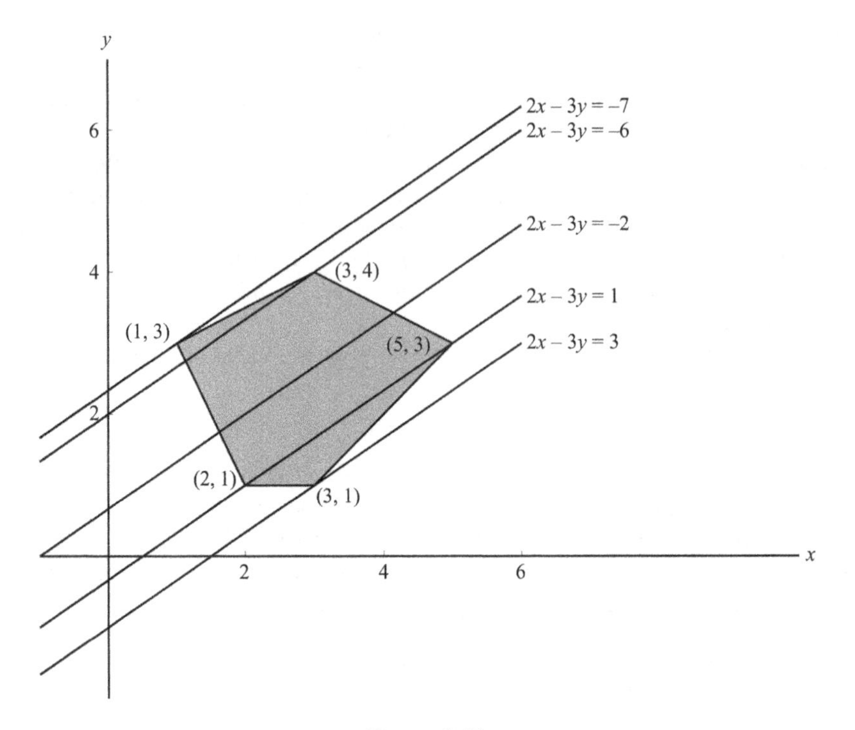

Figure 5.19

Consequently, the maximum and minimum values of the objective function are 3 and -7, respectively. Figure 5.19 displays both the polygon and the behavior of the objective function relative to this polygon. Of course, the Vertex Principle obviates the need for such diagrams.

Example 5.2.3. *Find the maximum and minimum values of the objective function $x + 2y$ over the pentagon with vertices (1, 3), (3, 4), (5, 3), (3, 1), and (2, 1).*

To find the answer we tabulate the values of $x + 2y$ at the given vertices:

Vertex	$x + 2y$
(1, 3)	$1 + 2 \cdot 3 = 1 + 6 = 7$
(3, 4)	$3 + 2 \cdot 4 = 3 + 8 = 11$
(5, 3)	$5 + 2 \cdot 3 = 5 + 6 = 11$
(3, 1)	$3 + 2 \cdot 1 = 3 + 2 = 5$
(2, 1)	$2 + 2 \cdot 1 = 2 + 2 = 4$

Consequently, the maximum and minimum values of the objective function are 11 and 4, respectively. Note that the maximum occurs at two points. Figure 5.20 displays both the polygon and the behavior of the objective function relative

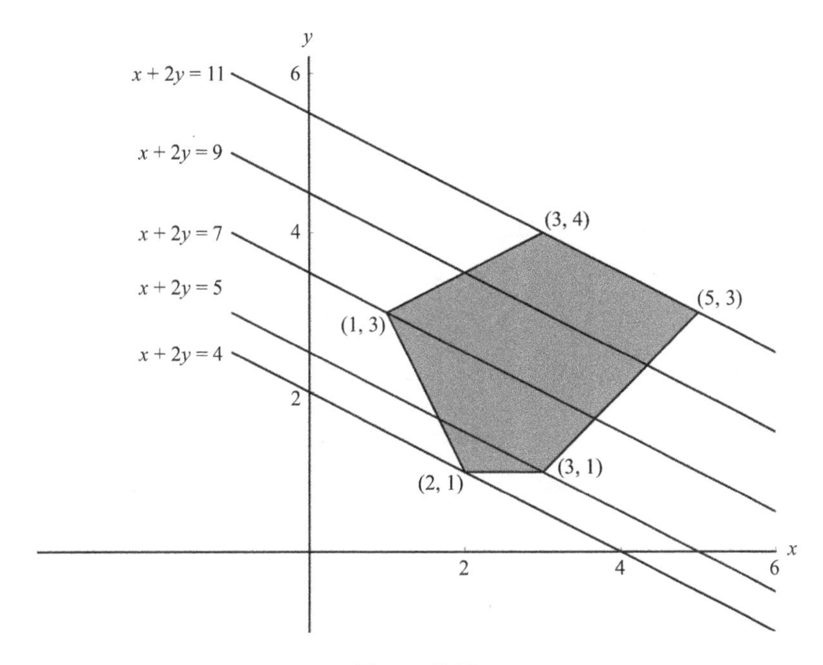

Figure 5.20

to this polygon. This diagram makes it clear that the maximum value of $x + 2y$ is actually assumed at every point on the side of the polygon that joins the vertices (3, 4) and (5, 3).

It was seen in Example 5.1.9 that the graphs of systems of inequalities can have infinite extents. In such cases objective functions may not have either maxima or minima. For instance, Figure 5.21 demonstrates that the objective function $x - y$ has neither a maximum nor a minimum over the graph of the system $x \geq 0$, $y \geq 0$ which is the first quadrant. The reason for this is that for every value of c, the straight line $x - y = c$, which has a southwest to northeast direction, is bound to intersect the first quadrant at some point.

The precise explanation of which objective functions have maxima or minima over infinitely extended regions falls beyond the scope of this text. For the purpose of the subsequent applications the following rule will suffice:

If an infinitely extended polygon lies completely in the first quadrant and A and B are positive numbers, then the objective function $Ax + By$ has a minimum and this minimum is assumed at one of the polygon's vertices.

Example 5.2.4. *Find the minimum value of $2x + 5y$ over the region depicted in Figure 5.22.*

According to the above rule such a minimum value exists, and it is assumed at one of the specified vertices. Straightforward substitution yields the table:

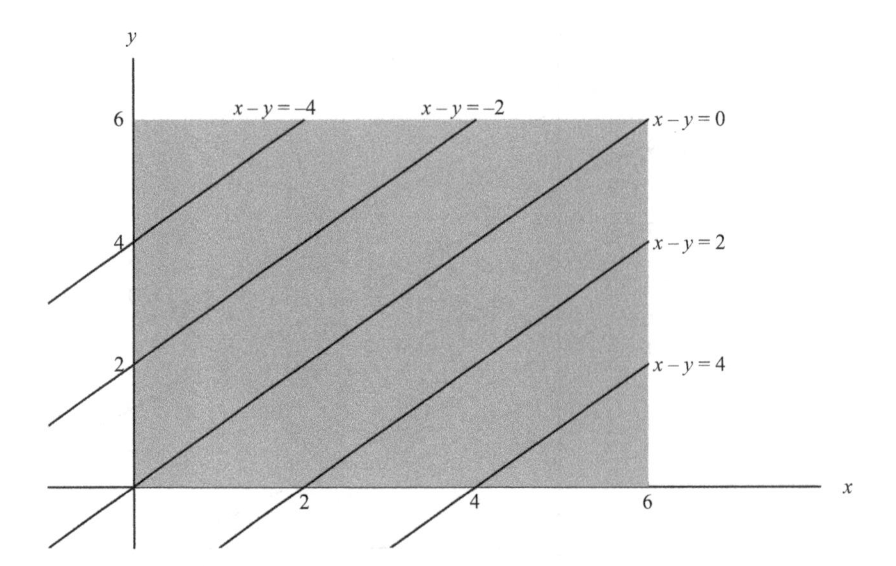

Figure 5.21

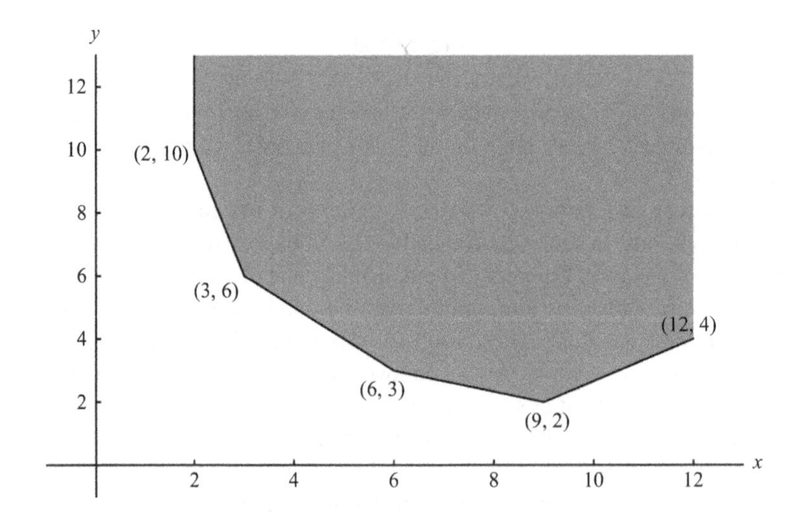

Figure 5.22

Vertex	$2x + 5y$
(2, 10)	$2 \cdot 2 + 5 \cdot 10 = 4 + 50 = 54$
(3, 6)	$2 \cdot 3 + 5 \cdot 6 = 6 + 30 = 36$
(6, 3)	$2 \cdot 6 + 5 \cdot 3 = 12 + 15 = 27$
(9, 2)	$2 \cdot 9 + 5 \cdot 2 = 18 + 10 = 28$
(12, 4)	$2 \cdot 12 + 5 \cdot 4 = 24 + 20 = 44$

Hence, the minimum value of this objective function is 27.

We now turn to the solution of some fairly realistic word problems.

Example 5.2.5. *Recall the small furniture factory of Example 5.1.1 whose constraints were graphed in Example 5.1.7 and whose objective function is $45x + 80y$. These lead to the table:*

Vertex	$45x + 80y$
$(0, 0)$	$45 \cdot 0 + 80 \cdot 0 = 0$
$(20, 0)$	$45 \cdot 20 + 80 \cdot 0 = 900$
$(5, 9)$	$45 \cdot 5 + 80 \cdot 9 = 945$
$(0, \frac{87}{8})$	$45 \cdot 0 + 80 \cdot \frac{87}{8} = 870$

Since 945 is the largest of the computed values of the objective function it follows that the company will optimize its profits by manufacturing five chairs and nine sofas weekly.

Example 5.2.6. *A dietitian must prepare a mixture of two foods I and II. Each unit of food I costs \$3 and contains 2 ounces of protein and 4 ounces of carbohydrates. Each unit of food II costs \$4 and contains 4 ounces of protein and 5 ounces of carbohydrates. What is the least expensive mixture that can be obtained, subject to the constraint that it must contain at least 26 ounces of protein and 40 ounces of carbohydrates?*

The objective function is the cost of the mixture. Since this cost depends on the amounts of foods I and II that are used in the mixture, these amounts are denoted by x and y, respectively, and in terms of these variables the objective function is:

$$\text{Objective function:} \quad 3x + 4y$$

In order to obtain the constraints it is helpful to organize the information in the problem in terms of a tableau that lists on the left those elements of the problem on which the objective function depends directly. Each of these will in turn depend on some other resources which should be listed at the top of the tableau. The actual dependencies become the entries of the array. The tableau's bottom row should list the various constraints on the resources.

Tableau:	Protein	Carbohydrates
Food I (x)	2	4
Food II (y)	4	5
Constraint	≥ 26	≥ 40

The constraint inequalities are then easily read off the tableau.

$$\underline{\text{Constraints:}} \quad \begin{aligned} 2x + 4y &\geq 26 \\ 4x + 5y &\geq 40 \\ x \geq 0,\ y &\geq 0. \end{aligned}$$

This yields the preliminary diagram of Figure 5.23. To find the missing vertex we solve simultaneously:

$$\begin{array}{lll} 2x + 4y = 26 & 4x + 8y = 52 & 10x + 20y = 130 \\ \quad\text{or} & & \quad\text{or} \\ 4x + 5y = 40 & 4x + 5y = 40 & 16x + 20y = 160 \\ \hline & 3y = 12 & -6x = -30 \\ & y = 12/3 = 4 & x = (-30)/(-6) = 5 \end{array}$$

and obtain the vertex coordinates (5, 4). The feasible region is depicted in Figure 5.24.

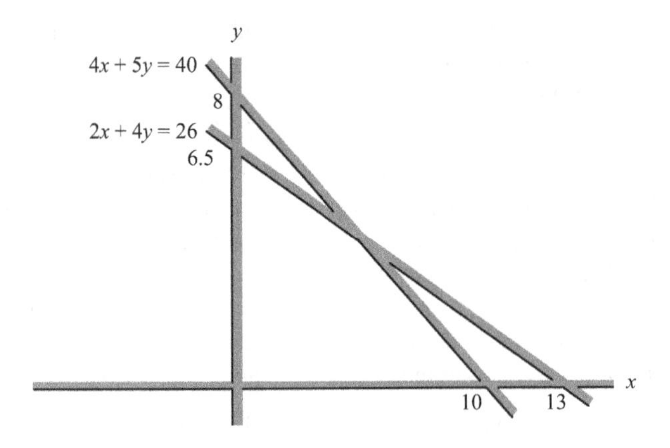

Figure 5.23

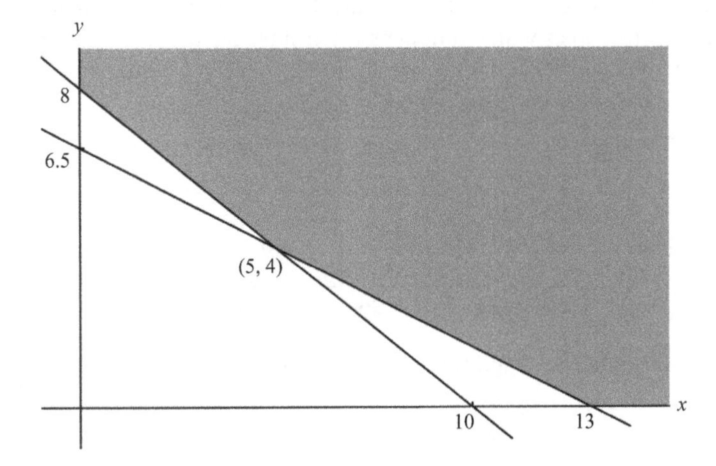

Figure 5.24

The evaluation of the objective function at the three vertices of the region yields the table:

Vertex	$3x + 4y$
(0, 8)	$3 \cdot 0 + 4 \cdot 8 = 32$
(5, 4)	$3 \cdot 5 + 4 \cdot 4 = 31$
(13, 0)	$3 \cdot 13 + 4 \cdot 0 = 39$

Hence the minimal cost of the dietitian's mixture is \$31, and it is obtained by mixing 5 units of food I with 4 units of food II.

Example 5.2.7. *Suppose Example 5.2.6 is modified by stipulating that the cost of each unit of food I is \$5 rather than \$3. How does this affect the problem's solution?*

Since the constraints are unchanged, Figure 5.24 still depicts the feasible region. On the other hand, the objective function is now $3x + 5y$ which, when evaluated at the vertices, yields the table:

Vertex	$5x + 4y$
(0, 8)	$5 \cdot 0 + 4 \cdot 8 = 32$
(5, 4)	$5 \cdot 5 + 4 \cdot 4 = 41$
(13, 0)	$5 \cdot 13 + 4 \cdot 0 = 65$

The minimum cost is therefore \$32, which is realized by using food II exclusively. Evidently, food I has become too expensive.

Example 5.2.8. *A nut company that can obtain from its suppliers 2800 lb of peanuts, 2600 lb of walnuts and 800 lb of cashews each month produces two mixes. Mix A uses 2 parts peanuts to 4 parts walnuts and 1 part cashews, whereas mix B uses 4 parts peanuts to 2 parts walnuts and 1 part cashews. If the mixes sell at \$3 and \$2 per pound, respectively, how much of each mix should the company prepare each month in order to maximize its revenue?*

The objective function is the revenue $3x + 2y$ where x denotes the amount of mix A and y the amount of mix B. The associated tableau is:

	Peanuts	Walnuts	Cashews
Mix A (x) uses	$\frac{2}{7}$	$\frac{4}{7}$	$\frac{1}{7}$
Mix B (y) uses	$\frac{4}{7}$	$\frac{2}{7}$	$\frac{1}{7}$
Constraint	≤ 2800	≤ 2600	≤ 800

These constraints translate to the inequalities

$$\frac{2}{7}x + \frac{4}{7}y \leq 2{,}800 \qquad \text{(Peanuts)}$$

$$\frac{4}{7}x + \frac{2}{7}y \leq 2{,}600 \qquad \text{(Walnuts)}$$

$$\frac{1}{7}x + \frac{1}{7}y \leq 800 \qquad \text{(Cashews)}$$

$$x \geq 0,\ y \geq 0.$$

When the fractions are cleared we obtain

$$2x + 4y \leq 19{,}600$$

$$4x + 2y \leq 18{,}200$$

$$x + y \leq 5{,}600$$

$$x \geq 0,\ y \geq 0.$$

The preliminary diagram is given in Figure 5.25. However, the intersection of the walnuts and peanuts does not lie on the boundary of the feasible region and is therefore irrelevant to the solution. For the other two vertices we solve:

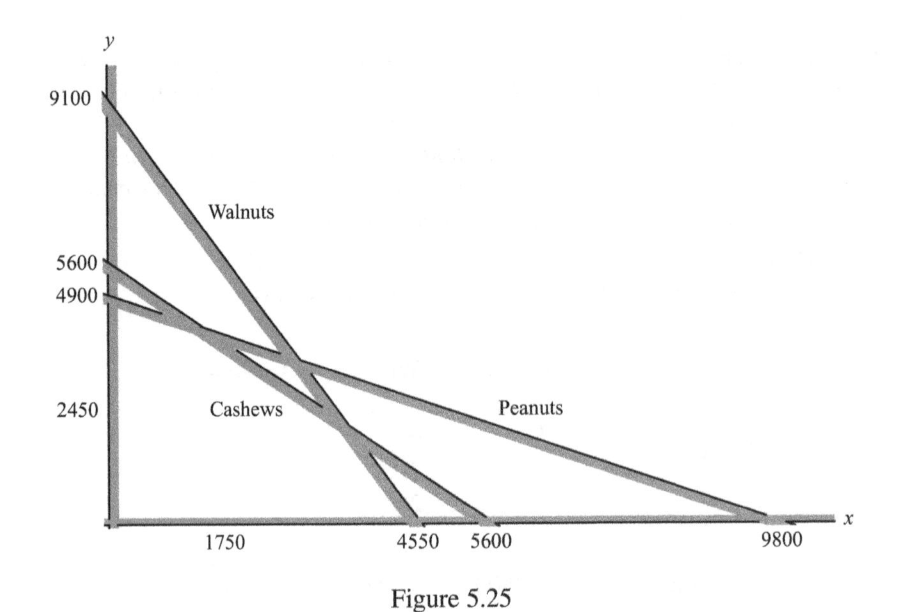

Figure 5.25

Cashews and peanuts:

$$2x + 4y = 19{,}600 \quad \text{or} \quad 2x + 4y = 19{,}600 \quad \text{or} \quad 2x + 4y = 19{,}600$$
$$x + y = 5{,}600 \qquad 2x + 2y = 11{,}200 \qquad 4x + 4y = 22{,}400$$

$$2y = 8{,}400 \qquad -2x = -2{,}800$$
$$y = 8{,}400/2 \qquad x = (-2{,}800)/(-2)$$
$$= 4{,}200 \qquad = 1{,}400$$

Cashews and walnuts:

$$4x + 2y = 18{,}200 \quad \text{or} \quad 4x + 2y = 18{,}200 \quad \text{or} \quad 4x + 2y = 18{,}200$$
$$x + y = 5{,}600 \qquad 4x + 4y = 22{,}400 \qquad 2x + 2y = 11{,}200$$

$$-2y = -4{,}200 \qquad 2x = 7{,}000$$
$$y = (-4{,}200)/(-2) \qquad x = 7{,}000/2$$
$$= 2{,}100 \qquad = 3{,}500$$

The feasible region is presented in Figure 5.26.
The values of the objective function at the vertices of this feasible region are:

Vertex	$3x + 2y$
(0, 0)	$3 \cdot 0 + 2 \cdot 0 = 0$
(0, 4900)	$3 \cdot 0 + 2 \cdot 4{,}900 = 9{,}800$
(1400, 4200)	$3 \cdot 1{,}400 + 2 \cdot 4{,}200 = 12{,}600$
(3500, 2100)	$3 \cdot 3{,}500 + 2 \cdot 2{,}100 = 14{,}700$
(4550, 0)	$3 \cdot 4{,}550 + 2 \cdot 0 = 13{,}650$

Since the largest value assumed by the objective function is \$14,700, we conclude that the company should prepare 3500 lb of mix A and 2100 lb of mix B.

Example 5.2.9. *Suppose the company of the previous example finds that mix A is popular and that it can therefore afford to raise its price from \$3 to \$3.50 per lb. Should the company alter its production?*

The constraints and feasible region remain unchanged, but the objective function is now $3.5x + 2y$. The table of its values at the vertices is:

Vertex	$3.5x + 2y$
(0, 0)	$3.5 \cdot 0 + 2 \cdot 0 = 0$
(0, 4900)	$3.5 \cdot 0 + 2 \cdot 4{,}900 = 9{,}800$
(1400, 4200)	$3.5 \cdot 1{,}400 + 2 \cdot 4{,}200 = 13{,}300$

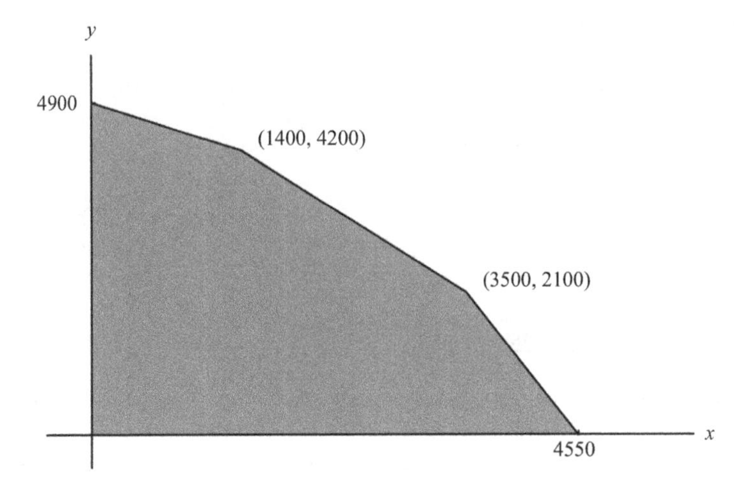

Figure 5.26

(3500, 2100)	$3.5 \cdot 3{,}500 + 2 \cdot 2{,}100 = 16{,}450$
(4550, 0)	$3.5 \cdot 4{,}550 + 2 \cdot 0 = 15{,}925$

Since the new maximum value of \$16,450 is assumed at the same vertex (3,500, 2,100), the company has no reason to change its production policy.

Example 5.2.10. *Suppose the company of the previous example finds that mix A is very popular and that it can therefore afford to raise its price from \$3 to \$5 per lb. Should the company alter its production?*

The constraints and feasible region remain unchanged, but the objective function is now $5x + 2y$. The table of its values at the vertices is:

Vertex	$5x + 2y$
(0, 0)	$5 \cdot 0 + 2 \cdot 0 = 0$
(0, 4900)	$5 \cdot 0 + 2 \cdot 4{,}900 = 9{,}800$
(1,400, 4,200)	$5 \cdot 1{,}400 + 2 \cdot 4{,}200 = 15{,}400$
(3,500, 2,100)	$5 \cdot 3{,}500 + 2 \cdot 2{,}100 = 22{,}050$
(4,550, 0)	$5 \cdot 4{,}550 + 2 \cdot 0 = 22{,}750$

Since the new maximum value of \$22,750 is assumed at the new vertex (4,550, 0), the company should stop producing mix B and focus on mix A exclusively.

It is interesting to note that the response of the company to the demand for mix A is discontinuous. As long as this price remains lower that \$4 the company should stick to the production schedule of (3,500, 2,100). Once the price rises above \$4 the company should switch to the vertex (4,550, 0).

Example 5.2.11. *Suppose mixes A and B of Example 5.2.8 are produced at different plants and that the company has a new agreement with its workers that specifies that at least 1,750 lb of mix A and at least 2,450 lb of mix B will be produced each month. How will this affect the production schedule?*

We now have a new set of constraints:

$$\frac{2}{7}x + \frac{4}{7}y \leq 2{,}800 \qquad \text{(Peanuts)}$$
$$\frac{4}{7}x + \frac{2}{7}y \leq 2{,}600 \qquad \text{(Walnuts)}$$
$$\frac{1}{7}x + \frac{1}{7}y \leq 800 \qquad \text{(Cashews)}$$
$$x \geq 1{,}750,\ y \geq 2{,}450.$$

When the inequalities are simplified they become

$$2x + 4y \leq 19{,}600 \qquad \text{(Peanuts)}$$
$$4x + 2y \leq 18{,}400 \qquad \text{(Walnuts)}$$
$$x + y \leq 5{,}600 \qquad \text{(Cashews)}$$
$$x \geq 1{,}750,\ y \geq 2{,}450.$$

The feasible region is therefore that triangle of Figure 5.27 whose sides have equations

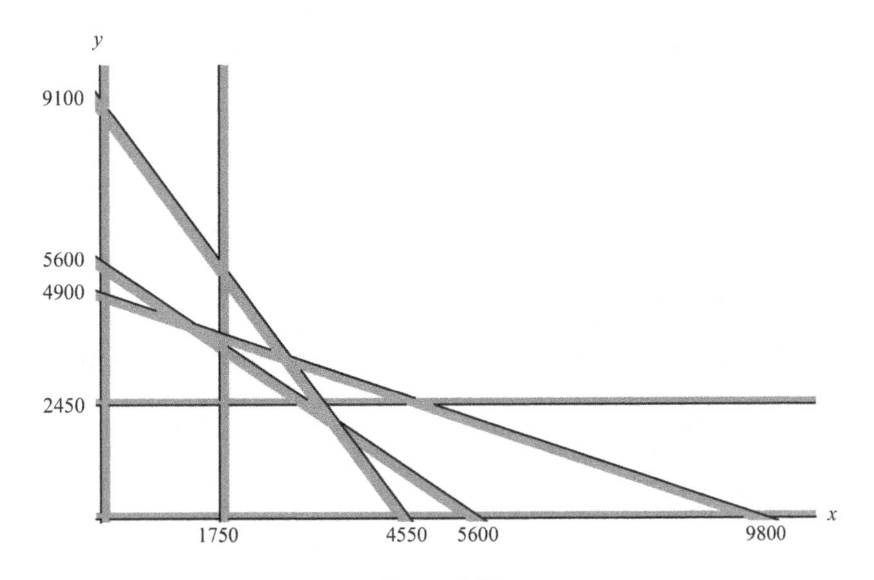

Figure 5.27

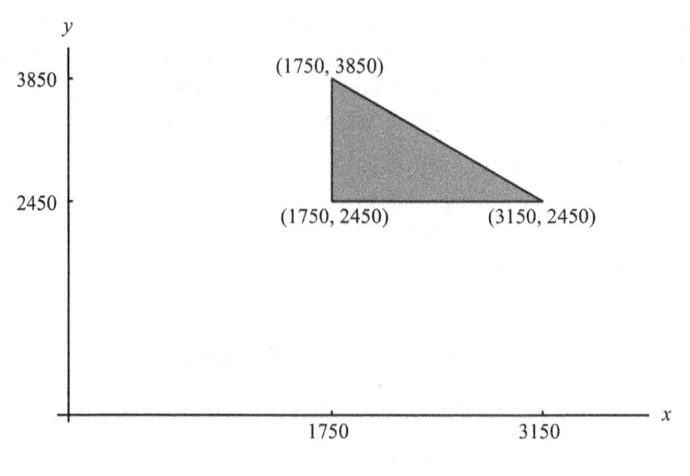

Figure 5.28

$x = 1{,}750$, $y = 2{,}450$, and $x + y = 5{,}600$, respectively. This triangle has vertices (1,750, 2,450), (1,750, 3,850), and (3,150, 2,450) and is depicted in Figure 5.28.

The objective function assumes the values:

Vertex	$3x + 2y$
(1,750, 2,450)	$3 \cdot 1{,}750 + 2 \cdot 2{,}450 = 10{,}150$
(1,750, 3,850)	$3 \cdot 1{,}750 + 2 \cdot 3{,}850 = 12{,}950$
(3,150, 2,450)	$3 \cdot 3{,}150 + 2 \cdot 2{,}450 = 14{,}350$

Thus, the company should prepare 3,150 lb of mix A and 2,450 lb of mix B.

Exercises 5.2

1. Find the maximum and minimum values of $x + 10y$ over the quadrilateral with vertices (2, 0), (0, 2), (1, 10), (9, 2).
2. Find the maximum and minimum values of $2x - 3y$ over the quadrilateral with vertices (2, 0), (0, 2), (1, 10), (9, 2).
3. Find the maximum and minimum values of $10x + y$ over the quadrilateral with vertices (2, 0), (0, 2), (1, 10), (9, 2).
4. Find the maximum and minimum values of $3x - 2y$ over the quadrilateral with vertices (2, 0), (0, 2), (1, 10), (9, 2).
5. A dietitian must prepare a mixture of two foods I and II. Each unit of food I contains 3 ounces of protein and 5 ounces of carbohydrates. Each unit of food II contains 5 ounces of protein and 4 ounces of carbohydrates. Determine the least expensive mixture that can be obtained, subject to the constraint that it must contain at least 27 ounces of protein

and 32 ounces of carbohydrates, and if the unit costs of food I and food II are, respectively, (a) \$4 and \$2 (b) \$3 and \$3 (c) \$2 and \$4.

6. A nut company that can obtain from its suppliers 2100 lb of peanuts, 6000 lb of walnuts and 5400 lb of cashews each month produces two mixes. Mix A uses 1 part peanuts to 3 parts walnuts and 2 parts cashews, whereas mix B uses 1 part peanuts to 2 parts walnuts and 3 parts cashews. How much of each mix should the company prepare each month in order to maximize its revenue if the mixes sell at the following respective prices per lb? (a) \$5 and \$3 (b) \$4 and \$3 (c) \$2 and \$3
7. Redo Exercise 6 under the additional assumption that at least 1,000 lb of each mix must be produced each month.
8. A company produces two products, floor lamps and table lamps. Production of one floor lamp requires 75 minutes of labor and materials that cost \$25. Production of one table lamp requires 50 minutes of labor and materials that cost \$20. Each employee is to work no more than 40 hours per week and to spend no more than \$900 for materials each week. If the profit is \$40 per floor lamp and \$30 per table lamp, how many of each product should each employee produce per week in order to maximize profits? What is the maximum weekly profit?
9. A baking company has five employees who make bread and cakes. Each loaf of bread requires 50 minutes of labor and ingredients costing \$0.90 and can be sold for \$1.20 profit. Each cake requires 30 minutes of labor and ingredients costing \$1.50 and can be sold for \$4.00 profit. No one is to work more than 8 hours a day and the company cannot spend more than \$96 per day on ingredients. How many loaves of bread and how many cakes should the company produce each day in order to maximize the profits? What is the maximum profit? How would the answer vary if each loaf yielded as much profit as each cake?
10. Charlie's Coffees sells two blends of coffee beans, Morning blend and South American blend. Morning blend is one-third Mexican beans and two-thirds Colombian beans. South American blend is two-thirds Mexican beans and one-third Colombian beans. Profit on the Morning blend is \$4 per pound while profit on the South American blend is \$3.50 per pound. Each day the shop can obtain 100 pounds of Mexican beans and 80 pounds of Colombian beans. How many pounds of Morning blend and how many pounds of South American blend should Charlie's Coffees prepare each day in order to maximize profits? What is the maximum daily profit?
11. A company sells bread to supermarkets. Shopgood Stores needs at least 15,000 loaves each week, and Rollie's Markets needs at least 20,000 loaves each week. The company can ship at most 45,000 loaves to these two stores each week if it wishes to satisfy its other customers' needs. If shipping costs an average of \$7 per loaf to Shopgood Stores and \$8 per loaf to Rollie's Markets, how many loaves should the

company allot to Shopgood and Rollie's each week in order to minimize the shipping costs? What are the minimum weekly shipping costs?

12. A company manufactures computer chips. Its two main customers, HAL Computers and Box Computers, just submitted orders that must be filled immediately. HAL needs at least 130 cases of chips, and Box needs at least 150 cases. Due to a limited supply of silicon, the company cannot send more than a total of 300 cases. If shipping costs $110 per case for shipments to HAL and $100 per case for shipments to Box, how many cases should the company send to each customer in order to minimize shipping costs? What shipping costs would this entail?
13. World Air Lines has contracted with a tour group to transport a minimum of 1,600 first-class passengers and 4,800 economy-class passengers from New York to London during a 6-month period. World Air Lines has two types of airplanes, the Orville 606 and the Wilbur W-1112. The Orville 606 carries 20 first-class passengers and 80 economy-class passengers, and costs $12,000 to operate. The Wilbur W-1112 carries 80 first-class passengers and 120 economy-class passengers and costs $20,000 to operate. During the time period involved, World Air Lines can schedule no more than 52 flights on Orville 606s and no more than 30 flights on Wilbur W-1112s. How should World Air Lines schedule its flights so as to minimize operating costs? What operating costs would this entail? What if the Wilburs only cost $16,000 to operate?
14. Compucraft sells personal computers and printers made by World Computers. The computers come in 12-cubic-foot boxes, and the printers come in 8-cubic-foot boxes. Compucraft's owner estimates that at least 30 computers can be sold each month and that the number of computers sold will be at least 50% more than the number of printers. The computers cost Compucraft $1,000 each and can be sold at $1,000 profit each, while the printers cost $300 each and can be sold at $350 profit each. Compucraft has 1,000 cubic feet of storage available for the World personal computers and printers and sufficient financing to spend $70,000 each month on computers and printers. How many computers and printers should Compucraft order from World each month in order to maximize profits? What is that maximum profit?
15. World Motors manufactures quarter-ton, half-ton, and three-quarter-ton panel trucks. Express Delivery Service has placed an order for at least 300 quarter-ton, 450 half-ton, and 450 three-quarter-ton panel trucks. World Motors builds the trucks at two plants, one in Detroit and one in Los Angeles. The Detroit plant produces 30 quarter-ton trucks, 60 half-ton trucks, and 90 three-quarter-ton trucks each week at a total cost of $540,000. The Los Angeles plant produces 60 quarter-ton trucks, 45 half-ton trucks, and 30 three-quarter-ton trucks each week, at a total cost of $360,000. How should World Motors schedule its two plants so that it can fill this order at minimum cost? What is

that minimum cost? How would the answer change if each plant costs \$500,000 per week?

16. Barton's Chocolates produces semisweet chocolate chips and milk chocolate chips at its plants in Bay City and Eustacia. The Bay City plant produces 3,000 pounds of semisweet chips and 2,000 pounds of milk chocolate chips each day at a cost of \$1,100, while the Eustacia plant produces 1,000 pounds of semisweet chips and 6,000 pounds of milk chocolate chips each day at a cost of \$1,800. Barton's has an order from SuperConvenience Stores for at least 30,000 pounds of semisweet chips and at least 60,000 pounds of milk chocolate chips. How should it schedule its production so that it can fulfill the order at minimum cost? What is that minimum cost?

Chapter 6

Planar Symmetries

The word symmetry originally meant a proper or excellent proportion and it is in this sense that it is used in Blake's poem *The Tiger*:

What immortal hand or eye
Could frame thy fearful symmetry?

Today this word refers to a quality of visual regularity. An object exhibits, or possesses, symmetry if it is composed of several replicas of the same part. This is the sense in which the butterfly and the starfish are symmetrical, each in its own way. Here we examine the various different kinds of symmetries that can be formed by planar configurations.

6.1 Introduction

The use of symmetry as an aspect of art goes back several millennia and some early samples of artistic creations into which symmetry was consciously incorporated are displayed in Figure 6.1. The Greek mathematician Euclid (circa 300 BC), whose opus *The Elements* codified geometry for the following two thousand years, included in this series of thirteen books many propositions regarding both two- and three-dimensional symmetrical figures. Interestingly, the first proposition of the first book concerns the construction of equilateral triangles and the last book is devoted to the description of the five regular solids (see Figure 6.1), all highly symmetrical

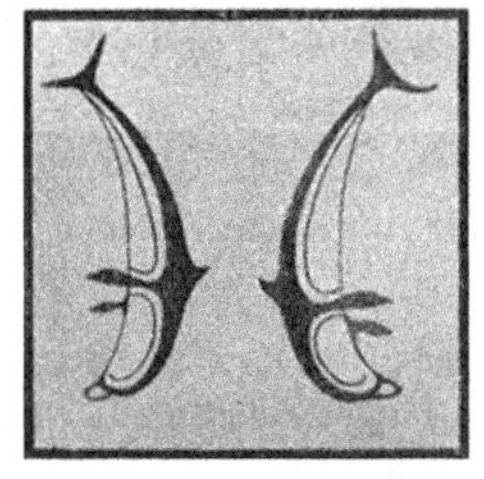

Figure 6.1: Symmetry in ancient art.

geometrical objects. This endowed Euclid's work with a symmetry of its own. Given that he could have developed the subject matter in many other ways, it is arguable that this symmetrical format was a result of a conscious decision on his part.

Euclid and his successors viewed symmetries as merely aesthetically pleasing aspects of geometrical objects rather than the subjects of mathematical investigations. In other words the symmetries of these figures attracted their attention, but their technical investigations were restricted to the mathematical consequences of these symmetries rather than the symmetries themselves. It was not until the eighteenth century that mathematicians realized that symmetries per se could and should be subject to mathematical research. Since then the mathematical study of symmetry, better known as *group theory*, has evolved into a deep and essential branch of mathematics with important applications to the physical sciences. This chapter is devoted to the description of the symmetries of plane figures.

6.2 Rigid Motions of the Plane

The Oxford English Dictionary's definition of geometrical symmetry is

> *Exact correspondence in position of the several points or parts of a figure or body with reference to a dividing line, plane, or a point.*

The mathematical understanding is different: a geometrical symmetry is a potential transformation of an object that would leave it in the same position. This requires a fair amount of clarification and we begin with the notion of transformation or, more technically, a rigid motion.

A rigid motion of the plane takes place when the plane is moved through space, without distortions, so that the plane as a whole ends up occupying its initial position, even though individual points and figures are deposited in new locations. While the actual movement through space is a useful tool for visualizing a rigid motion, it is its final position that is of interest to us here and consequently any two rigid motions that leave the points and figures of the plane in the same final positions are considered as identical, even though in their intermediary stages they may move the same points along different paths through space. Such is the case with the clockwise 90° and counterclockwise 270° turns: both leave us facing to the right relative to our original position. In general, motions will be visualized in the most efficient, or economical, manner possible. Moreover, rather then describe them as an actual movement through space, a difficult task within the confines of a textbook, the rigid motions will be described by means of their final effect on some figures.

The first type of rigid motions to be described are the *translations*. These are the rigid motions that keep all line segments parallel to their initial positions. The horizontal 3 inches shift to the right is a translation that carries the collection A of Figure 6.2 to location B. Of course, the horizontal 3 inches shift to the left is the translation that carries B to A. Similarly, there is a north-easterly translation that carries the mushroom M of Figure 6.3 to N.

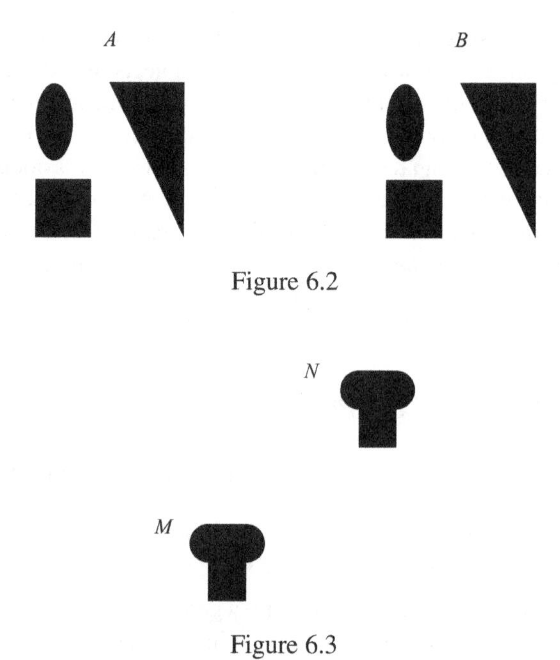

Figure 6.2

Figure 6.3

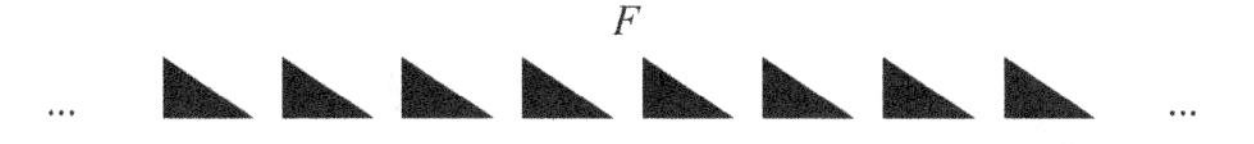

Figure 6.4: A translation.

Figure 6.5: A translation.

Before going on to discuss other types of rigid motions, it is necessary to make some more general definitions. If a rigid motion picks up a figure F and moves it to a new position G, it will be said that G is the *image* of F under the action of that motion. Thus figure B above is the image of A under the horizontal 3 inch rightward shift, whereas A is the image of B under the horizontal 3 inch leftward shift.

Given a figure F and a rigid motion m, the collection of all the images generated from F by the iteration of both that motion and its reversal constitute the *orbit* of F generated by m. Figure 6.4 above depicts an orbit generated by a horizontal translation.

Figure 6.5 above depicts an orbit generated by a north-easterly translation. The orbits generated by any rigid motion and its reversal are, of course, identical, provided that they begin with the same figures.

The second class of rigid motions are the *rotations*. These motions turn the plane through a certain signed angle about some pivot point. When the specified angle is positive the rotation moves counterclockwise from the point of view of the reader; when the angle is negative the motion is clockwise. Although in general any angle can serve as an angle of rotation, the subsequent discussion will be restricted to rotations whose angles measure 60°, 90°, 120°, and 180°. Typical orbits of such rotations appear in Figure 6.2. Their orbits are said to be 6-fold, 4-fold, 3-fold, and 2-fold, respectively. Because we are only interested in the question of where figures land, rather than how they got there, rotations by angles of 180° and $-180°$ with the same pivot points are equal to each other.

Yet another class of rigid motions is the *reflection,* so called because their effect on geometrical figures is akin to that of mirror reflections. They can be visualized as the effect of a 180° spatial rotation about a straight line, the *axis*, that lies in the rotated plane. The orbit of any figure generated by a reflection consists of two copies of that figure.

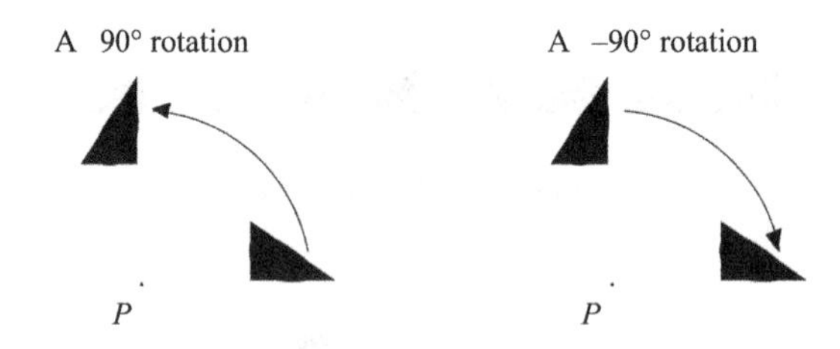

Figure 6.6: Rotations about a pivot point P.

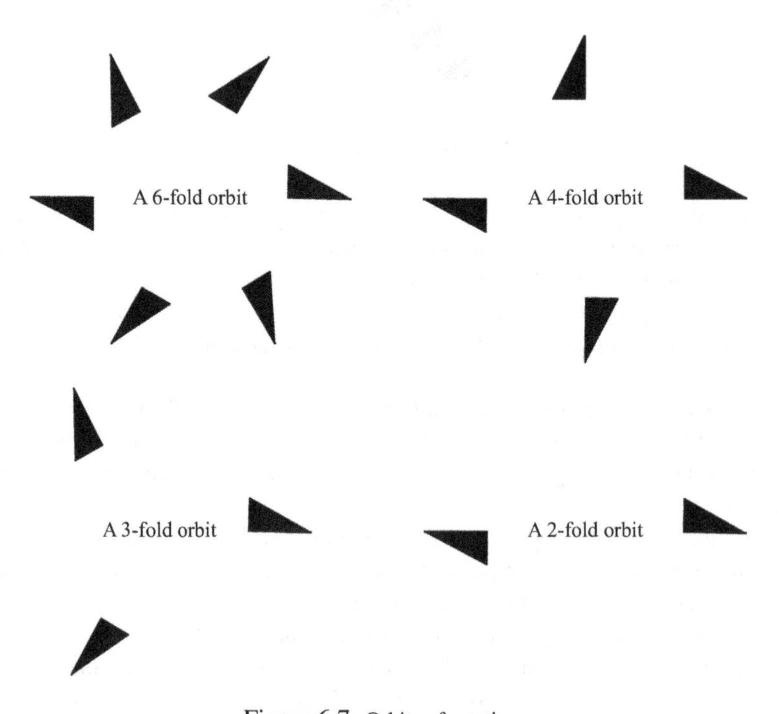

Figure 6.7: Orbits of rotations.

Since only the final position of the moving figure is of concern here, the fact that this rotation switches the "sides" of the plane is immaterial.

The last of the types of the rigid motion is the *glide reflection*. As its name implies, this is a combination of a translation and a reflection whose axis is parallel to the direction of the translation. The orbit of a figure generated by a glide reflection consists of an infinite number of copies of the figure that are situated alternately on the two sides of the axis. As was the case for translations, the orbit of a figure generated by a glide reflection also includes the copies obtained by applying the reversal of the glide reflection, and so this orbit extends infinitely far in two directions.

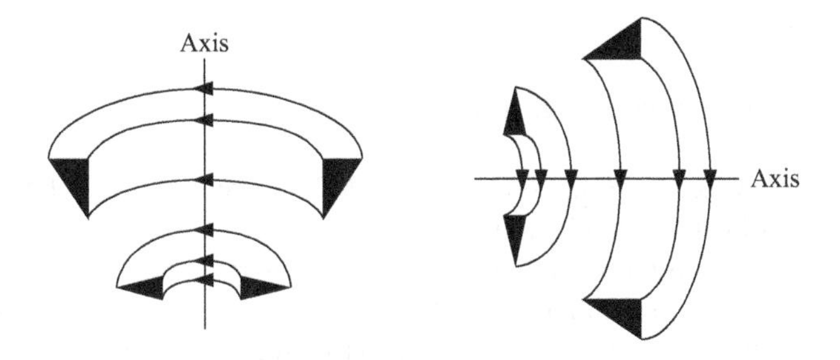

Figure 6.8: Orbits of reflections.

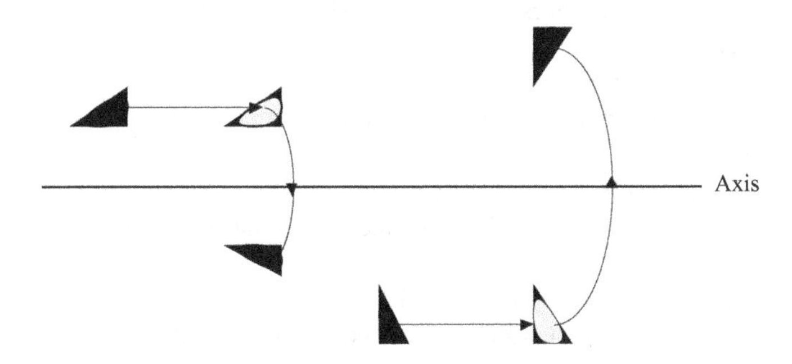

Figure 6.9: A glide reflection.

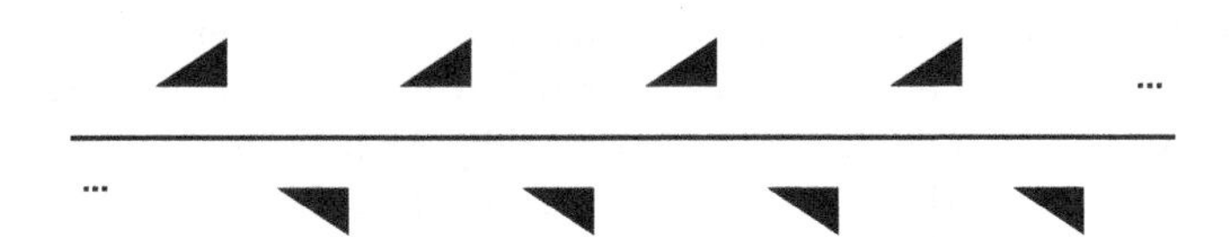

Figure 6.10: The orbit of a glide reflection.

The glide reflections complete the roster of all the rigid motions of the plane. Surprisingly, no new rigid motions are created by combining other rigid motions. This is not easy to visualize, but the combination of a translation with a rotation yields another rotation, and the combination of a rotation with a reflection will generally yield a glide reflection. The following theorem is due to M. Chasles (1793-1880).

Theorem 6.2.1. *The rigid motions of the plane consist of translations, rotations, reflections, and glide reflections.*

Just as 0 turns out to be a convenient number, so is it useful to consider the "motion" that does not move anything as a motion which is called the *identity*. As this motion does not change the distances between any two points it qualifies as a rigid motion. The identity motion is denoted by *Id*.

Exercises 6.2

Exercises 1 – 13 all refer to Figure 6.11.

1. Sketch six figures in the orbit of ΔABC generated by a 1 inch translation in a direction parallel to the straight line m.
2. Sketch six figures in the orbit of ΔABC generated by a 1 inch translation in a direction parallel to the straight line n.
3. Sketch the orbit of ΔABC generated by a rotation that is pivoted at A and has angle
(a) 60°, (b) 90°, (c) 120°, (d) 180°.
4. Sketch the orbit of ΔABC generated by a rotation that is pivoted at B and has angle
(a) 60°, (b) 90°, (c) 120°, (d) 180°.
5. Sketch the orbit of ΔABC generated by a rotation that is pivoted at P and has angle
(a) 60°, (b) 90°, (c) 120°, (d) 180°.
6. Sketch the orbit of ΔABC generated by a rotation that is pivoted at Q and has angle
(a) 60°, (b) 90°, (c) 120°, (d) 180°.
7. Sketch the orbit of ΔABC generated by a rotation that is pivoted at R and has angle
(a) 60°, (b) 90°, (c) 120°, (d) 180°.
8. Sketch the orbit of ΔABC generated by the reflection with axis m.
9. Sketch the orbit of ΔABC generated by the reflection with axis n.
10. Sketch the orbit of ΔABC generated by the reflection with axis AB.
11. Sketch the orbit of ΔABC generated by the reflection with axis BC.
12. Sketch the orbit of ΔABC generated by the reflection with axis QR.
13. Sketch the orbit of ΔABC generated by the reflection with axis RS.
14. Sketch six figures in the orbit of ΔABC generated by a 1 inch glide reflection with axis m.

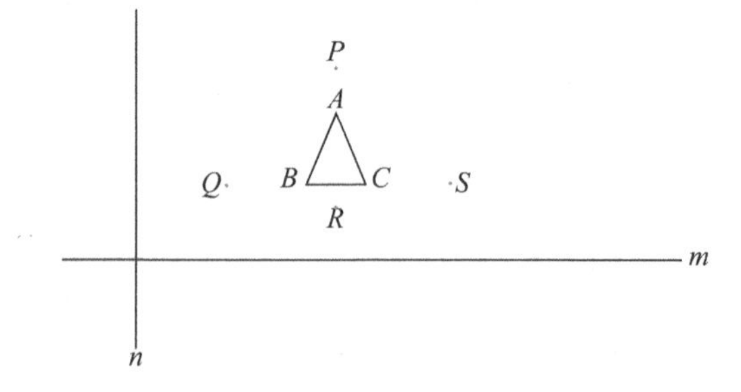

Figure 6.11

15. Sketch six figures in the orbit of $\Delta\ ABC$ generated by a 1 inch glide reflection with axis n.
16. Explain why the 360° rotation about a point P equals the identity motion.

6.3 Symmetries of Polygons

A *(mathematical) symmetry* of a figure is a rigid motion that carries that figure onto itself. Alternately, a rigid motion is a symmetry of a figure provided this figure constitutes the entire orbit generated by the motion. Thus, the reflections with axes d, m, e, n are all symmetries of the square below and they are denoted by ρ_d, ρ_m, ρ_e, and ρ_n, respectively.

These reflections do not constitute all of the square's symmetries. The 90°, 180°, and 270° (or −90°) rotations about the center of the square are also such symmetries. If C denotes this center, then $R_{C,90°}$, $R_{C,180°}$, $R_{C,270°}$ denote these respective rotations. The identity rigid motion Id is another such symmetry. The set of all the symmetries of a figure is called its *symmetry group* or just *group*. Thus, the symmetry group of the square is

$$\{Id,\ \rho_d,\ \rho_e,\ \rho_m,\ \rho_n\ ,R_{C,90°},\ R_{C,180°},\ R_{C,270°}\}.$$

By definition, every plane figure F has a symmetry group that contains at least the identity motion Id. The isosceles triangle below has $\{Id,\ \rho_v\}$ as its symmetry group whereas that of the equilateral triangle is $\{Id,\ \rho_d,\ \rho_e,\ \rho_f,\ R_{C,120°},\ R_{C,240°}\}$.

Since the orbits of translations and glide reflections are infinite in extent it follows that no finite figure can have symmetries of either of these types.

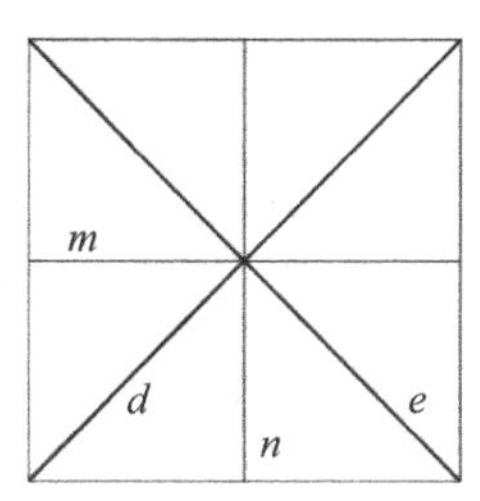

Figure 6.12: Some symmetries of the square.

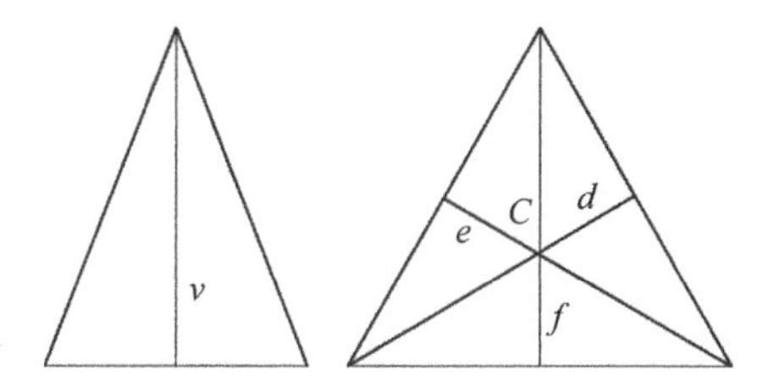

Figure 6.13: An isosceles and an equilateral triangle.

Infinitely extended figures, however, can have such symmetries. We first show how to create such figures and then examine the question of what their symmetry groups are like.

Exercises 6.3

1. Write down the symmetry groups of the following figures.
   a) The rectangle with unequal sides
   b) The regular pentagon
   c) The regular hexagon
   d) The regular heptagon
   e) The regular octagon

6.4 Frieze Patterns

Orbits of finite figures generated by translations are called *frieze patterns.* These frieze patterns are the mathematical idealization of such decorative designs as borders used to accent wallpapers and trim sewn or printed around a cloth (see Figure 6.14). Unlike their physical manifestations, frieze patterns are understood to extend indefinitely in both directions, just like a straight line.

The frieze pattern of a figure F necessarily has its generating translation as a symmetry and in addition it inherits all of the symmetries of F. This observation,

Figure 6.14: Chinese frieze patterns.

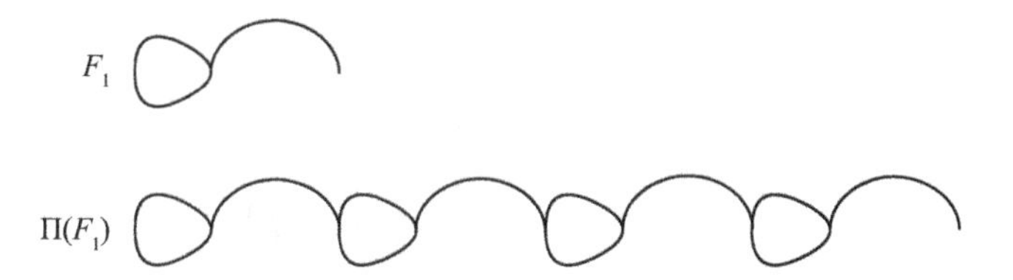

Figure 6.15: A frieze pattern with symmetry group $\Gamma_1 = < \tau >$.

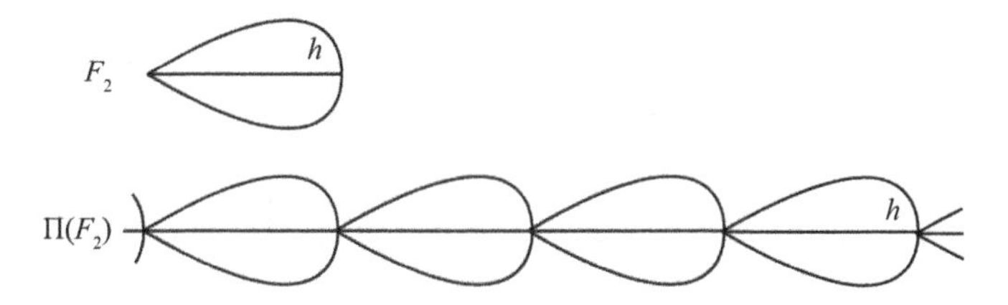

Figure 6.16: A frieze pattern with symmetry group $\Gamma_2 = < \tau, \rho_h, \gamma >$.

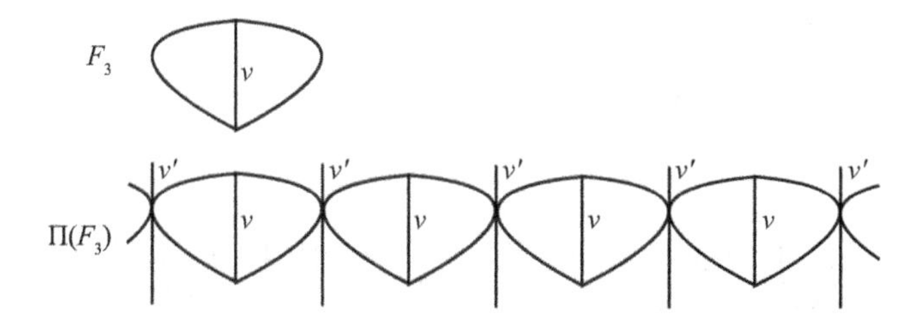

Figure 6.17: A frieze pattern with symmetry group $\Gamma_3 = < \tau, \rho_v >$.

however, does not account for all the symmetries of the frieze pattern. In the case of the figure F_1, there are no other symmetries, and so the pattern's symmetry group is denoted by $\Gamma_1 = < \tau >$, where τ denotes the horizontal translation that generates the infinite frieze pattern $\Pi(F_1)$. This is not meant to imply that τ is the only symmetry of $\Pi(F_1)$. It is clear that the translation 2τ that denotes two consecutive applications of τ is also a symmetry of $\Pi(F_1)$ and similarly, for each positive integer k, the translation $k\tau$, which consists of k iterations of τ, is also a symmetry of $\Pi(F_1)$. Frieze patterns always have infinite symmetry groups and instead of listing their elements it is customary simply to list the types of symmetries they contain.

Figure F_2 possesses the symmetry ρ_h (h for horizontal) which is of course also a symmetry of its frieze pattern $\Pi(F_2)$. In addition, this pattern also necessarily possesses the glide reflection γ obtained by combining t with ρ_h as a symmetry. This frieze's symmetry group is denoted by $\Gamma_2 = < \tau, \rho_h, \gamma >$. The symmetry ρ_v of F_3 results in a multitude of symmetries of the frieze $\Pi(F_3)$ which are all essentially identical. However, this frieze pattern possesses an additional symmetry, namely

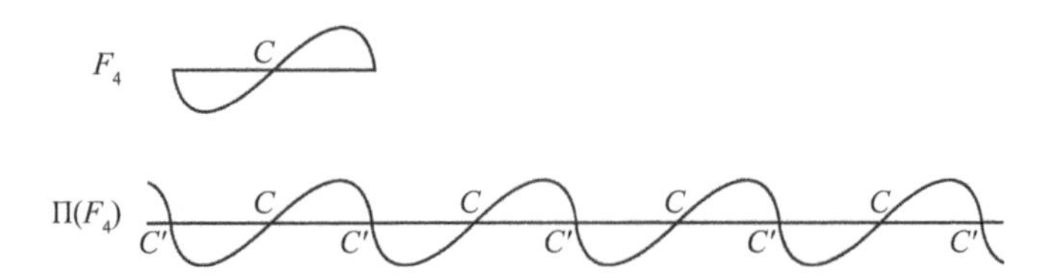

Figure 6.18: A frieze pattern with symmetry group $\Gamma_4 = < \tau, R >$.

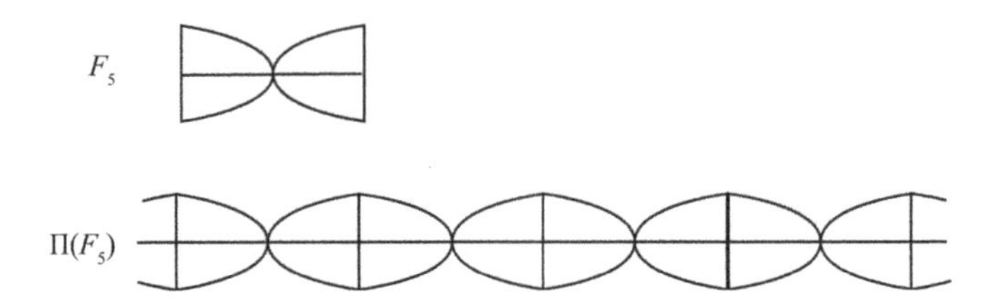

Figure 6.19: A frieze pattern with symmetry group $\Gamma_5 = < \tau, \rho_h, \rho_v, R, \gamma >$.

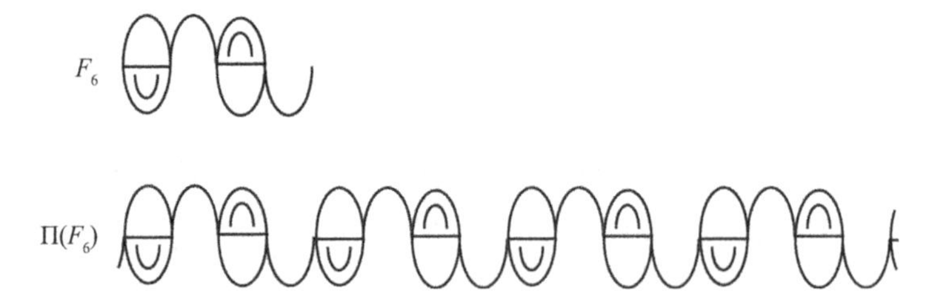

Figure 6.20: A frieze pattern with symmetry group $\Gamma_6 = < \tau, \gamma >$.

the reflection $\rho_{v'}$, which has no counterpart in the generating figure F_3. Because of its similarity to ρ_v, the symmetry $\rho_{v'}$ is not listed in the symmetry group $\Gamma_3 = < \tau, \rho_v >$ of this frieze pattern. Such an additional reflection could not have appeared with the horizontal reflection of F_2, but similar "accidental" symmetries can arise in other cases, as will be seen below. The symmetry $R_{C,180°}$ of the figure F_4 results in the symmetries $R = R_{C,180°}$ of the frieze pattern. Once again the frieze pattern has the additional symmetry $R_{C',180°}$. This frieze pattern's symmetry group is $\Gamma_1 = < \tau, R >$. The figure F_5 possesses all three of the above symmetries, as does the generated frieze pattern. Its symmetry group is $\Gamma_5 = < \tau, \rho_h, \rho_v, R, \gamma >$. This pattern of course also has the additional symmetries described for Γ_3 and Γ_4. Just like Γ_2, the next two patterns have symmetry groups $\Gamma_6 = < \tau, \gamma >$ and $\Gamma_7 = < \tau, \gamma, R >$ that contain glide reflections.

Unlike the glide reflection of Γ_2, those of Γ_6 and Γ_7 do not have their component translation and reflection in the group.

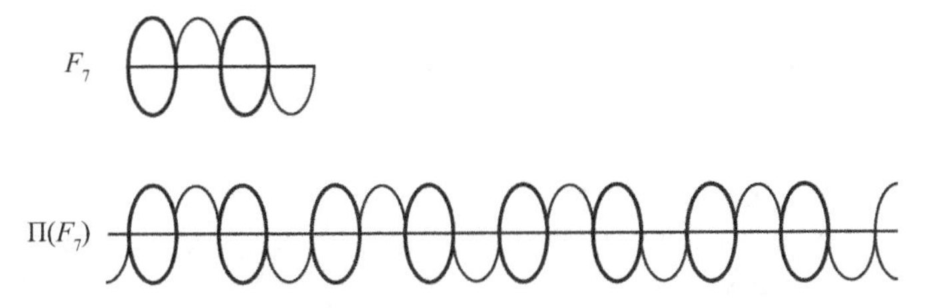

Figure 6.21: A frieze pattern with symmetry group $\Gamma_7 = <\tau, \gamma, R>$.

The following theorem, proved by P. Niggli in 1926, states that these are all the possible symmetry groups that frieze patterns can possess.

Theorem 6.4.1 *Every frieze pattern has a symmetry group that is identical with one of the groups* $\Gamma_1 = <\tau>$, $\Gamma_2 = <\tau, \rho_h, \gamma>$, $\Gamma_3 = <\tau, \rho_v>$, $\Gamma_4 = <\tau, R>$, $\Gamma_5 = <\tau, \rho_h, \rho_v, R, \gamma>$, $\Gamma_6 = <\tau, \gamma>$, $\Gamma_7 = <\tau, \gamma, R>$.

Exercise 6.4

Identify the groups of the following frieze patterns.

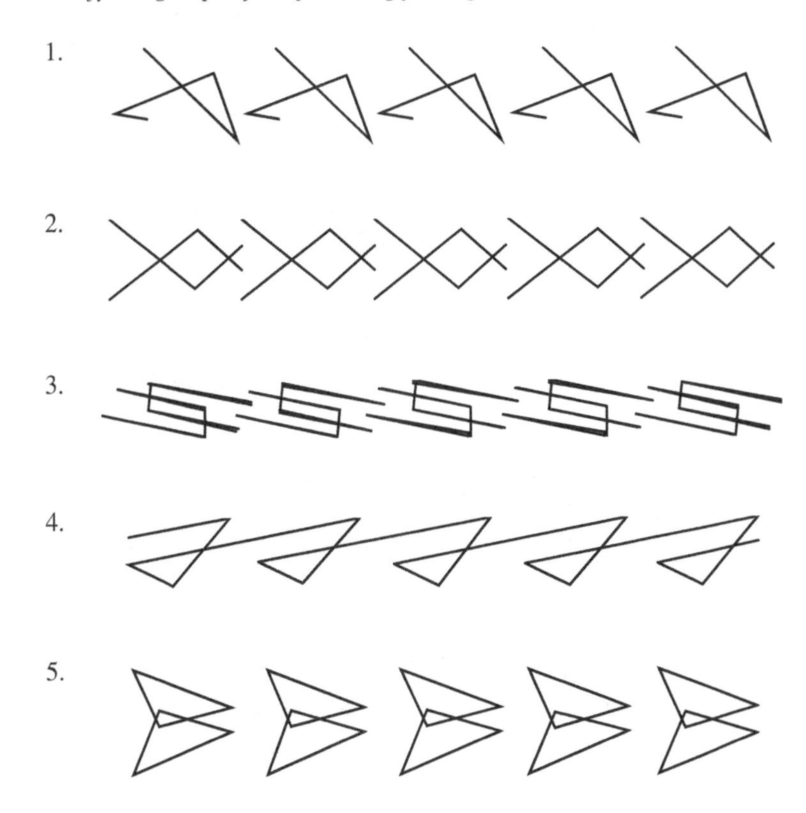

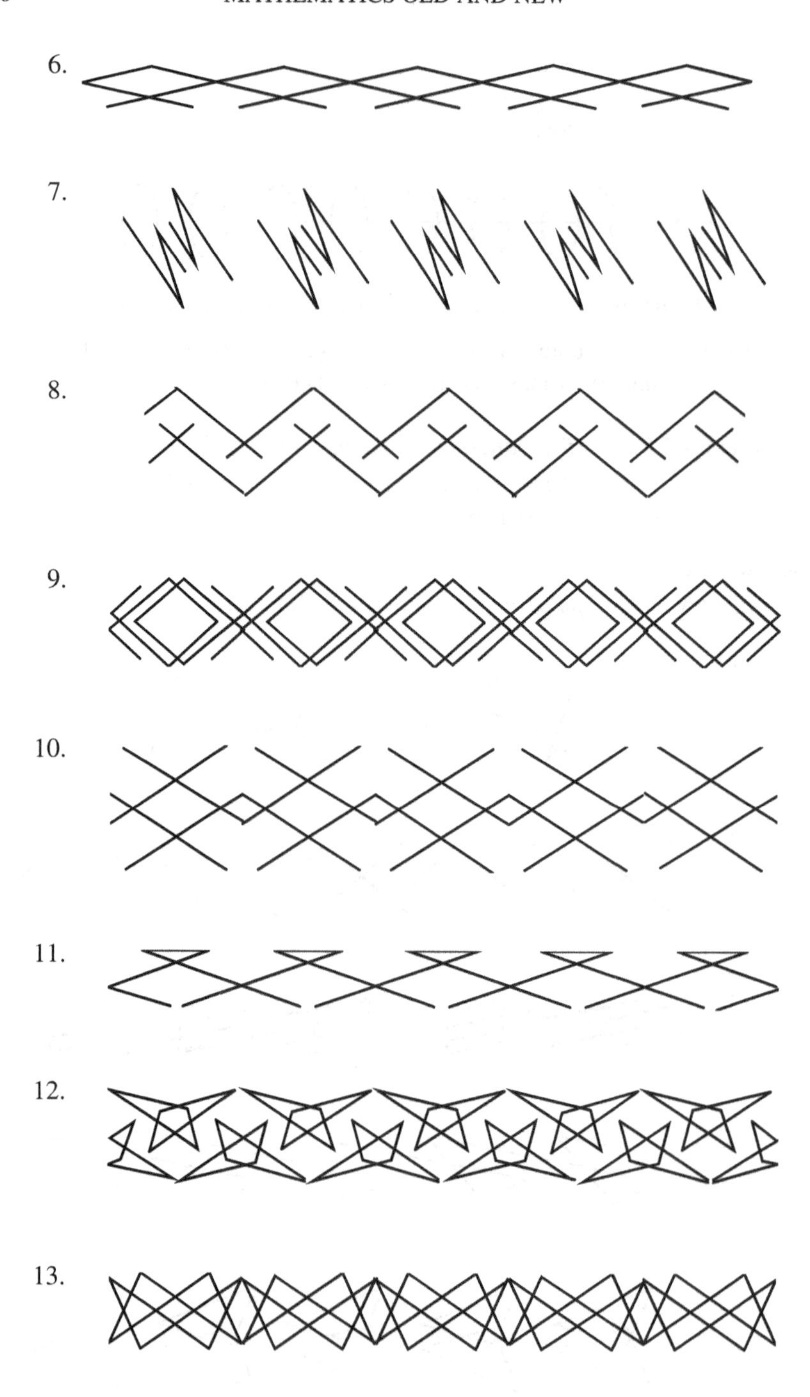
6.
7.
8.
9.
10.
11.
12.
13.

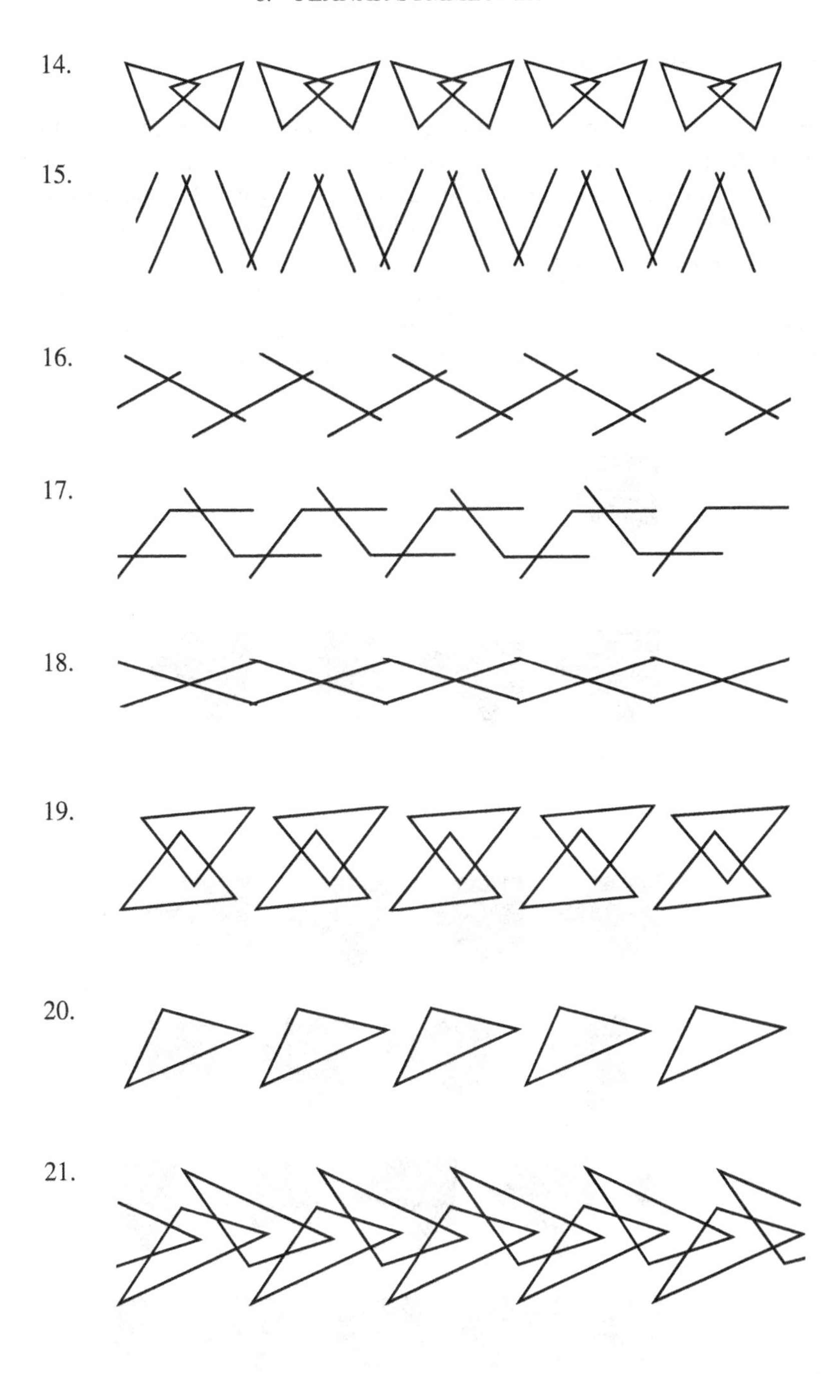
14.
15.
16.
17.
18.
19.
20.
21.

6.5 Wallpaper Designs

Wallpaper designs are the two-dimensional analogs of frieze patterns. More precisely, let $\Pi(F)$ denote the frieze pattern generated by a figure F and a translation τ. If τ^* is another translation whose direction is not parallel to that of τ, then the orbit $\Omega(F) = \Pi(\Pi(F))$ of τ^* generated by $\Pi(F)$ is a *wallpaper design*. In other words, a wallpaper design is an orbit of an orbit.

It is clear that both τ and τ^* are symmetries of the wallpaper design they generate, and an illustration of this appears in Figures 6.22 and 6.23. As was the case for frieze patterns, the generated design $\Pi(F)$ may possess further

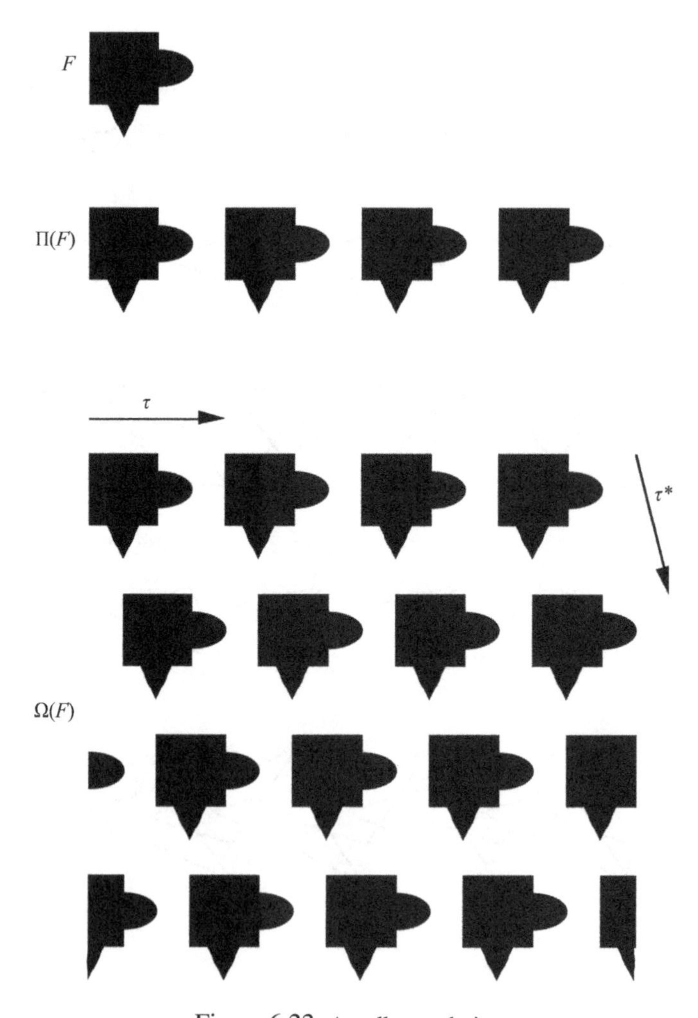

Figure 6.22: A wallpaper design.

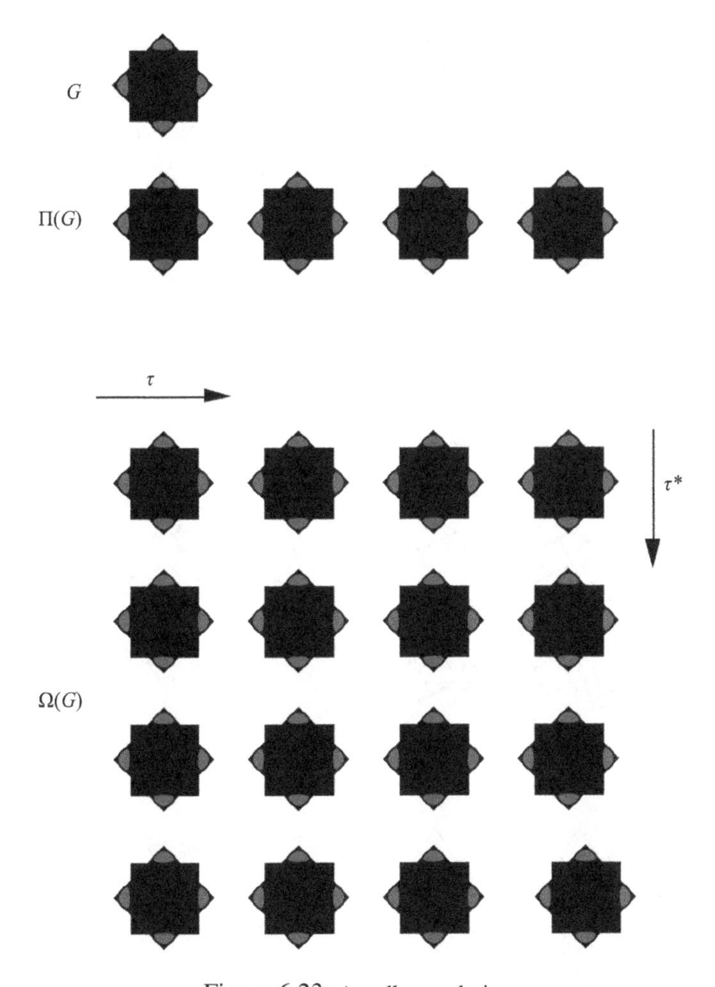

Figure 6.23: A wallpaper design.

symmetries that are not present in F. In contrast with the seven different groups of symmetries of frieze patterns, there are seventeen different possibilities for the symmetry groups of wallpaper designs. Wallpaper designs exemplifying all of these are exhibited in Figures 6.24 – 6.26. In these diagrams the presence of reflections and glide-reflections is denoted by a dashed line with a label of either ρ (for reflection) or γ (for glide translation). The centers of rotational symmetries are denoted by the symbols ◇ (180°), Δ (120°), □ (90°), ⬡ (60°). The following table lists the salient symmetry characteristics of each design. A rotation through an angle of $360°/n$ is said to have *order n*. A glide-reflection is said to be *non-trivial* if its component translation and reflection are not symmetries of the pattern.

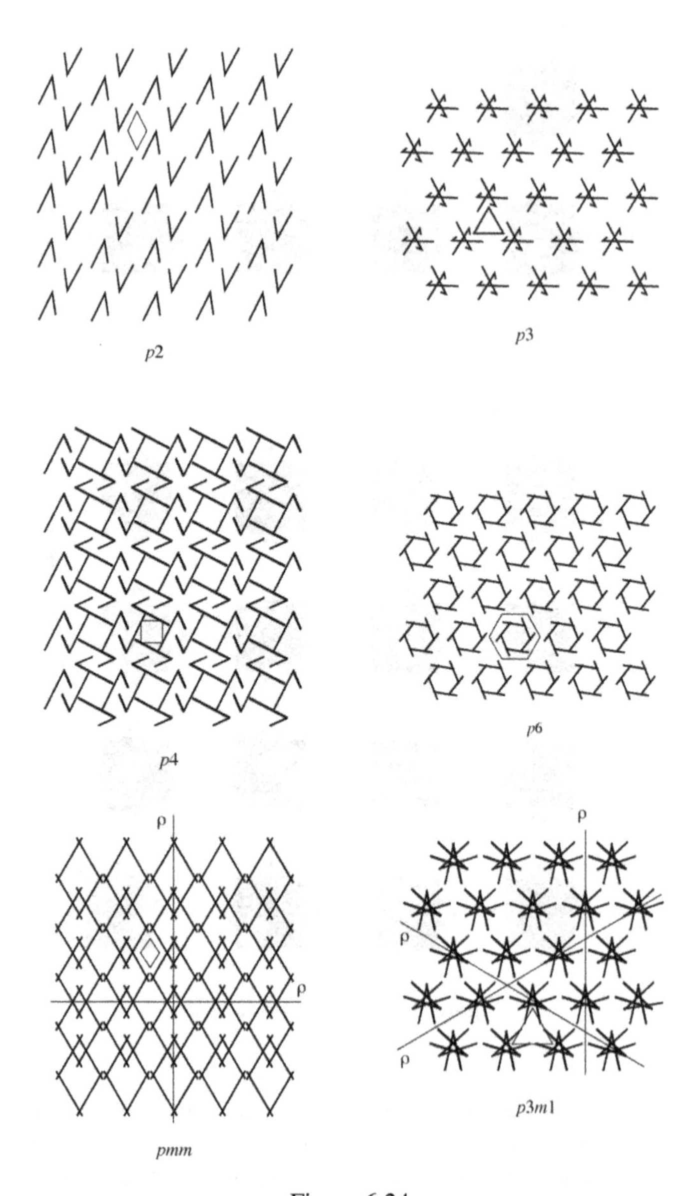

Figure 6.24

The symbols *p1, pgg, p31m,* ... are used to denote both a type of wallpaper design and its symmetry group. They are known as the crystallographic notation for the symmetry groups. If the second character in this symbol is an integer, it is the highest order of all the rotations in that group. The significance of the other characters is too technical to explain here.

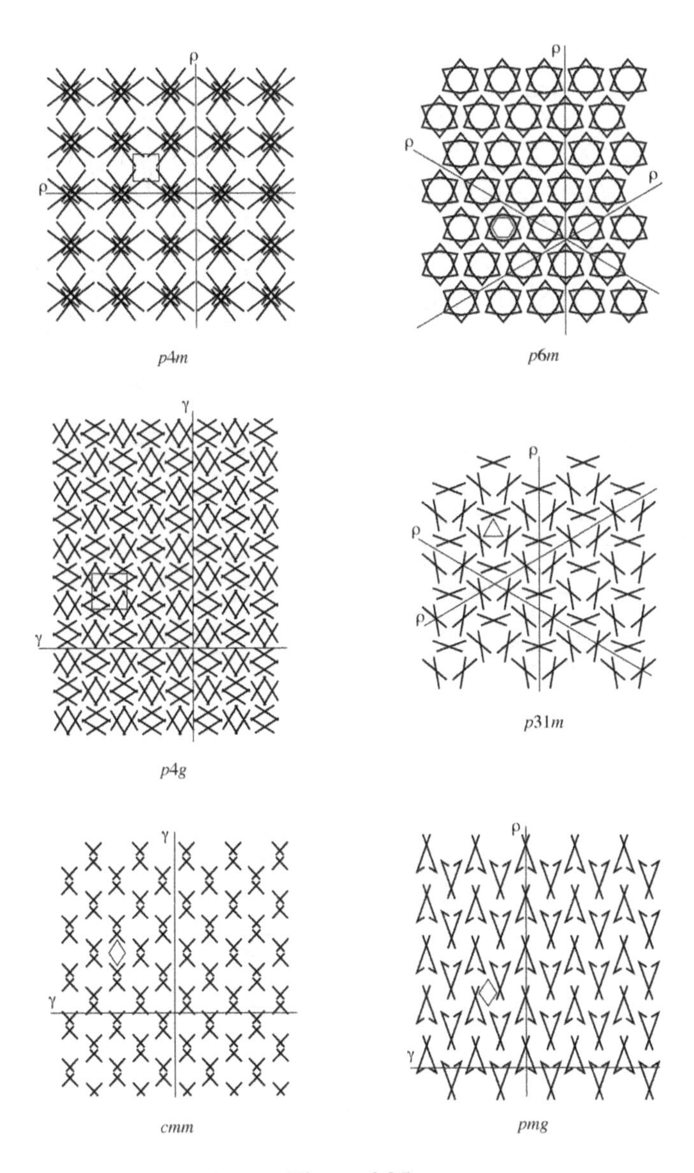

Figure 6.25

Theorem 6.5.1 There are exactly 17 wallpaper groups.

The characteristics of the designs that correspond to these groups are displayed in the table below. The table itself is then followed by Figures 6.26–8 which display one design for each of these groups.

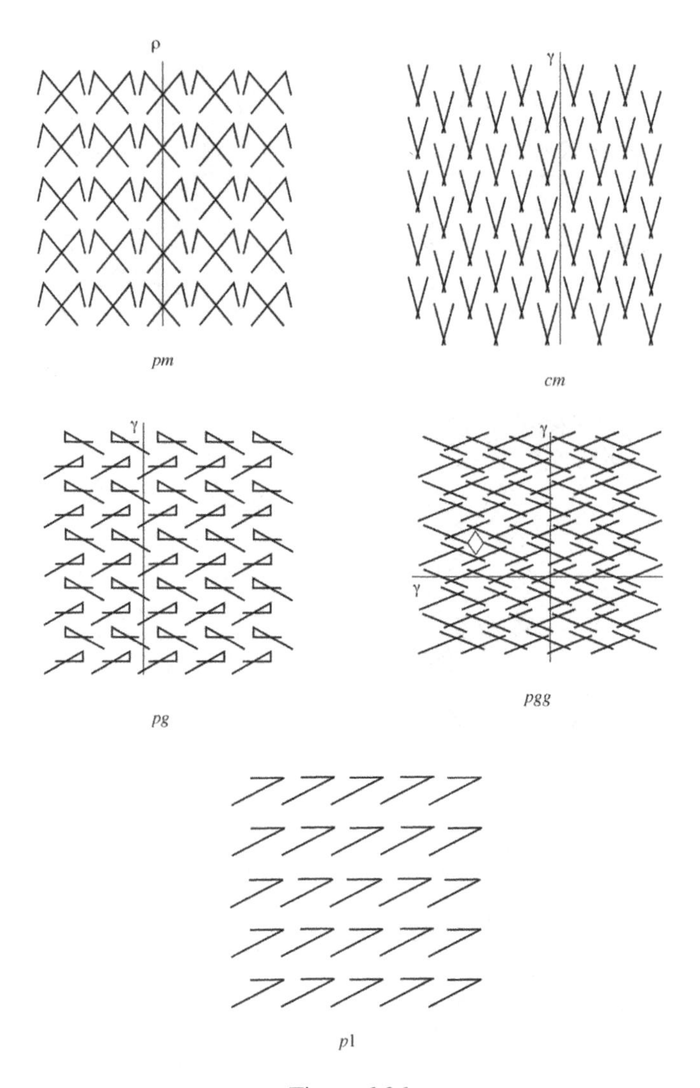

Figure 6.26

Theorem 6.5.1 was first discovered in 1891 by E. S. Fedorov, thirty-five years before Niggli stated and proved its one-dimensional analog on frieze groups. Curiously, this work had been preceded by Fedorov's and Arthur Schönflies's (1853-1928) independent classifications of the 230 crystallographic groups, these being the three-dimensional analogs of the wallpaper groups. It has since then been established that there are exactly 4,783 classes of such groups in four-dimensional space. For spaces of more than four dimensions it is only known that the number of such symmetry groups is finite.

Recognition Chart for Plane Periodic Patterns (From Doris Schattschneider's article ...)				
Type	Highest Order of Rotations	Reflection	Non-Trivial Glide Reflections	Helpful Distinguishing Properties
p1	1	no	no	
p2	2	no	no	
pm	1	yes	no	
pg	1	no	yes	
cm	1	yes	yes	
pmm	2	yes	no	
pmg	2	yes	yes	parallel reflection axes
pgg	2	no	yes	
cmm	2	yes	yes	perpendicular reflection axes
p4	4	no	no	
p4m	4	yes	yes	4-fold centers on reflection axes
p4g	4	yes	yes	4-fold centers not on reflection axes
p3	3	no	no	
p3m1	3	yes	yes	all 3-fold centers on reflection axes
p31m	3	yes	yes	not all 3-fold centers on reflection axes
p6	6	no	no	
p6m	6	yes	yes	

Exercises 6.5

Determine the crystallographic symbol of each of the following wallpaper designs. In each case

(a) display a rotation of the highest order;
(b) denote the presence of a glide reflection by drawing its axis with an accompanying γ;
(c) denote the presence of a reflection by drawing its axis with an accompanying ρ;
(d) avoid redundancy by only drawing only one axis in any direction;
(e) in case you have to choose between a ρ and a γ, display the γ.

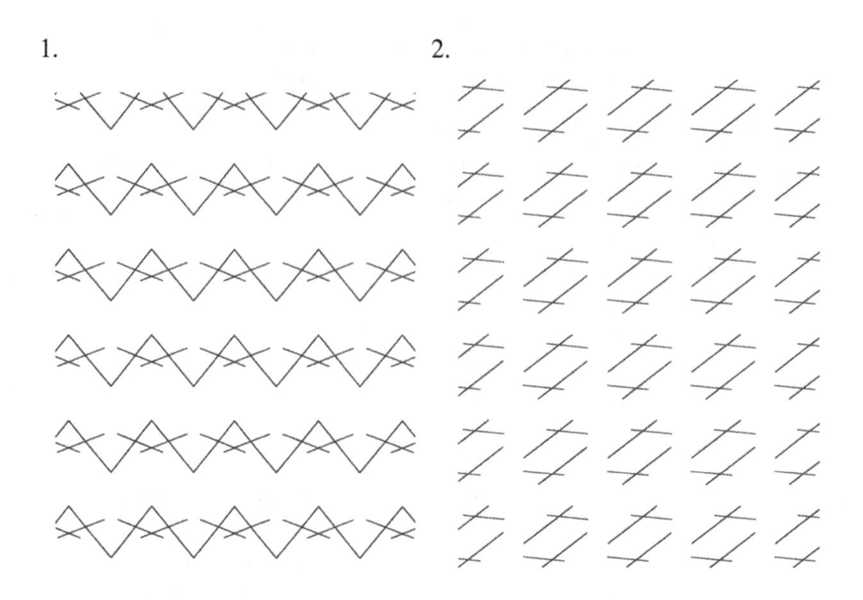
1.
2.

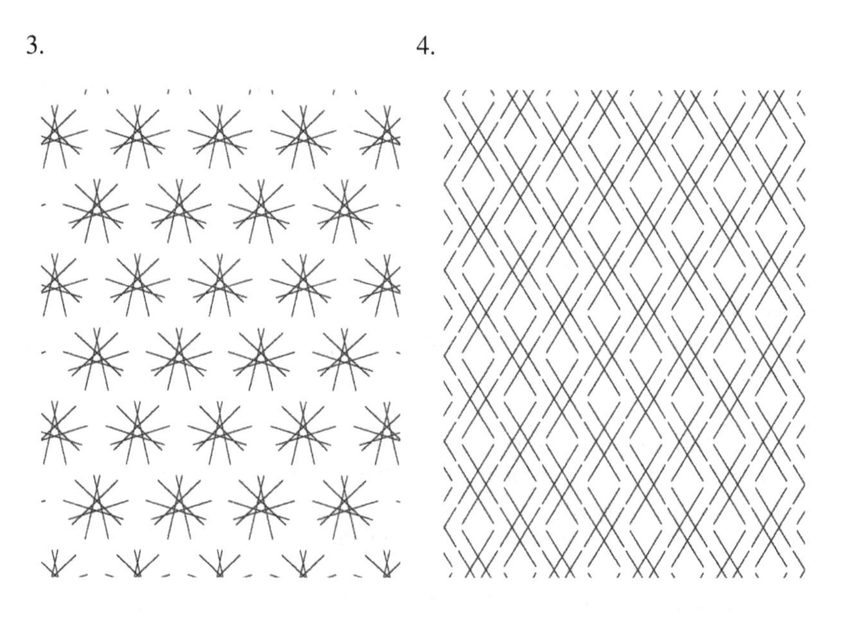
3.
4.

5. 6.

7. 8.

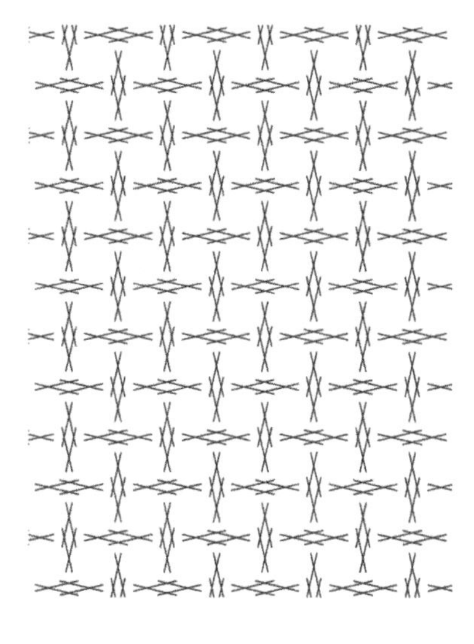

9.

10.

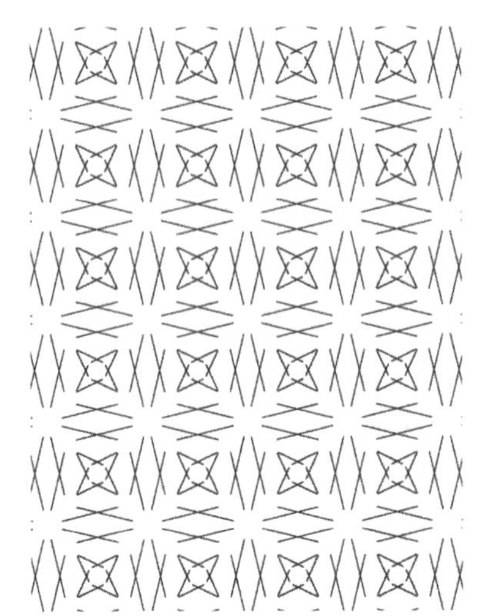

11.

12.

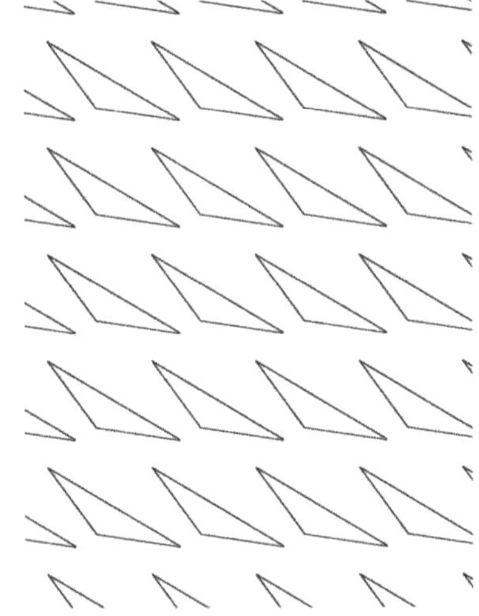

13.

14.

15.

16.

17. 18.

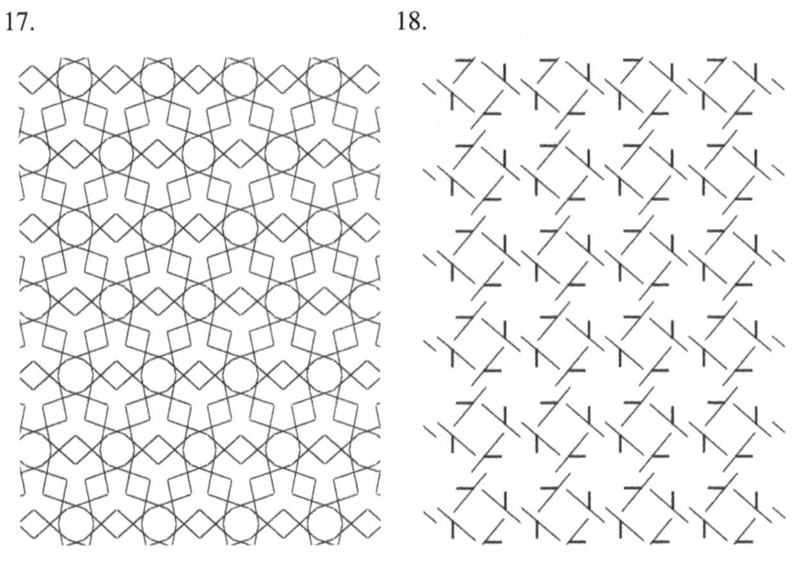

19. 20.

21. 22.

23. 24.

25.

26.

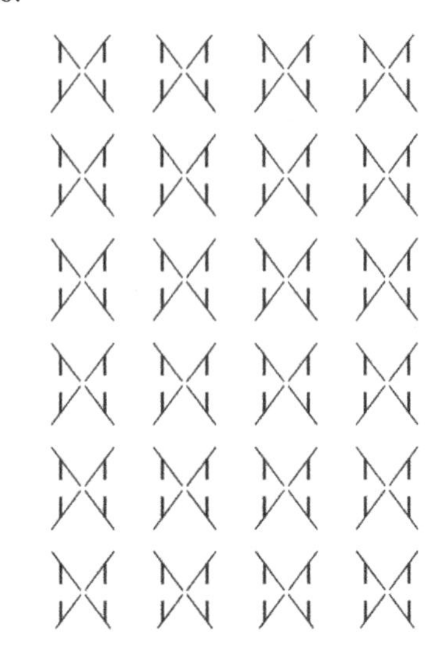

27.

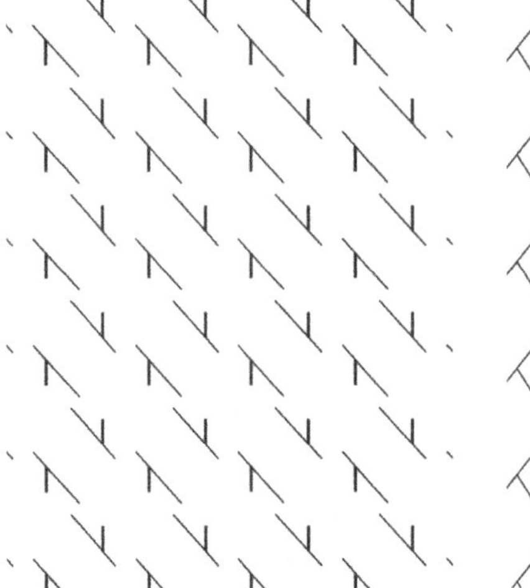

28.

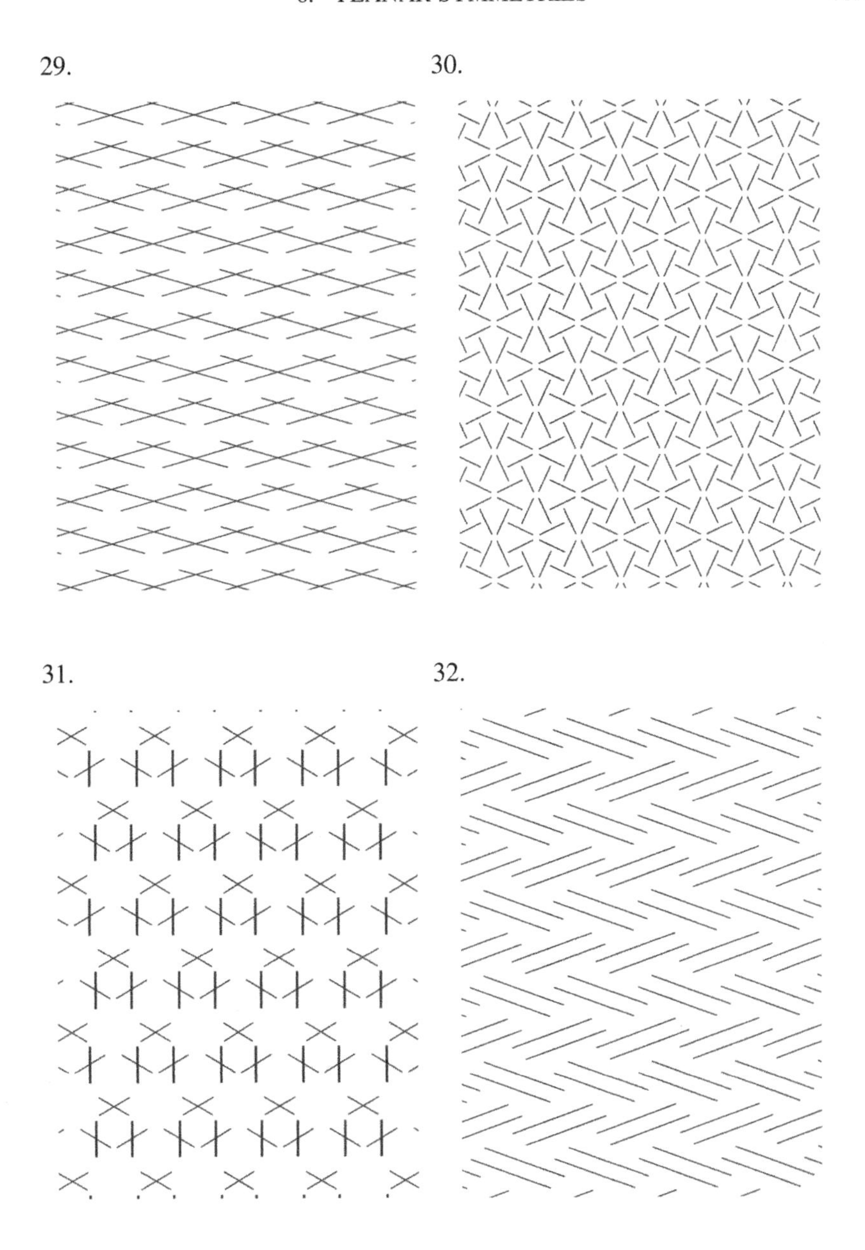
29.
30.
31.
32.

33. 34.

Chapter 7

Spatial Symmetries

The focus now shifts to a discussion of three-dimensional symmetries. Attention is restricted mostly to the symmetries of the five regular solids. The chapter and text conclude with a discussion of some recent discoveries in group theory.

7.1 Regular and Semiregular Solids

As was mentioned above, the last of the thirteen books that comprise Euclid's famed *The Elements* is devoted to the three-dimensional analogs of the regular polygons. These highly symmetric solids are aesthetically pleasing to the eye; they have been used as alternative dice in popular games and also as decorations. They have also proved useful to engineers and scientists.

Our discussion must begin with some definitions. A *polyhedron* is a solid body of finite extent whose surface consists of several polygons, called *faces*. The sides and vertices of these polygonal faces are respectively the *edges* and *vertices* of the polyhedron. The vertices, edges, and faces of a polyhedron are collectively referred to as its *cells*.

A *regular polyhedron* is a polyhedron whose cells satisfy the following constraints:

1. *All the faces are the same regular polygon;*
2. *All the vertices are equivalent in the sense that for any two vertices in positions* u *and* v *there is a rotation of the solid that replaces the vertex*

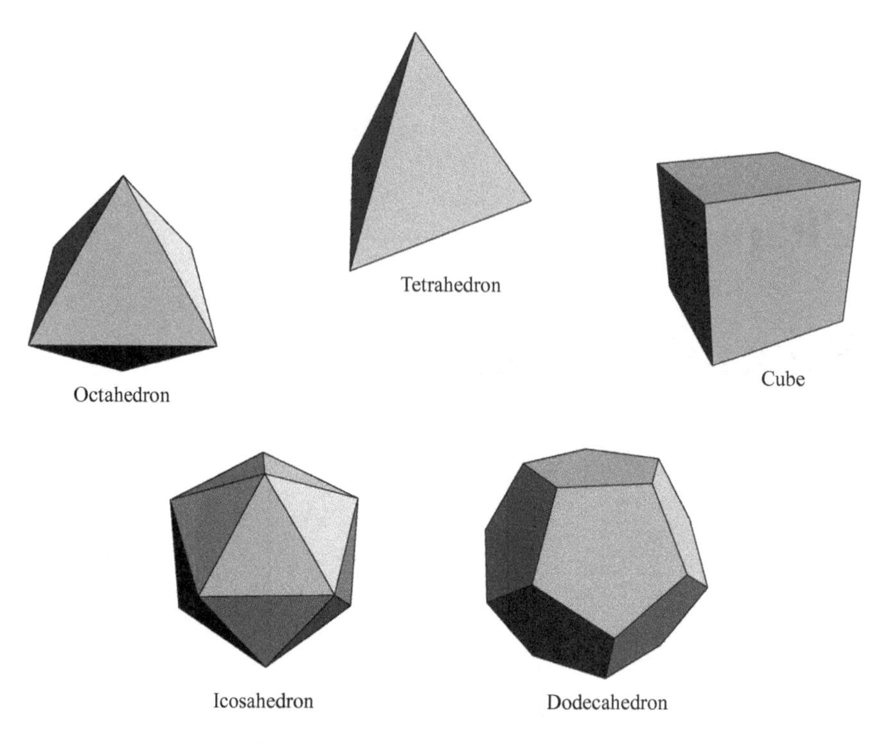

Figure 7.1: The Platonic or regular solids.

at u *with the vertex at* v *and also replaces all the edges emanating from* u *with the edges emanating from* v.

As proved by Euclid in the third century BC, there are five regular polyhedra (see Figure 7.1). However, it is commonly accepted that the Pythagorean school which was active in Greece in the fifth century BC was already aware of all five regular solids two hundred years earlier. The cube, tetrahedron, and octahedron are natural structures and were probably independently constructed in many cultures several millennia ago. The dodecahedron and icosahedron are much less obvious. Theaetetus (415?–369? BC) is credited with being the first mathematician to formally prove their existence.

The easiest regular solid to visualize is of course the *cube* whose faces consist of six congruent squares. It has 12 edges and 8 vertices. Almost as immediate as the cube is the *tetrahedron*, a triangle-based pyramid, whose faces consist of four equilateral triangles. It has 6 edges and 4 vertices. The *octahedron*, a double square-based pyramid, has 8 equilateral triangles as its faces. It has 12 edges and 6 vertices. The *dodecahedron* has 12 regular pentagons as its faces, 30 edges and 20 vertices. The *icosahedron* has 20 equilateral triangles as its faces, 30 edges and 12 vertices. These counts are tallied in Table 7.1.

	v = vertices	e = edges	f = faces	$v - e + f$
Cube	8	12	6	$8 - 12 + 6 = 2$
Tetrahedron	4	6	4	$4 - 6 + 4 = 2$
Octahedron	6	12	8	$6 - 12 + 8 = 2$
Dodecahedron	20	30	12	$20 - 30 + 12 = 2$
Icosahedron	12	30	20	$12 - 30 + 20 = 2$

Table 7.1: Cell counts for the Platonic solids.

The rightmost column of this table indicates that these counts are subject to a very simple and surprising relationship which actually holds for all polyhedra and not just for the regular ones.

Theorem 7.1.1 (Euler's Equation). *For any polyhedron*

$$v - e + f = 2.$$

This equation is named after its discoverer, the Swiss mathematician Leonhard Euler (1707–1783), who was the most prolific mathematician of all time. The discovery of this equation initiated a flourishing branch of mathematics now known as *topology*.

The Greeks also discovered another class of highly symmetrical and intriguing polyhedra. A polyhedron is said to be *semiregular* if

1. *All the faces are regular polygons*;
2. *All the vertices are equivalent in the sense that for any two vertices in positions* u *and* v *there is a rotation of the solid that replaces the vertex at* u *with the vertex at* v *and also replaces all the edges emanating from* u *with the edges emanating from* v.

The definition of a semiregular solid differs from that of a regular one only in that the faces need not all be the same regular polygon. These solids can be divided into three classes of which two contain an infinite number, while the third has only 13 members. These are called, respectively, the prisms (Figure 7.2a), the antiprisms (Figure 7.2b), and the Archimedean solids (Figures 7.3, 7.4).

While some semiregular polyhedra were mentioned by Plato (427-347 BC), their first serious study is commonly attributed to Archimedes (287-212 BC) who is said to have found all of those displayed in Figures 7.2–7.4 and to have demonstrated that no others exist. His work on this topic was lost and it was Johannes Kepler (1571–1630) who reconstructed all the Archimedean polyhedra and discussed their relation to the regular ones. It turns out that some of the Archimedean polyhedra can be derived by truncating the corners of the regular solids. This process is demonstrated here for the cube. In this description the cells of the original cube are referred to as the *old* cells and to those of the derived solid as *new* ones.

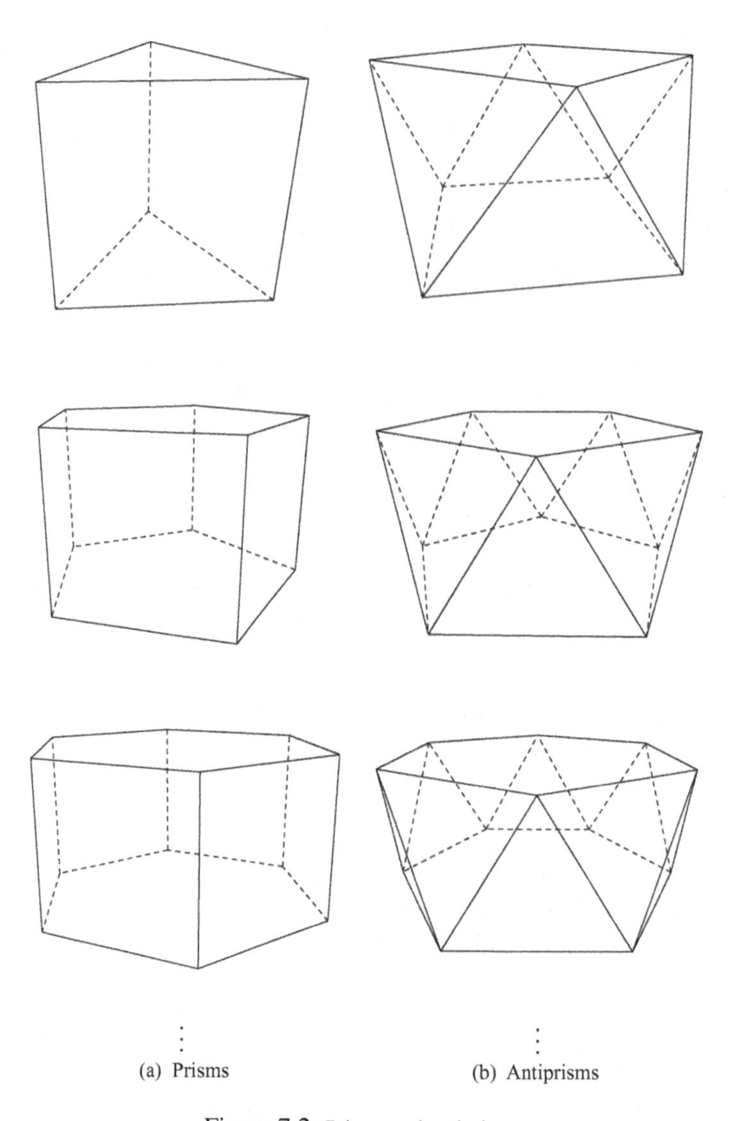

Figure 7.2: Prisms and antiprisms.

Truncated Cube I. All the corners of the cube are cut so that the (central) portion of the old edge that remains has a length which equals that of the edge of the new triangular face created by the truncation process (Figure 7.5). There are 24 new vertices, two for each of the old edges. Each of the 8 old vertices contributes 3 new edges, and there are also the 12 remnants of the old edges. These add up to a total of 36 new edges. Each of the 6 old square faces has been trimmed down to an octagon and the truncation of each of the 8 of the cube's corners has left a

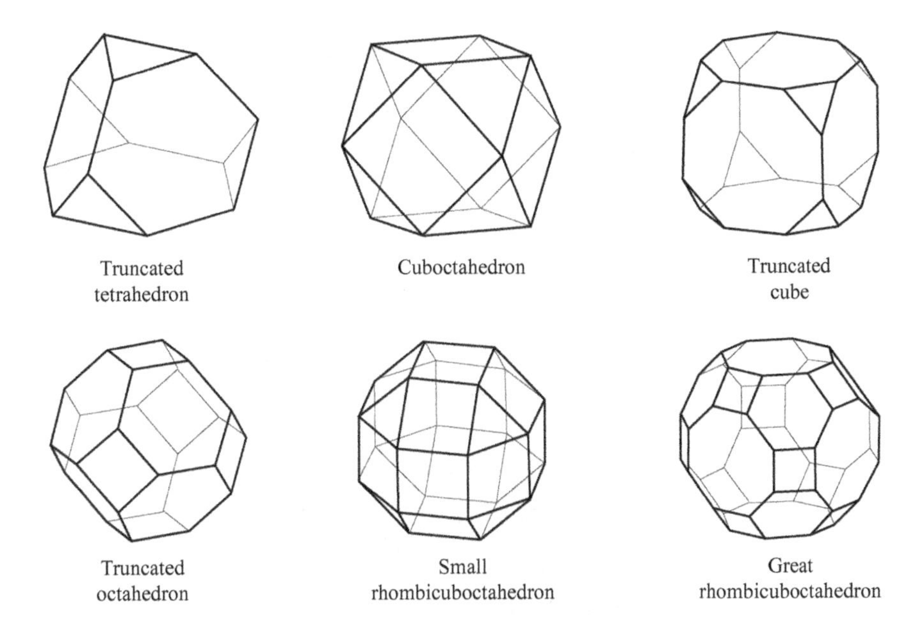

Figure 7.3: Six of the Archimedean solids.

triangular face. Hence this solid has a total of $6 + 8 = 14$ faces. Note that, as predicted by Euler's Theorem,

$$24 - 36 + 14 = 2.$$

Truncated Cube II. All the corners of the cube are cut off in such a manner that the cutting planes meet at the midpoints of the old edges (Figure 7.6). There are 12 new vertices, one for each of the old edges. Each of the 8 old vertices contributes 3 new edges, for a total of 24. Each of the 6 old square faces has been trimmed down to a smaller new square face and each of the 8 truncated corners has left a new triangular face, for a total of $6 + 8 = 14$ new faces. Once again,

$$12 - 24 + 14 = 2,$$

in agreement with Euler's Theorem.

The prisms and antiprisms of Figure 7.2 are offered as further examples of non-regular solids for which Euler's equation holds. Their cell counts satisfy the respective equations

$$6 - 9 + 5 = 2 \qquad 8 - 16 + 10 = 2$$

$$10 - 15 + 7 = 2 \qquad 10 - 20 + 12 = 2$$

$$12 - 18 + 8 = 2 \qquad 12 - 24 + 14 = 2.$$

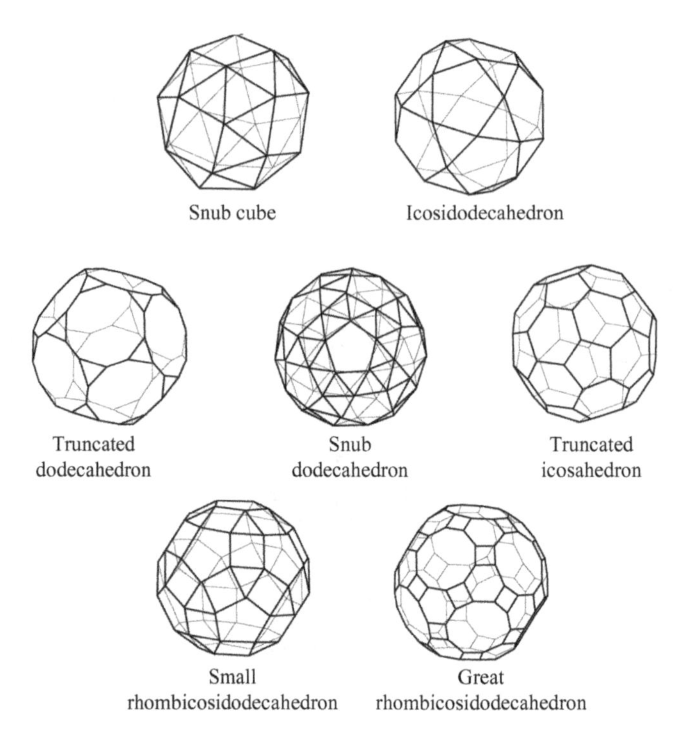

Figure 7.4: Seven of the Archimedean solids.

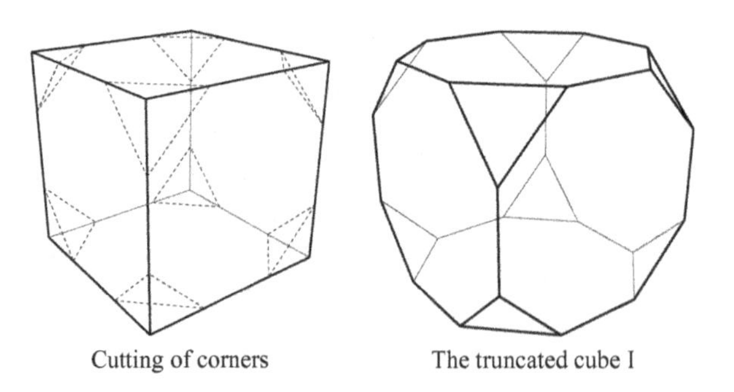

Figure 7.5: One way to truncate a cube.

On the other hand, the double cube of Figure 7.7 has 15 vertices, 24 edges, and 12 faces which yield

$$15 - 24 + 12 = 3 \neq 2.$$

The reason for this is of course that this is not a proper polyhedron. Other exceptions can be easily constructed but they all involve some feature that could call their "solidity" into question.

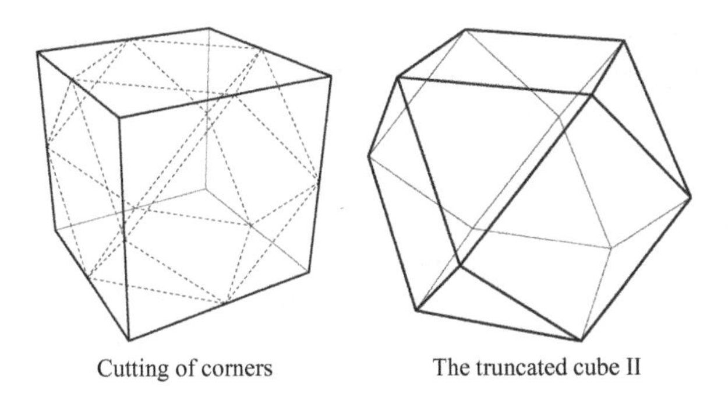

Figure 7.6: Another way to truncate a cube.

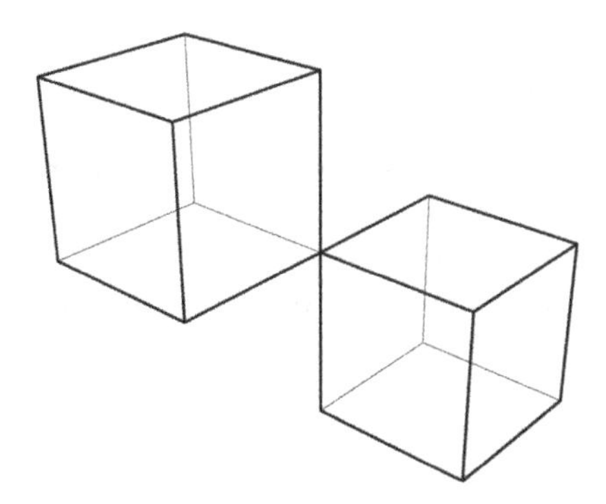

Figure 7.7: A counterexample to Euler's equation.

Exercises 7.1

1. Answer the following questions for each of the solids obtained from the octahedron by the first truncation methods described in Figure 7.5 (parts a, b, c are to be answered without reference to Euler's equation):
   (a) How many vertices does it have?
   (b) How many edges does it have?
   (c) What regular polygons appear as its faces and how many times?
   (d) Identify the truncated solid in Figure 7.3 or in Figure 7.1.
   (e) Verify that the cells of this solid satisfy Euler's equation.
2. Repeat Exercise 1 for the tetrahedron.
3. Repeat Exercise 1 for the dodecahedron.
4. Repeat Exercise 1 for the icosahedron.
5. Answer the following questions for each of the solids obtained from the octahedron by the second truncation methods described in Figure 7.6 (parts a, b, c are to be answered without reference to Euler's equation):
   (a) How many vertices does it have?
   (b) How many edges does it have?

   (c) What regular polygons appear as its faces and how many times?
   (d) Identify the truncated solid in Figures 7.3, 7.4 or in Figure 7.1.
   (e) Verify that the cells of this solid satisfy Euler's equation.
6. Repeat Exercise 5 for the tetrahedron.
7. Repeat Exercise 5 for the dodecahedron.
8. Repeat Exercise 5 for the icosahedron.
9. The truncation procedure that produced the truncated cube I can be applied to arbitrary solids so as to obtain new solids. Without using Euler's equation find the number of vertices, edges, and faces of the solids obtained by applying this procedure to each of the solids below. Also verify Euler's equation for each derived solid.
   (a) the two truncated cubes
   (b) the two truncated tetrahedra
   (c) the two truncated octahedra
   (d) the two truncated dodecahedra
   (e) the two truncated icosahedra
   (f) a solid with v vertices, e edges, and f faces, in which each vertex is incident to d edge
10. The truncation procedure that produced the truncated cube II can be applied to arbitrary solids so as to obtain new solids. Find the number of vertices, edges, and faces of the solids obtained by applying this procedure to each of:
   (a) the two truncated cubes
   (b) the two truncated tetrahedra
   (c) the two truncated octahedra
   (d) the two truncated dodecahedra
   (e) the two truncated icosahedra
   (f) a solid with v vertices, e edges, and f faces, in which each vertex is incident to d edges.
11. Show that the cell counts of the solid of Figure 7.8 do not satisfy Euler's equation. Explain why this is not a counterexample to this equation.

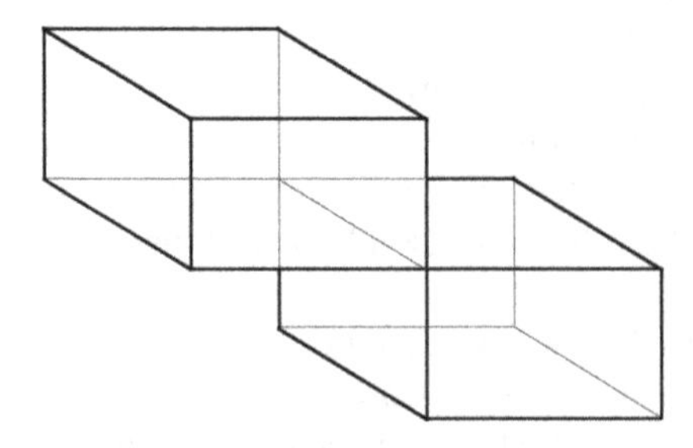

Figure 7.8: This is not a polyhedron.

12. Show that the cell counts of the solid of Figure 7.9 do not satisfy Euler's equation. Explain why this is not a counterexample to this equation.

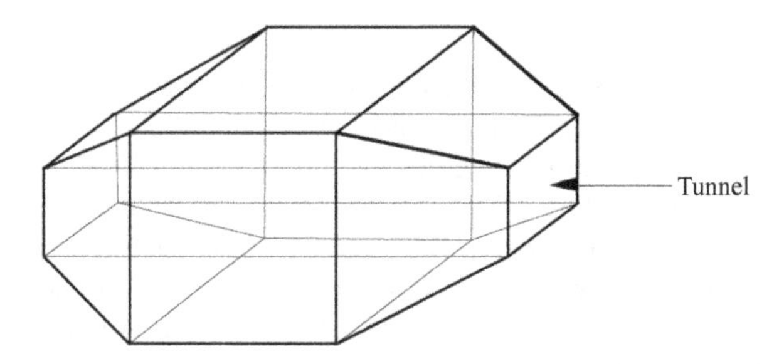

Figure 7.9: This is not a polyhedron.

13. Show that the cell counts of the solid of Figure 7.10 do not satisfy Euler's equation. Explain why this is not a counterexample to this equation.

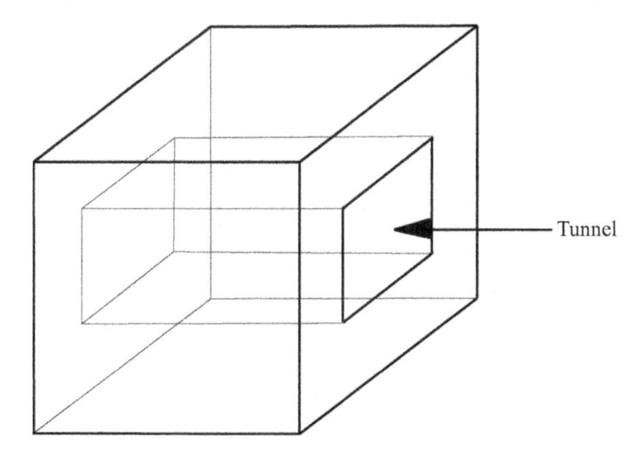

Figure 7.10: This is not a polyhedron.

14. Show that the cell counts of the solid of Figure 7.11 do not satisfy Euler's equation. Explain why this is not a counterexample to this equation.

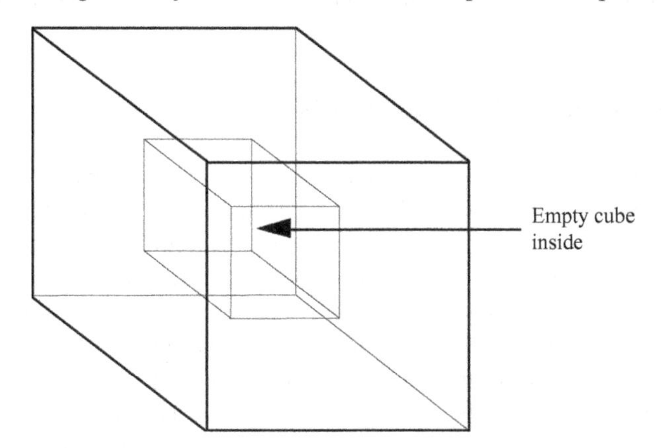

Figure 7.11: This is not a polyhedron.

15. Construct a cube using the medium of your choice.
16. Construct a tetrahedron using the medium of your choice.
17. Construct an octahedron using the medium of your choice.
18. Construct a dodecahedron using the medium of your choice.
19. Construct an icosahedron using the medium of your choice.
20. A paper model of the dodecahedron can be constructed from thirty square sheets of paper (8.5″ by 8.5″ is easy to work with). Each piece should be folded in half and then each half is to be folded in half again, accordion fashion. Next, fold each piece along the dashed lines indicated below, where the two corners are isosceles right triangles. These last three folds should all bend toward you. The pieces are to be tucked into each other as indicated in Figure 7.12 so that short dashed lines fall along long dashed lines.

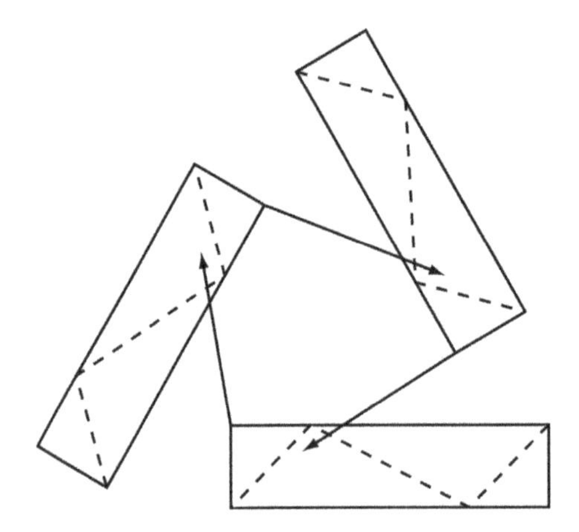

Figure 7.12: A corner of a dodecahedron.

21. Construct all thirteen Archimedean polyhedra using the medium of your choice.

7.2 Symmetries of Regular Solids

A *symmetry* of a polyhedron is a rotation of that solid's ambient space that ends with the solid in its exact original position. During the rotation the polyhedron may very well pass through nearby parts of space that it did not occupy initially, but once the rotation is completely carried out the solid's position must coincide exactly with its initial position. Every such rotation of course has an axis and an (oriented) angle of rotation. The axes of those rotations that constitute symmetries of some polyhedron are constrained by the fact that they must pass through a vertex, the midpoint of an edge, or else the center of some face. This observation is used to label the rotations. Thus, $R_{3,\bullet}$ denotes a symmetry of the cube whose axis passes through the vertex at 3, $R_{23,\bullet}$ denotes a symmetry of the cube whose axis passes

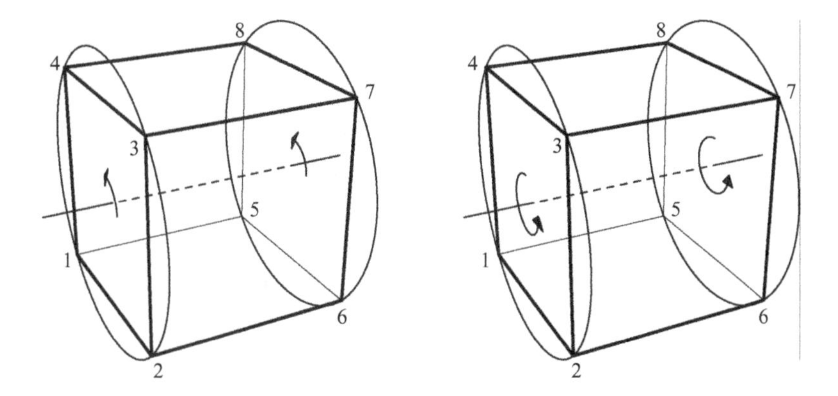

Figure 7.13: $R_{1234,90^\circ} = R_{5678,-90^\circ}$ and $R_{1234,270^\circ} = R_{5678,-270^\circ}$.

through the midpoint of the edge 23, and $R_{3487,\bullet}$ denotes a symmetry of the cube whose axis passes through the center of the square 3487. This notation is subject to some redundancy. Thus, for the cube, the symmetry $R_{3,\bullet}$ can also be written as $R_{5,\bullet}$, the symmetry $R_{23,\bullet}$ can also be written as $R_{58,\bullet}$, and the symmetry $R_{3487,\bullet}$ can also be written as $R_{1265,\bullet}$.

The angle of a rotation is also incorporated into its symbol. Thus $R_{A,\alpha}$ denotes a rotation by the oriented angle α where the orientation is understood to be determined by an observer positioned outside the solid near A. Thus, $R_{1234,90^\circ}$ denotes the 90° rotation of the cube about the axis that passes through the centers of the squares 1234 and 5678, counterclockwise from the point of view of an observer situated outside the cube near the face 1234 (Figure 7.13). Note that since -90° denotes a clockwise rotation it follows that $R_{1234,90^\circ} = R_{5678,-90^\circ}$. The circles in the illustrations are meant to help visualize the rotation; they are the "tracks" in which the vertices move to their new positions.

From Rotations to Permutations. It is convenient to represent rotations by their effect on vertices. Thus, since the rotation $R_{1234,90^\circ}$, cycles the vertices in positions 1, 2, 3, 4 and also the vertices in positions 5, 6, 7, 8, it has the *permutation representation* (1 2 3 4)(5 6 7 8) as well as

(4 1 2 3)(7 8 5 6), (7 8 5 6)(3 4 1 2), (2 3 4 1)(8 5 6 7)

and many others, all of which are considered as the same permutation.

Similarly, the 180° rotation $R_{1234,180^\circ}$ has, amongst others, the following permutation representations:

(1 3)(2 4)(5 7)(8 6), (8 6)(7 5)(1 3)(2 4), (5 7)(6 8)(2 4)(3 1).

A qualitatively different symmetry of the cube is obtained by a 180° rotation about an axis that passes through the midpoints of two diametrically opposite edges of the cube. Such, for example, is the rotation $R_{23,180^\circ} = R_{58,180^\circ}$ of Figure 7.14. It has the permutation representation (1 7)(4 6)(2 3)(5 8). While this permutation

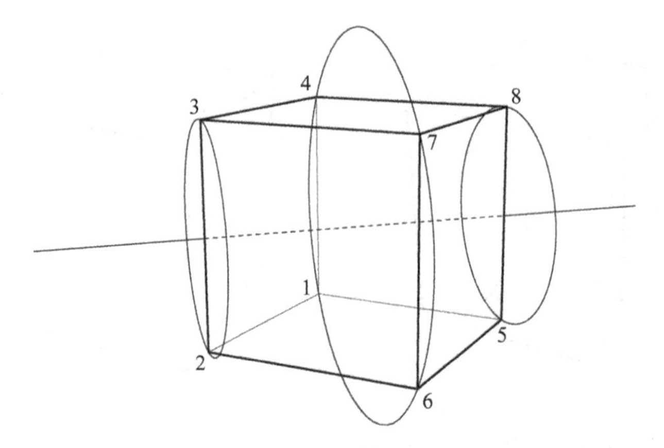

Figure 7.14: $R_{23,180°} = R_{58,180°}$.

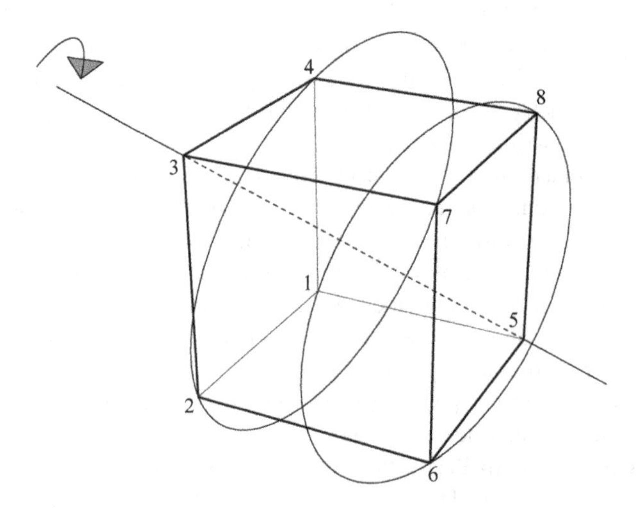

Figure 7.15: $R_{3,120°} = R_{5,240°}$.

looks very much like that of $R_{1265,180°}$ above, there is a significant geometrical difference between them. The permutation representation (1 7)(4 6)(2 3)(5 8) of $R_{23,180°}$ has cycles that are in fact cells of the cube, namely (2 3) and (5 8). On the other hands, none of the cycles of the permutation representation (1 3)(2 4)(5 7)(6 8) of $R_{1234,180°}$ are cells of the cube.

Yet another kind of symmetry is obtained by using a diagonal of the cube as axis. The rotation $R_{3,120°}$ (Figure 7.15) has the diagonal 35 as its axis and the angle of the rotation is 120°. A permutation representation of this rotation is (1 6 8)(2 7 4)(3)(5).

The set of symmetries of a polyhedron is called its *symmetry group* and the number of such symmetries is the *order* of this symmetry group. Note that the

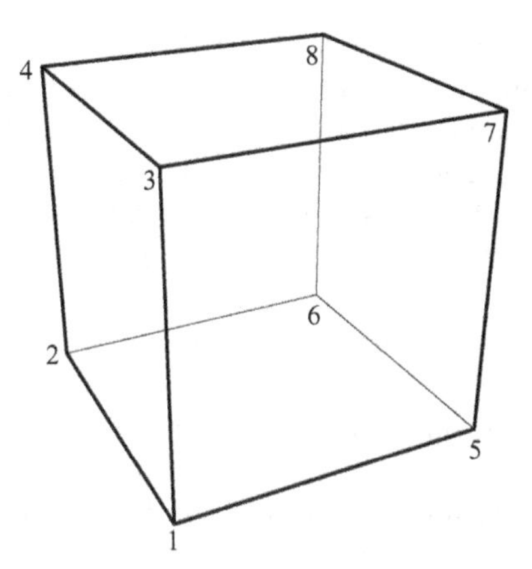

Figure 7.16: A cube.

identity is trivially a symmetry of every polyhedron and so the order of every symmetry group is at least one. The information garnered above about the symmetry group of the cube is now summarized.

Proposition 7.2.1. *The symmetry group of the cube has order 24 and its symmetries are classified as:*

Id
4 symmetries of each of the types $R_{vertex,120^\circ}$ *and* $R_{vertex,-120^\circ}$
6 symmetries of the type $R_{edge,180^\circ}$
3 symmetries of each of the types $R_{face,90^\circ}$, $R_{face,180^\circ}$, $R_{face,-90^\circ}$

From Permutations to Rotations. It will soon be necessary to work in the reverse direction and to identify a rotation from its permutation representation. This is an easy matter if one keeps the following observation in mind:

Proposition 7.2.2. *If a cycle of a permutation representation of a rotation is also a cell of the polyhedron, then the axis of the rotation passes through the center of that cell.*

Consider, for example, the permutation (1 5)(2 8)(3 7)(4 6) of the cube of Figure 7.16. Note that the cycles (1 5) and (3 7) are actually cells of this cube whereas that is not the case for the cycles (2 8) and (4 6). It follows from the above observation that the axis of the corresponding rotation passes through the centers of (1 5) and (3 7) and so it is of the type $R_{15,\bullet}$ or $R_{37,\bullet}$. The angle of the rotation must be 180° because this symmetry must interchange 1 and 5 as well as 3 and 7. This symmetry is therefore $R_{15,180^\circ} = R_{37,180^\circ}$.

Similarly, the permutation (1 3 8)(2 7 5)(4)(6) has the cycles (4) and (6) which correspond to the vertices 4 and 6 of the cube and so its rotation must be of the form $R_{4,\bullet}$ or $R_{6,\bullet}$. As noted in Proposition 7.2.1, the angle of such a rotation is either 120° or −120°. To determine the angle we examine the rotation from the point of view of the vertex 6 since it is easier for us to see things from this point of view rather than that of 4. The neighbors of 6, i.e., the vertices connected to 6 by edges, are rotated according to the cycle (2 7 5), which from the point of view of 6, is a counterclockwise rotation. This permutation therefore represents the rotation $R_{6,120°}$. Alternately, this rotation moves the three neighbors of 4 according to the cycle (1 3 8), which, from the point of view of 4, is a clockwise rotation (this is somewhat harder to visualize, which is the reason it was suggested that we work with the vertex 6 instead). Hence the same rotation can be written as $R_{4,-120°}$.

The only type of rotation for which Proposition 7.2.2 fails to yield the axis directly is that of $R_{1234,180°}$ = (1 3)(2 4)(5 7)(6 8). However, this uniqueness makes it easy to identify rotations of this type. For example, none of the cycles of the permutation (1 8)(2 7)(3 6)(4 5) is a cell of the cube of Figure 7.16. Consequently, this is a rotation of type $R_{abcd,180°}$. Since this rotation interchanges vertices 1 and 8, it must be $R_{1485,180°}$. On the other hand, the permutation (1 7)(2 6)(3 5)(4 8) has two cycles (2 6) and (4 8) which are also cells and hence it is the rotation $R_{26,180°}$.

The Composition of Rotations. The interaction of symmetries is of particular interest to both mathematicians and physicists. These interactions are easily described by means of the permutation representations of the symmetries in question. For this purpose we digress here to discuss the composition of permutations in general. If τ and σ are two permutations (or rotations) then $\tau_o\sigma$ denotes their combined effect then they are applied successively, first τ and then σ, to the same solid. For example, working with $\tau = R_{1265,90°}$ = (1 2 6 5)(4 3 7 8) and $\sigma = R_{2376,180°}$ = (2 7) (3 6)(4 5)(1 8) we have

$$\tau_o\sigma = (1\ 2\ 6\ 5)(3\ 7\ 8\ 4)\text{o}(2\ 7)(3\ 6)(4\ 5)(1\ 8) = (1\ 7)(2\ 3)(4\ 6)(5\ 8)$$

where the computation goes as follows:

τ rotates	σ rotates	so $\tau_o\sigma$ takes
1 to 2	2 to 7	1 to 7
7 to 8	8 to 1	7 to 1
2 to 6	6 to 3	2 to 3
3 to 7	7 to 2	3 to 2
6 to 5	5 to 4	6 to 4
4 to 3	3 to 6	4 to 6
5 to 1	1 to 8	5 to 8
8 to 4	4 to 5	8 to 5

The first two rows of the array above tell us that (1 7) is a cycle of $\tau_o\sigma$. The next six rows, taken two at a time, show that (2 3), (4 6), (5 8) are also cycles of $\tau_o\sigma$. Consequently,

$$\tau_o\sigma = (1\ 7)(2\ 3)(4\ 6)(5\ 8) = R_{23,180^\circ},$$

or

$$(R_{1265,90^\circ})\ \text{o}\ (R_{2376,180^\circ}) = R_{23,180^\circ}$$

Similarly,

$$\begin{aligned}(R_{7,120^\circ})\text{o}(R_{1265,90^\circ}) &= (1)(2\ 4\ 5)(3\ 8\ 6)(7)\text{o}(1\ 2\ 6\ 5)(3\ 7\ 8\ 4)\\ &= (1\ 2\ 3\ 4)(5\ 6\ 7\ 8)\\ &= R_{1234,90^\circ}.\end{aligned}$$

Or,

$$\begin{aligned}(R_{26,180^\circ})\text{o}(R_{34,180^\circ}) &= (1\ 7)(2\ 6)(3\ 5)(4\ 8)\text{o}(1\ 7)(2\ 8)(3\ 4)(5\ 6)\\ &= (1)(2\ 5\ 4)(3\ 6\ 8)(7)\\ &= R_{7,-120^\circ}.\end{aligned}$$

Whereas the axis of any symmetry of the cube joins the midpoints of cells of the same dimension, the tetrahedron presents us with a new alternative. The axis of the symmetry $R_{3,-120^\circ}$ joins the vertex 1 to the center of the triangular face with vertices 1, 2, and 4. This symmetry has the permutation representation (1 4 2)(3). The only qualitatively different symmetry of the tetrahedron is the 180° rotation about the line joining the midpoints of two opposite edges. Such, for example, is $R_{24,180^\circ} = (1\ 3)(2\ 4)$. Note that

$$(R_{3,-120^\circ})\text{o}(R_{24,180^\circ}) = (1\ 4\ 2)(3)\text{o}(1\ 3)(2\ 4) = (1\ 2\ 3)(4) = R_{4,-120^\circ}.$$

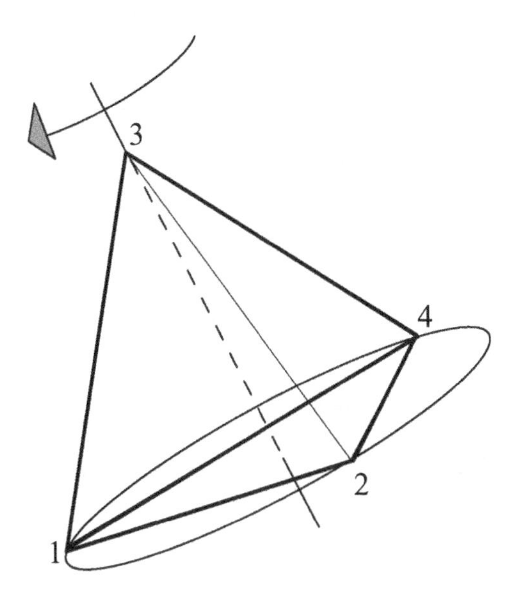

Figure 7.17: $R_{3,240^\circ}$.

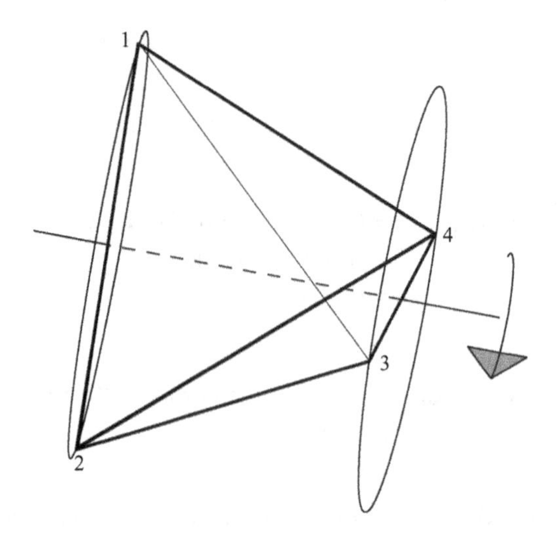

Figure 7.18: $R_{13,180°}$.

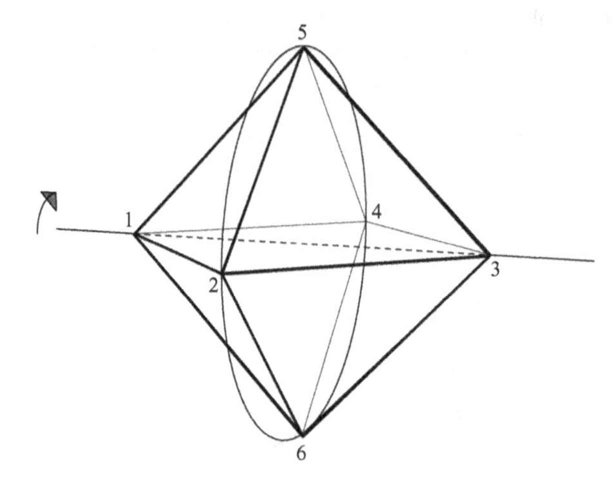

Figure 7.19: The symmetry $R_{1,90°} = (1)(2\ 5\ 4\ 6)(3)$ of the octahedron.

The symmetries of the tetrahedron are illustrated in Figures 7.17–7.18 and hereby summarized.

Proposition 7.2.3. *The symmetry group of the tetrahedron has order 12 and its symmetries are classified as:*

Id
4 symmetries of each of the types $R_{vertex,120°}$ *and* $R_{vertex,240°}$
3 symmetries of the type $R_{edge,180°}$

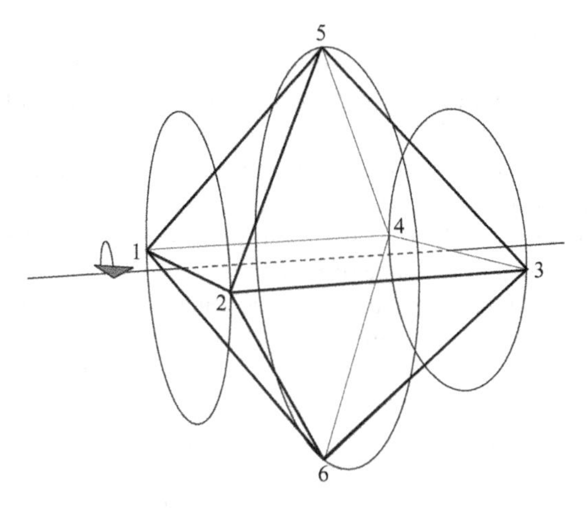

Figure 7.20: The symmetry $R_{12,180°} = (1\ 2)(3\ 4)(5\ 6)$ of the octahedron.

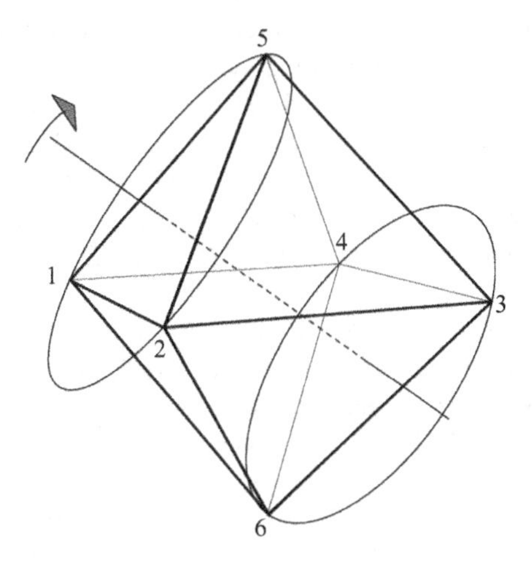

Figure 7.21: The symmetry $R_{152,120°} = (1\ 5\ 2)(3\ 6\ 4)$ of the octahedron.

We close this section with a display of the three types of symmetries of the octahedron (Figures 7.19–7.21).

Exercises 7.2

1. The symmetry group of the octahedron has order 24 and its symmetries are classified as:
   (a) Id
   (b) 3 symmetries of each of the types $R_{\text{vertex},90°}$, $R_{\text{vertex},180°}$, $R_{\text{vertex},-90°}$

(c) 6 symmetries of the type $R_{\text{edge},180°}$
(d) 4 symmetries of each of the types $R_{\text{face},90°}$, $R_{\text{face},180°}$, $R_{\text{face},-90°}$
(e) Using Figures 7.18–7.20, display all the rotations of the octahedron as permutations.

2. Classify the symmetries of the dodecahedron of Figure 7.22.
3. Classify the symmetries of the icosahedron of Figure 7.23.
4. Suppose $A = R_{26,180°}$, $B = R_{1,120°}$, $C = R_{2376,90°}$, $D = \mathrm{R}_{1234,180°}$ are symmetries of the cube of Figure 7.16.
   (a) Find the permutation representations of *A, B, C, D*.
   (b) Identify the following symmetries:

i)	$A \circ A$	ii)	$A \circ B$	iii)	$A \circ C$	iv)	$A \circ D$
v)	$B \circ A$	vi)	$B \circ B$	vii)	$B \circ C$	viii)	$B \circ D$
ix)	$C \circ A$	x)	$C \circ B$	xi)	$C \circ C$	xii)	$C \circ D$
xiii)	$D \circ A$	xiv)	$D \circ B$	xv)	$D \circ C$	xvi)	$D \circ D$

5. Repeat Exercise 4 with $A = R_{15,180°}$, $B, = R_{2,-120°}$ $C = R_{3487,270°}$, $D = R_{5678,180°}$.
6. Suppose $A = R_{24,180°}$, $B = R_{1,120°}$, $C = R_{2,240°}$, $D = R_{14,180°}$ are symmetries of the tetrahedron of Figure 7.17.
   (a) Find the permutation representations of *A, B, C, D*.
   (b) Identify the following symmetries:

i)	$A \circ A$	ii)	$A \circ B$	iii)	$A \circ C$	iv)	$A \circ D$
v)	$B \circ A$	vi)	$B \circ B$	vii)	$B \circ C$	viii)	$B \circ D$
ix)	$C \circ A$	x)	$C \circ B$	xi)	$C \circ C$	xii)	$C \circ D$
xiii)	$D \circ A$	xiv)	$D \circ B$	xv)	$D \circ C$	xvi)	$D \circ D$

7. Repeat Exercise 6 with $A = R_{23,180°}$, $B = R_{3,120°}$, $C = R_{3,240°}$, $D = R_{12,180°}$.
8. Suppose $A = R_{25,180°}$, $B = R_{1,90°}$, $= R_{1,90°}$, $C = R_{235,240°}$, $D = R_{4,180°}$ are symmetries of the octahedron of Figure 7.19.
   (a) Find the permutation representations of *A, B, C, D*.
   (b) Identify the following symmetries:

i)	$A \circ A$	ii)	$A \circ B$	iii)	$A \circ C$	iv)	$A \circ D$
v)	$B \circ A$	vi)	$B \circ B$	vii)	$B \circ C$	viii)	$B \circ D$
ix)	$C \circ A$	x)	$C \circ B$	xi)	$C \circ C$	xii)	$C \circ D$
xiii)	$D \circ A$	xiv)	$D \circ B$	xv)	$D \circ C$	xvi)	$D \circ D$

9. Repeat Exercise 8 with $A = R_{46,180°}$, $B = R_{1,180°}$, $C = \mathrm{R}_{345,120°}$, $D = \mathrm{R}_{6,90°}$.
10. Suppose $A = R_{57,120°}$, $B = R_{5,120°}$, $C = R_{57jbf,72°}$, $D = R_{1d5f2,144°}$ are symmetries of the dodecahedron of Figure 7.22.
   (a) Find the permutation representations of *A, B, C, D*.
   (b) Identify the following symmetries:

i)	$A \circ A$	ii)	$A \circ B$	iii)	$A \circ C$	iv)	$A \circ D$
v)	$B \circ A$	vi)	$B \circ B$	vii)	$B \circ C$	viii)	$B \circ D$
ix)	$C \circ A$	x)	$C \circ B$	xi)	$C \circ C$	xii)	$C \circ D$
xiii)	$D \circ A$	xiv)	$D \circ B$	xv)	$D \circ C$	xvi)	$D \circ D$

11. Repeat Exercise 10 with $A = R_{1,240^\circ}$, $B = R_{bj,180^\circ}$, $C = R_{68iae,144^\circ}$, $D = R_{2fbch,72^\circ}$.

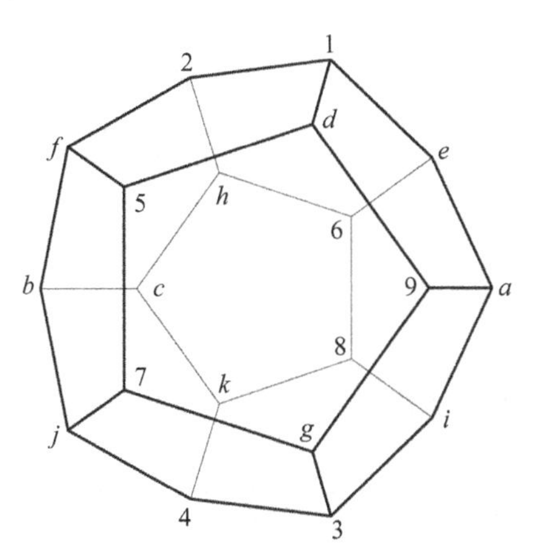

Figure 7.22: A labeled dodecahedron.

12. Suppose $A = R_{1,72^\circ}$, $B = R_{4a,180^\circ}$, $C = R_{349,240^\circ}$, $D = R_{4,144^\circ}$ are symmetries of the icosahedron of Figure 7.23.
   (a) Find the permutation representations of A, B, C, D.
   (b) Identify the following symmetries:

i)	$A \circ A$	ii)	$A \circ B$	iii)	$A \circ C$	iv)	$A \circ D$
v)	$B \circ A$	vi)	$B \circ B$	vii)	$B \circ C$	viii)	$B \circ D$
ix)	$C \circ A$	x)	$C \circ B$	xi)	$C \circ C$	xii)	$C \circ D$
xiii)	$D \circ A$	xiv)	$D \circ B$	xv)	$D \circ C$	xvi)	$D \circ D$

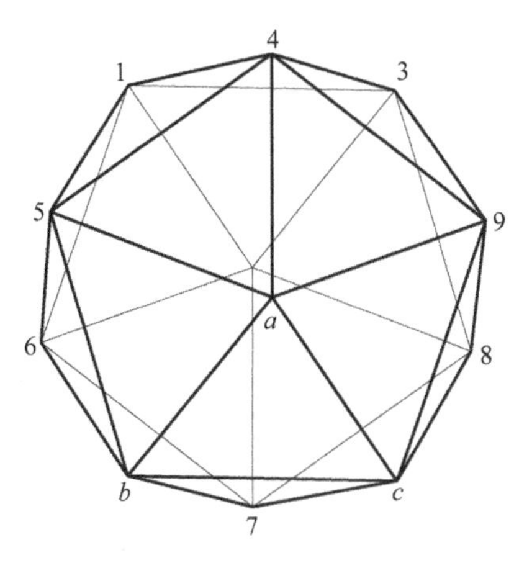

Figure 7.23: A labeled icosahedron.

13. Repeat Exercise 12 with $A = R_{1,144°}$, $B = R_{89,180°}$, $C = R_{126,120°}$, $D = R_{a,144°}$.
14. None of the faces of a rectangular box are square. How many symmetries does it have?
15. Two of the faces of a rectangular box are square. How many symmetries does it have?
16. How many symmetries does a triangular prism with an equilateral base have?
17. How many symmetries does a triangular prism have if its base has sides 6, 6, 4?
18. How many symmetries does a prism have if its base is a regular n-gon?

7.3 Monstrous Moonshine

The previous section described the symmetries of the cube and the octahedron separately. Those of the cube were listed in Proposition 7.2.1 whereas those of the octahedron were relegated to Exercise 7.2.1. Superficially, the symmetries of the cube are permutations of eight vertices whereas those of the octahedron are permutations of six vertices. Nevertheless, there is a sense in which these two groups are identical. Observe that in Figure 7.24 a cube has been placed inside an octahedron so that each of the vertices of the first is the center of a square face of the latter. The possibility of this placement implies that every $R_{\text{face},\bullet}$ symmetry of the cube is also an $R_{\text{vertex},\bullet}$ symmetry of the octahedron. Similarly, every $R_{\text{edge},\bullet}$ symmetry of the cube is also an $R_{\text{edge},\bullet}$ symmetry of the octahedron and every $R_{\text{vertex},\bullet}$ symmetry of the cube is an $R_{\text{face},\bullet}$ symmetry of the octahedron. Thus, Figure 7.24 reveals that the symmetry groups of the cube and the octahedron are identical. Technically, they are said to be *isomorphic*. It is clear that isomorphic groups must have the same order and so the symmetry groups of the cube and the tetrahedron are non-isomorphic, as they have groups of orders 24 and 12, respectively. Groups of the same order need not be isomorphic either. This is demonstrated by the symmetry

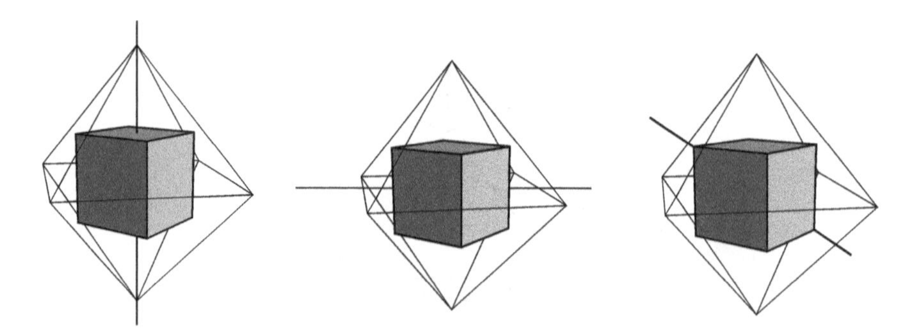

Figure 7.24: An isomorphism of symmetry groups.

group of the regular 12-gon. Just like the cube, this polygon has 24 symmetries. However, the 30° rotation of this polygon is a symmetry of order 12 and no symmetry of the cube has such an order. Hence the symmetry groups of the regular 12-gon and the cube are non-isomorphic, even though they have the same orders.

Group theory, the mathematical theory of symmetry, has its origins in the work of Joseph Louis Lagrange (1736–1813) on the theory of equations. It was later used by Niels Henrik Abel (1802–1829) and Évariste Galois (1811–1832) to settle the question of which equations could be solved by explicit algebraic formulas and which could only be solved by means of successive approximations. It was the investigations of Felix Klein (1849–1925) and Henri Poincaré (1854–1912) that pointed out the central role that symmetry plays in geometry.

One of the main targets of group theory is the classification of all groups up to isomorphism. While there is no expectation that this goal will be achieved in the foreseeable future, a significant milestone was passed less than twenty years ago when the *finite simple groups* were completely classified. There is nothing simple about the simple groups, nor is it possible to characterize them in this text. The symmetries of the icosahedron (and the dodecahedron) constitute a simple group whereas the groups of the cube and tetrahedron are not simple, but this difference does not have a geometrical interpretation. Algebraically, though, the difference is extremely important. The simplicity of the dodecahedral group turns out to be responsible for the non-existence of a formulaic solution for the general fifth degree equation

$$ax^5 + bx^4 + cx^3 + dx^2 + ex + f = 0.$$

Conversely, the non-simplicity of the rotation groups of the octahedron, cube and tetrahedron implies the existence of such a formula for the fourth degree equation

$$ax^4 + bx^3 + cx^2 + dx + e = 0.$$

The classification of the finite simple groups constitutes the most monumental task ever accomplished by mathematicians. Its proof is spread over 500 articles comprising more than 14,000 journal pages written by hundreds of researchers. This classification asserts that the finite simple groups fall into two categories: several infinite families of groups that possess clear patterns and 26 exceptional groups, known as the *sporadic groups*, for which no general pattern has been found.

The first of the sporadic groups was discovered in 1861 and the last two almost simultaneously in 1980. The largest of these was nicknamed MONSTER because of its order which is

808, 017, 424, 794, 512, 875, 886, 459, 904, 961, 710, 757, 005, 754, 368, 000, 000, 000.

MONSTER, discovered by Bernd Fischer and Robert L. Griess, is the group of symmetries of a (non-regular) solid in 196,883 dimensions. When word of this discovery reached John McKay, he pointed out the remarkable coincidence that the number 196,884 plays an important role in the theory of the patterns that underlie non-Euclidean wallpaper designs. As mathematicians were at a loss to explain this conjunction they dubbed the following equation as McKay's Formula:

$$196{,}884 = 1 + 196{,}883.$$

John H. Conway assigned the name Moonshine to this and other related unexplained phenomena in 1979 "... intending the word to convey our feelings that they are seen in a dim light, and that the whole subject is rather vaguely illicit." It should be remembered that at that time the existence of MONSTER had only been conjectured so that even the number 196,883 was questionable, not to mention its purported relation with non-Euclidean geometry.

The existence of MONSTER was conclusively demonstrated in 1980 by R. Griess who tried, unsuccessfully, to change its name to *The Friendly Giant.* Monstrous Moonshine Mathematics was finally explained by Richard E. Borcherds who found the connection in the theory of Vertex Algebras, a discipline developed recently for the purpose of providing a mathematical foundation to the new and controversial Superstring Theory of physics. For this work Borcherds received the 1998 Fields Medal, the most prestigious award bestowed by the mathematical community.

7.4 Activities/Websites

1. Construct dodecahedra in class as described in Exercise 7.1.20. Have each student prepare a dozen folded strips before class and let them work in groups of three.
2. Use the Triangle Tiler at the Geometry Center at the ScienceU.com website to draw the five regular solids.
3. Use the Triangle Tiler at the Geometry Center at the ScienceU.com website to draw all the semiregular solids that can be obtained from the regular ones by truncation.

Chapter 8

Bell-Shaped Curve

Statistics is the most widely applied of all mathematical disciplines and at the center of statistics lies the normal distribution, known to millions of people as the bell curve, or the bell-shaped curve. This is actually a two-parameter family of curves that are graphs of the equation

$$y = \frac{1}{\sqrt{2\pi}\sigma} e^{-\frac{1}{2}\left(\frac{x-\mu}{\sigma}\right)^2}. \tag{1}$$

Several of these curves appear in Figure 8.1. Not only is the bell curve familiar to these millions, but they also know of its main use: to describe the general, or idealized, shape of graphs of data. It has, of course, many other uses and plays as significant a role in the social sciences as differentiation does in the natural sciences. As is the case with many important mathematical concepts, the rise of this curve to its current prominence makes for a tale that is both instructive and amusing. Actually, there are two tales here: the invention of the curve as a tool for computing probabilities and the recognition of its utility in describing data sets.

8.1 An Approximation Tool

The origins of the mathematical theory of probability are justly attributed to the famous correspondence between Fermat and Pascal, which was instigated in

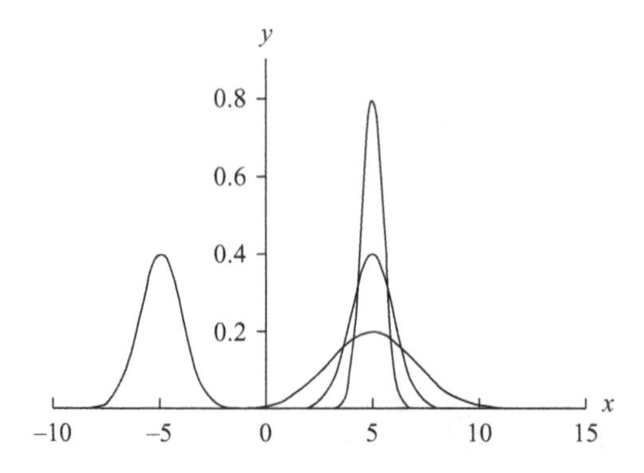

Figure 8.1: Bell-shaped curves.

1654 by the queries of the gambling Chevalier de Mèrè. Among the various types of problems they considered were binomial distributions which today would be described by such sums as

$$\sum_{k=i}^{j} \binom{n}{k} p^k (1-p)^{n-k} \tag{2}$$

that denotes the likelihood of between i and j successes in trials of probability p. As the examples they worked out involved only small values of n, they were not concerned with the computational challenge that the evaluation of sums of this type presents in general. However, more complicated computations were not long in coming. For example, in 1712 the Dutch mathematician 's Gravesande tested the hypothesis that male and female births are equally likely against the actual births in London over the 82 years 1629-1710 [14, 16]. He noted that the relative number of male births varies from a low of 7765/15,448 = 0.5027 in 1703 to a high of 4748/8855 = 0.5362 in 1661. 's Gravesande multiplied these ratios by 11,429 – the average number of births over this 82 year span. These gave him nominal bounds of 5745 and 6128 on the number of male births in each year. Consequently, the probability that the observed excess of male births is due to randomness alone is the 82nd power of

$$\Pr\left[5745 \leq x \leq 6128 \middle| p = \frac{1}{2}\right] = \sum_{x=5745}^{6128} \binom{11{,}429}{x} \left(\frac{1}{2}\right)^{11{,}429}$$

$$\approx \frac{3{,}849{,}150}{12{,}196{,}800} \approx 0.292.$$

's Gravesande did make use of the recursion

$$\binom{n}{x+1} = \binom{n}{x}\frac{n-x}{x+1}$$

suggested by Newton for similar purposes, but even so this is clearly an onerous task. Since the probability of this difference in birth rates recurring 82 years in a row is the extremely small number 0.292^{82}, 's Gravesande drew the conclusion that the higher male birth rates were due to divine intervention.

In the following year, James and Nicholas Bernoulli found estimates for binomial sums of the type of (2). These estimates, however, did not involve the exponential function e^x. De Moivre began his search for such approximations in 1721. In 1733, he proved that

$$\binom{n}{\frac{n}{2}+d}\left(\frac{1}{2}\right)^n \approx \frac{2}{\sqrt{2\pi n}}e^{-2d^2/n} \tag{3}$$

and

$$\sum_{|x-n/2|\leq d}\binom{n}{x}\left(\frac{1}{2}\right)^n \approx \frac{4}{\sqrt{2\pi}}\int_0^{d/\sqrt{n}} e^{-2y^2}dy. \tag{4}$$

De Moivre also asserted that the above (4) could be generalized to a similar asymmetrical context. This is, of course, quite clear, and a proof is required only to clarify the precision of the approximation. Figure 8.2 demonstrates how the binomial probabilities associated with 50 independent repetitions of a Bernoulli trial with probability $p = 0.3$ of success are approximated by such an exponential curve. (A Bernoulli trial is the most elementary of all random experiments. It has two outcomes, usually termed *success* and *failure*.) De Moivre's discovery is

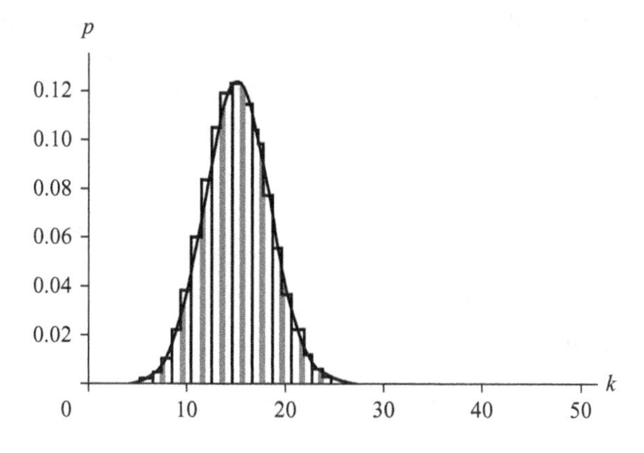

Figure 8.2: An approximation of binomial probabilities.

standard fare in all introductory statistics courses where it is called the normal approximation to the binomial and rephrased as

$$\sum_{i}^{j}\binom{p}{k}p^{k}(1-p)^{n-k}\approx N\left(\frac{j-np}{\sqrt{np(1-p)}}\right)-N\left(\frac{i-np}{\sqrt{np(1-p)}}\right)$$

where

$$N(z)=\frac{1}{\sqrt{2\pi}}\int_{-\infty}^{z}e^{-x^{2}/2}dx.$$

Since the above integral is easily evaluated by numerical methods and quite economically described by tables, it does indeed provide a very practical approximation for cumulative binomial probabilities.

8.2 The Search for an Error Curve

Astronomy was the first science to call for accurate measurements. Consequently, it was also the first science to be troubled by measurement errors and to face the question of how to proceed in the presence of several distinct observations of the same quantity. In the 2nd century BCE Hipparchus seems to have favored the midrange. Ptolemy, in the 2nd century CE, when faced with several discrepant estimates of the length of a year, may have decided to work with the observation that best fit his theory [29]. Toward the end of the sixteenth century, Tycho Brahe incorporated the repetition of measurements into the methodology of astronomy. Curiously, he failed to specify how these repeated observations should be converted into a single number. Consequently, astronomers devised their own, often ad hoc, methods for extracting a mean, or *data representative*, out of their observations. Sometimes they averaged, sometimes they used the median, sometimes they grouped their data and resorted to both averages and medians. Sometimes they explained their procedures, but often they did not. Consider, for example, the following excerpt which comes from Kepler [19] and whose observations were, in fact, made by Brahe himself:

On 1600 January 13/23 at 11^h 50^m the right ascension of Mars was:

	°	''	'
Using the bright foot of Gemini:	134	23	39
Using Cor Leonis:	134	27	37
Using Pollux:	134	23	18
At 12^h 17^m, using the third in the wing of Virgo:	134	29	48
The mean, treating the observations impartially:	134	24	33

Kepler's choice of data representative is baffling. Note that

$$\textit{Average}: \quad 134^\circ \quad 26' \quad 5.5''$$

$$\textit{Median}: \quad 134^\circ \quad 25' \quad 38''$$

and it is difficult to believe that an astronomer who recorded angles to the nearest second could fail to notice a discrepancy larger than a minute. The consensus is that the chosen mean could not have been the result of an error but must have been derived by some calculations. The literature contains at least two attempts to reconstruct these calculations but the author finds neither convincing, since both explanations are ad hoc, there is no evidence of either ever having been used elsewhere, and both result in estimates that differ from Kepler's by at least five seconds.

To the extent that they recorded their computation of data representatives, the astronomers of the time seem to be using improvised procedures that had both averages and medians as their components [7, 29, 30]. The median versus average controversy lasted for several centuries and now appears to have been resolved in favor of the latter, particularly in scientific circles. As will be seen from the excerpts below, this decision had a strong bearing on the evolution of the normal distribution.

The first scientist to note in print that measurement errors are deserving of a systematic and scientific treatment was Galileo in his famous *Dialogue Concerning the Two Chief Systems of the World - Ptolemaic and Copernican*, published in 1632. His informal analysis of the properties of random errors inherent in the observations of celestial phenomena is summarized by S.M. Stigler, as:

1. There is only one number which gives the distance of the star from the center of the earth, the true distance.
2. All observations are encumbered with errors, due to the observer, the instruments, and the other observational conditions.
3. The observations are distributed symmetrically about the true value; that is the errors are distributed symmetrically about zero.
4. Small errors occur more frequently than large errors.
5. The calculated distance is a function of the direct angular observations such that small adjustments of the observations may result in a large adjustment of the distance.

Unfortunately, Galileo did not address the question of how the true distance should be estimated. He did, however, assert that: "... it is plausible that the observers are more likely to have erred little than much." It is therefore not unreasonable to attribute to him the belief that the most likely true value is that which minimizes the sum of its deviations from the observed values. (That Galileo believed in the straightforward addition of deviations is supported by his calculations.)

In other words, faced with the observed values $x_1, x_2, \ldots, x_n$, Galileo would probably have agreed that the most likely true value is the x that minimizes the function

$$f(x) = \sum_{n=1}^{n} |x - x_i|. \tag{5}$$

As it happens, this minimum is well known to be the median of $x_1, x_2, \ldots, x_n$ and not their average, a fact that Galileo was likely to have found quite interesting.

This is easily demonstrated by an inductive argument that is based on the observation that if these values are re-indexed so that $x_1 < x_2 < \ldots < x_n$, then

$$\sum_{i=1}^{n} |x - x_i| = \sum_{i=2}^{n-1} |x - x_i| + (x_n - x_1) \quad \text{if } x \in [x_1, x_n]$$

whereas

$$\sum_{i=1}^{n} |x - x_i| > \sum_{i=2}^{n-1} |x - x_i| + (x_n - x_1) \quad \text{if } x \notin [x_1, x_n].$$

It took hundreds of years for the average to assume the near universality that it now possesses and its slow evolution is quite interesting. Circa 1660 we find Robert Boyle, later president of the Royal Society, arguing eloquently against the whole idea of repeated experiments:

> *... experiments ought to be estimated by their value, not their number; ... a single experiment ... may as well deserve an entire treatise ... . As one of those large and orient pearls ... may outvalue a very great number of those little ... pearls, that are to be bought by the ounce ...*

In an article that was published posthumously in 1722, Roger Cotes made the following suggestion:

> *Let p be the place of some object defined by observation, q, r, s, the places of the same object from subsequent observations. Let there also be weights P, Q, R, S reciprocally proportional to the displacements which may arise from the errors in the single observations, and which are given from the given limits of error; and the weights P, Q, R, S are conceived as being placed at p, q, r, s, and their center of gravity Z is found: I say the point Z is the most probable place of the object, and may be safely had for its true place.*

Cotes apparently visualized the observations as tokens $x_1, x_2, \ldots, x_n (= p, q, r, s, \ldots)$ with respective physical weights $w_1, w_2, \ldots, w_n (= P, Q, R, S, \ldots)$ lined up

Figure 8.3: A well-balanced explanation of the average.

on a horizontal axis (Figure 8.3 displays a case where $n = 4$ and all the tokens have equal weight). Having done this, it was natural for him to suggest that the center of gravity Z of this system should be designated to represent the observations. After all, in physics too, a body's entire mass is assumed to be concentrated in its center of gravity and so it could be said that the totality of the body's points are represented by that single point. That Cotes's proposed center of gravity agrees with the weighted average can be argued as follows. By the definition of the center of gravity, if the axis is pivoted at Z it will balance and hence, by Archimedes's law of the lever,

$$\sum_{i=1}^{n} w_i(Z - x_i) = 0$$

or

$$Z = \frac{\sum_{i=1}^{n} w_i x_i}{\sum_{i=1}^{n} w_i}. \tag{6}$$

Of course, when the weights w_i are all equal, Z becomes the classical average

$$\overline{x} = \frac{1}{n}\sum_{i=1}^{n} x_i.$$

It has been suggested that this is an early appearance of the method of least squares. In this context this method proposes that we represent the data $x_1, x_2, \ldots, x_n$ by the x which minimizes the function

$$g(x) = \sum_{i=1}^{n} w_i(x - x_i)^2. \tag{7}$$

Differentiation with respect to x makes it clear that it is the Z of (6) that provides this minimum. Note that the median minimizes the function $f(x)$ of (5) whereas the (weighted) average minimizer the function $g(x)$ of (7). It is curious that each of the two foremost data representatives can be identified as the minimize of a nonobvious, though fairly natural, function. It is also frustrating that so little is known about the history of this observation.

Thomas Simpson's paper of 1756 is of interest here for two reasons. First comes his opening paragraph:

> *It is well known to your Lordship, that the method practiced by astronomers, in order to diminish the errors arising from the imperfections of instruments, and of the organs of sense, by taking the Mean of several observations, has not been generally received, but that some persons, of considerable note, have been of opinion, and even publickly maintained, that one single observation, taken with due care, was as much to be relied on as the Mean of a great number.*

Thus, even as late as the mid-18th century doubts persisted about the value of repetition of experiments. More important, however, was Simpson's experimentation with specific *error curves* — probability densities that model the distribution of random errors. In the two propositions, Simpson computed the probability that the error in the mean of several observations does not exceed a given bound when the individual errors take on the values

$$-v,\ldots,-3,-2,-1,0,1,2,3,\ldots,v$$

with probabilities that are proportional to either

$$r^{-v},\ldots,r^{-3},r^{-2},r^{-1},r^{0},r^{1},r^{2},r^{3},\ldots,r^{v}$$

or

$$r^{-v},2r^{1-v},3r^{2-v}\ldots,(v+1)r^{0}\ldots,3r^{v-2},2r^{v-1},r^{v}.$$

Simpson's choice of error curves may seem strange, but they were in all likelihood dictated by the state of the art of probability at that time. For $r = 1$ these two distributions yield the two top graphs of Figure 8.4. One year later, Simpson, while effectively inventing the notion of a continuous error distribution, dealt with similar problems in the context of the error curves described in the bottom of Figure 8.4.

In 1774, Laplace proposed the first of his error curves. Denoting this function by $\phi(x)$, he stipulated that it must be symmetric in x, monotone decreasing for $x > 0$ and:

> *Now, as we have no reason to suppose a different law for the ordinates than for their differences, it follows that we must, subject to the rules of probabilities, suppose the ratio of two infinitely small consecutive differences to be equal to that of the corresponding ordinates. We thus will have*

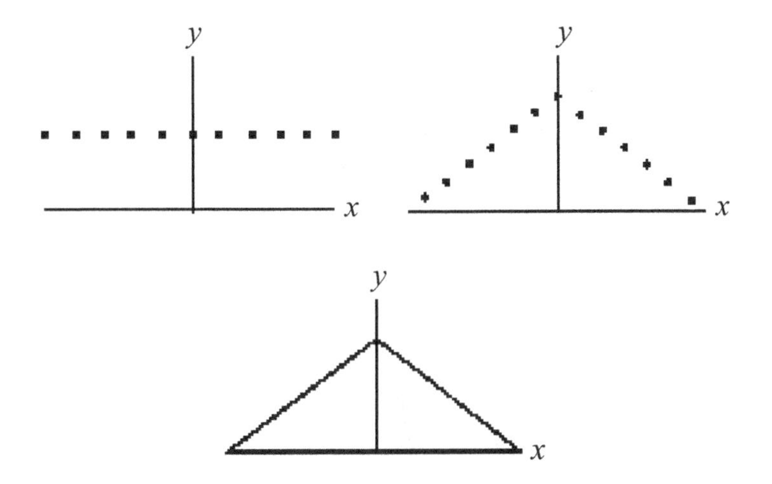

Figure 8.4: Simpson's error curves.

$$\frac{d\phi(x+dx)}{d\phi(x)} = \frac{\phi(x+dx)}{\phi(x)}$$

Therefore

$$\frac{d\phi(x)}{dx} = -m\phi(x)$$

... therefore

$$\phi(x) = \frac{m}{2}e^{-m|x|}.$$

Laplace's argument can be paraphrased as follows. Aside from their being symmetrical and descending (for $x > 0$), we know nothing about either $\phi(x)$ or $\phi'(x)$. Hence, presumably by Occam's razor, it must be assumed that they are proportional (the simpler assumption of equality leads to $\phi(x) = Ce^{|x|}$, which is impossible). The resulting differential equation is easily solved and the extracted error curve is displayed in Figure 8.5. There is no indication that Laplace was in any way disturbed by this curve's nondifferentiability at $x = 0$. We are about to see that he was perfectly willing to entertain even more drastic singularities.

Laplace must have been aware of the shortcomings of his rationale, for three short years later he proposed an alternative curve. Let a be the supremum of all the possible errors (in the context of a specific experiment) and let n be a positive integer.

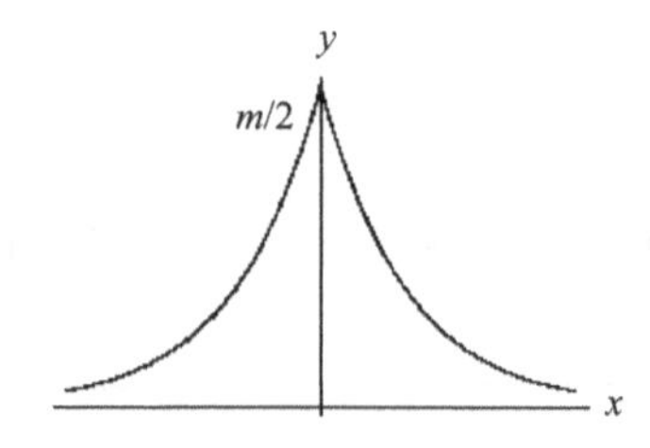

Figure 8.5: Laplace's first error curve.

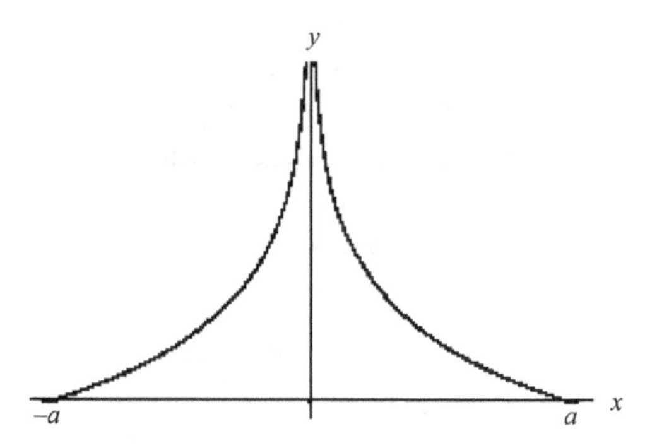

Figure 8.6: Laplace's second error curve.

Choose n points at random within the unit interval, thereby dividing it into $n + 1$ spacings. Order the spacings as:

$$d_1 > d_2 > \ldots > d_{n+1}, \quad d_1 + d_2 + \ldots + d_{n+1} = 1.$$

Let $\overline{d}_i$ be the expected value of d_i. Draw the points $(i/n, \overline{d}_i)$, $i = 1,2,\ldots, n + 1$ and let n become infinitely large. The limit configuration is a curve that is proportional to $\ln(a/x)$ on $(0, a]$. Symmetry and the requirement that the total probability must be 1 then yield Laplace's second candidate for the error curve (Fig. 8.6):

$$y = \frac{1}{2a}\ln\left(\frac{a}{|x|}\right) \quad -a \leq x \leq a.$$

This curve, with its infinite singularity at 0 and finite domain (a reversal of the properties of the error curve of Figure 8.5 and the bell-shaped curve), constitutes a step backward in the evolutionary process and one suspects that Laplace was seduced by the considerable mathematical intricacies of the curve's derivation. So much so that he felt compelled to comment on the curve's excessive complexity and to suggest that error analyses using this curve should be carried out only in "very delicate" investigations, such as the transit of Venus across the sun.

Shortly thereafter, in 1777, Daniel Bernoulli wrote:

> *Astronomers as a class are men of the most scrupulous sagacity; it is to them therefore that I choose to propound these doubts that I have sometimes entertained about the universally accepted rule for handling several slightly discrepant observations of the same event. By these rules the observations are added together and the sum divided by the number of observations; the quotient is then accepted as the true value of the required quantity, until better and more certain information is obtained. In this way, if the several observations can be considered as having, as it were, the same weight, the center of gravity is accepted as the true position of the object under investigation. This rule agrees with that used in the theory of probability when all errors of observation are considered equally likely. But is it right to hold that the several observations are of the same weight or moment or equally prone to any and every error? Are errors of some degrees as easy to make as others of as many minutes? Is there everywhere the same probability? Such an assertion would be quite absurd, which is undoubtedly the reason why astronomers prefer to reject completely observations which they judge to be too wide of the truth, while retaining the rest and, indeed, assigning to them the same reliability.*

It is interesting to note that Bernoulli acknowledged averaging to be universally accepted. As for the elusive error curve, he took it for granted that it should have a finite domain and he was explicit about the tangent being horizontal at the maximum point and almost vertical near the boundaries of the domain. He suggested the semi-ellipse as such a curve, which, following a scaling argument, he then replaced with a semicircle.

The next important step has its roots in a celestial event that occurred on January 1, 1801. On that day the Italian astronomer Giuseppe Piazzi sighted a heavenly body that he strongly suspected to be a new planet. He announced his discovery and named it Ceres. Unfortunately, six weeks later, before enough observations had been taken to make possible an accurate determination of its orbit, so as to ascertain that it was indeed a planet, Ceres disappeared behind the sun and was not expected to re-emerge for nearly a year. Interest in this possibly new planet was widespread and astronomers throughout Europe prepared themselves by compu-guessing the location where Ceres was most likely to reappear. The young Gauss, who had already made a name for himself as an extraordinary mathematician, proposed that an area of the sky be searched that was quite different from those suggested by the other astronomers and he turned out to be right.

Gauss explained that he used the least squares criterion to locate the orbit that best fit the observations. This criterion was justified by a theory of errors that was based on the following three assumptions:

1. Small errors are more likely than large errors.
2. For any real number ε, the likelihood of errors of magnitudes ε and $-\varepsilon$ are equal.
3. In the presence of several measurements of the same quantity, the most likely value of the quantity being measured is their average.

On the basis of these assumptions he concluded that the probability density for the error (that is, the error curve) is

$$\phi(x) = \frac{h}{\sqrt{\pi}} e^{-h^2x^2}$$

where h is a positive constant that Gauss thought of as the "precision of the measurement process." We recognize this as the bell curve determined by $\mu = 0$ and $\sigma = 1/\sqrt{2}h$.

Gauss's ingenious derivation of this error curve made use of only some basic probabilistic arguments and standard calculus facts. As it falls within the grasp of undergraduate mathematics majors with a course in calculus-based statistics, his proof is presented here with only minor modifications.

8.3 The Proof

Let p be the true (but unknown) value of the measured quantity, let n independent observations yield the estimates M_1, M_2,...M_n, and let $\phi(x)$ be the random error's probability density function. Gauss took it for granted that this function is differentiable. Assumption 1 above implies that $\phi(x)$ has a maximum at $x = 0$ whereas Assumption 2 means that $\phi(-x) = \phi(x)$. If we define

$$f(x) = \frac{\phi'(x)}{\phi(x)}$$

then

$$f(-x) = -f(x).$$

Note that $M_i - p$ denotes the error of the ith measurement and consequently, since these measurements (and errors) are assumed to be stochastically independent, it follows that

$$\Omega = \phi(M_1 - p)\phi(M_2 - p)...\phi(M_n - p)$$

is the joint density function for the n errors. Gauss interpreted Assumption 3 as saying, in modern terminology, that

$$\overline{M} = \frac{M_1 + M_2 + ... + M_n}{n}$$

is the maximum likelihood estimate of p. In other words, given the measurements $M_1, M_2, \ldots, M_n$ the choice $p = \overline{M}$ maximizes the value of Ω. Hence,

$$0 = \frac{\partial \Omega}{\partial p}_{|p=\overline{M}} = -\phi'(M_1 - \overline{M})\phi(M_2 - \overline{M})\ldots\phi(M_n - \overline{M})$$

$$-\phi(M_1 - \overline{M})\phi'(M_2 - \overline{M})\ldots\phi(M_n - \overline{M})$$

$$\ldots$$

$$-\phi(M_1 - \overline{M})\phi(M_2 - \overline{M})\ldots\phi'(M_n - \overline{M})$$

$$= -\left(\frac{\phi'(M_1 - \overline{M})}{\phi(M_1 - \overline{M})} + \frac{\phi'(M_2 - \overline{M})}{\phi(M_2 - \overline{M})} + \ldots + \frac{\phi'(M_n - \overline{M})}{\phi(M_n - \overline{M})}\right)\Omega.$$

It follows that

$$f(M_1 - \overline{M}) + f(M_2 - \overline{M}) + \ldots + f(M_n - \overline{M}) = 0. \tag{8}$$

Recall that the measurements M_i can assume arbitrary values and in particular, if M and N are arbitrary real numbers, we may use

$$M_1 = M, \quad M_2 = M_3 = \ldots = M_n = M - nN$$

for which set of measurements

$$\overline{M} = M - (n-1)N.$$

Substitution into (8) yields

$$f((n-1)N) + (n-1)f(-N) = 0$$

or

$$f((n-1)N) = (n-1)f(N).$$

It is a well known exercise that this homogeneity condition, when combined with the continuity of f, implies that $f(x) = kx$ for some real number k. This yields the differential equation

$$\frac{\phi'(x)}{\phi(x)} = kx.$$

Integration with respect to x yields

$$\ln \phi(x) = \frac{k}{2}x^2 + c \quad or \quad \phi(x) = Ae^{kx^2/2}.$$

In order for $\phi(x)$ to assume a maximum at $x = 0$, k must be negative and so we may set $k/2 = -h^2$. Finally, since

$$\int_{-\infty}^{\infty} e^{-h^2x^2}\,dx = \frac{\sqrt{\pi}}{h}$$

it follows that

$$\phi(x) = \frac{h}{\sqrt{\pi}}e^{-h^2x^2}.$$

Figure 8.7 displays a histogram of some measurements of the right ascension of Mars, together with an approximating exponential curve. The fit is certainly striking.

It was noted above that the average is in fact a least squares estimator of the data. This means that Gauss used a particular least squares estimation to justify his theory of errors which in turn was used to justify the general least squares criterion. There is an element of boot-strapping in this reasoning that has left later statisticians dissatisfied and may have had a similar effect on Gauss himself. He returned to the subject twice, twelve and thirty years later, to explain his error curve by means of different chains of reasoning.

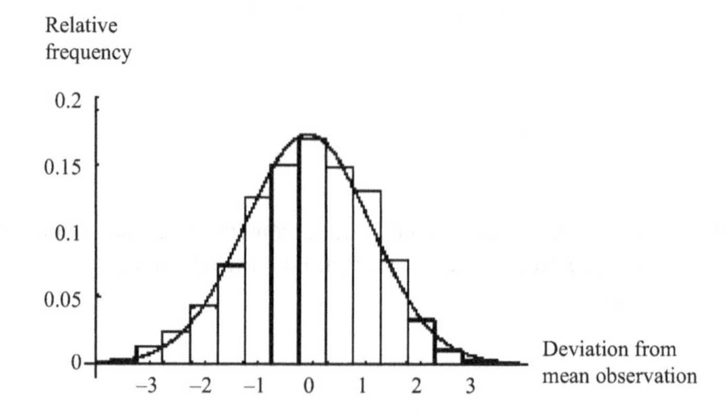

Figure 8.7: Normally distributed measurements.

Actually, a highly plausible explanation is implicit in the Central Limit Theorem published by Laplace in 1810. Laplace's translated and slightly paraphrased statement is:

> *... if it is assumed that for each observation the positive and negative errors are equally likely, the probability that the mean error of n observations will be contained within the bounds ±rh / n, equals*
>
> $$\frac{2}{\sqrt{\pi}} \cdot \sqrt{\frac{k}{2k'}} \cdot \int dr \cdot e^{-\frac{k}{2k'} \cdot r^2}$$
>
> where h is the interval within which the errors of each observation can fall. If the probability of error
>
> *$\pm x$ is designated by $\phi(x/h)$, then k is the integral $\int dx \cdot \phi(x/h)$ evaluated from $x = -\frac{1}{2}h$ to $x = \frac{1}{2}h$, and k' is the integral $\int \frac{x^2}{h^2} \cdot dx \cdot \phi(x/h)$ evaluated in the same interval.*

Loosely speaking, Laplace's theorem states that if the error curve of a single observation is symmetric, then the error curve of the sum of several observations is indeed approximated by one of the Gaussian curves. Hence if we take the further step of imagining that the error involved in an individual observation is the aggregate of a large number of "elementary" or "atomic" errors, then this theorem predicts that the random error that occurs in that individual observation is indeed controlled by De Moivre and Gauss's curve.

This assumption, promulgated by Hagen and Bessel, became known as the *hypothesis of elementary errors*. A supporting study had already been carried out by Daniel Bernoulli in 1780, albeit one of much narrower scope. Assuming a fixed error $\pm\alpha$ for each oscillation of a pendulum clock, Bernoulli concluded that the accumulated error over, say, a day would be, in modern terminology, approximately normally distributed.

This might be the time to recapitulate the average's rise to the prominence it now enjoys as the estimator of choice. Kepler's treatment of his observations shows that around 1600 there still was no standard procedure for summarizing multiple observations. Around 1660 Boyle still objected to the idea of *combining* several measurements into a single one. Half a century later, Cotes proposed the average as the best estimator. Simpson's article of 1756 indicates that the opponents of the process of averaging, while apparently a minority, had still not given up. Bernoulli's article of 1777 admitted that the custom of averaging had become universal. Finally, some time in the first decade of the 19th century, Gauss assumed the optimality of the average as an axiom for the purpose of determining the distribution of measurement errors.

8.4 Beyond Errors

The first mathematician to extend the provenance of the normal distribution beyond the distribution of measurement errors was Adolphe Quetelet (1796–1874). He began his career as an astronomer but then moved on to the social sciences. Consequently, he possessed an unusual combination of qualifications that placed him in just the right position for him to be able to make one of the most influential scientific observations of all times.

In his 1846 book *Letters addressed to H. R. H. the grand duke of Saxe Coburg and Gotha, on the Theory of Probabilities as Applied to the Moral and Political Sciences,* Quetelet extracted the contents of Table 1 from the *Edinburgh Medical and Surgical Journal* (1817) and contended that the pattern followed by the variety of its chest measurements was identical with that formed by the type of repeated measurements that are so common in astronomy. In modern terminology, Quetelet claimed that the chest measurements of Table 1 were normally distributed. The readers are left to draw their own conclusions regarding the closeness of the fit attempted in Figure 8. The more formal χ^2 normality test yields a χ^2_{Test} value of 47.1, which is much larger than the cutoff value of $\chi^2_{10,.05} = 18.3$, meaning that by modern standards these data cannot be viewed as being normally distributed. This discrepancy indicates that Quetelet's justification of his claim of the normality the chest measurements merits a substantial dose of skepticism. It appears here in translation:

Girth	Frequency
33	3
34	18
35	81
36	185
37	420
38	749
39	1,073
40	1,079
41	934
42	658
43	370
44	92
45	50
46	21
47	4
48	1
	5,738

Table 8.1: Chest measurements of Scottish soldiers.

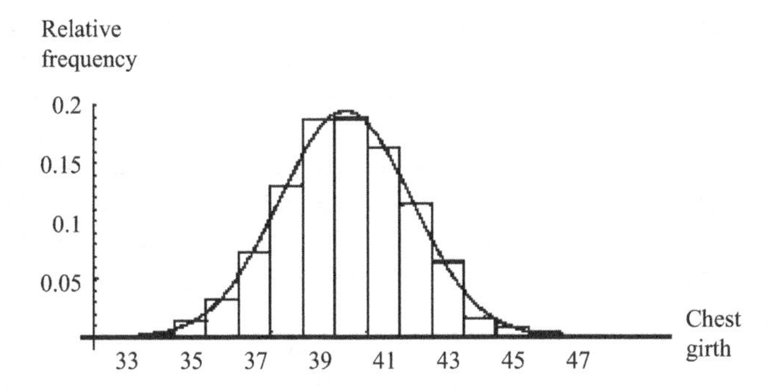

Figure 8.8: Is this data normally distributed?

I now ask if it would be exaggerating, to make an even wager, that a person little practiced in measuring the human body would make a mistake of an inch in measuring a chest of more than 40 inches in circumference? Well, admitting this probable error, 5,738 measurements made on one individual would certainly not group themselves with more regularity, as to the order of magnitude, than the 5,738 measurements made on the scotch [sic] soldiers; and if the two series were given to us without their being particularly designated, we should be much embarrassed to state which series was taken from 5,738 different soldiers, and which was obtained from one individual with less skill and ruder means of appreciation.

This argument, too, is unconvincing. It would have to be a strange person indeed who could produce results that diverge by 15" while measuring a chest of girth 40". Any idiosyncrasy or unusual conditions (fatigue, for example) that would produce such unreasonable girths is more than likely to skew the entire measurement process to the point that the data would fail to be normal.

It is of interest to note that Quetelet was the man who coined the phrase *the average man*. In fact, he went so far as to view this mythical being as an ideal form whose various corporeal manifestations were to be construed as measurements that are beset with errors:

If the average man were completely determined, we might ... consider him as the type of perfection; and everything differing from his proportions or condition, would constitute deformity and disease; everything found dissimilar, not only as regarded proportion or form, but as exceeding the observed limits, would constitute a monstrosity.

Quetelet was quite explicit about the application of this, now discredited, principle to the Scottish soldiers. He takes the liberty of viewing the measurements of the soldiers' chests as a repeated estimation of the chest of the average soldier:

I will astonish you in saying that the experiment has been done. Yes, truly, more than a thousand copies of a statue have been measured, and though I will not assert it to be that of the Gladiator, it differs, in any event, only slightly from it: these copies were even living ones, so that the measures were taken with all possible chances of error, I will add, moreover, that the copies were subject to deformity by a host of accidental causes. One may expect to find here a considerable probable error.

Finally, it should be noted that Table 1 contains substantial errors. The original data was split amongst tables for eleven local militias, cross-classified by height and chest girth, with no marginal totals, and Quetelet made numerous mistakes in extracting his data. The actual counts are displayed in Table 2 where they are compared to Quetelet's counts.

Quetelet's book was very favorably reviewed in 1850 by the eminent and eclectic British scientist John F. W. Herschel [18]. This extensive review contained the outline of a different derivation of Gauss's error curve, which begins with the following three assumptions:

1. ... the probability of the concurrence of two or more independent simple events is the product of the probabilities of its constituents considered singly;
2. ... the greater the error the less its probability ...
3. ... errors are equally probable if equal in numerical amount ...

Herschel's third postulate is much stronger than the superficially similar symmetry assumption of Galileo and Gauss. The latter is one-dimensional and is formalized as $\phi(\varepsilon)=\phi(-\varepsilon)$ whereas the former is multi-dimensional and is formalized as asserting the existence of a function ψ such that

Girth	Actual Frequency	Quetelet's Frequency
33	3	3
34	19	18
35	81	81
36	189	185
37	409	420
38	753	749
39	1,062	1,073
40	1,082	1,079
41	935	934
42	646	658
43	313	370
44	168	92
45	50	50
46	18	21
47	3	4
48	1	1
	5,738	5,732

Table 8.2: Chest measurements of Scottish soldiers.

$$\phi(x)\phi(y)...\phi(t) = \psi(x^2 + y^2 + ... + t^2).$$

Essentially the same derivation had already been published by the American R. Adrain in 1808, prior to the publication of but subsequent to the location of Ceres. In his 1860 paper on the kinetic theory of gases, the renowned British physicist J. C. Maxwell repeated the same argument and used it, in his words:

To find the average number of particles whose velocities lie between given limits, after a great number of collisions among a great number of equal particles.

The social sciences were not slow to realize the value of Quetelet's discovery to their respective fields. The American Benjamin A. Gould, the Italians M. L. Bodio and Luigi Perozzo, the Englishman Samuel Brown, and the German Wilhelm Lexis all endorsed it. Most notable amongst its proponents was the English gentleman and scholar Sir Francis Galton who continued to advocate it over the span of several decades. This aspect of his career began with his 1869 book *Hereditary Genius* in which he sought to prove that genius runs in families. As he was aware that exceptions to this rule abound, it had to be verified as a statistical, rather than absolute, truth. What was needed was an efficient quantitative tool for describing populations and that was provided by Quetelet whose claims of the wide ranging applicability of Gauss's error curve Galton had encountered and adopted in 1863.

As the description of the precise use that Galton made of the normal curve would take us too far afield, we shall only discuss his explanation for the ubiquity of the normal distribution. In his words:

Considering the importance of the results which admit of being derived whenever the law of frequency of error can be shown to apply, I will give some reasons why its applicability is more general than might have been expected from the highly artificial hypotheses upon which the law is based. It will be remembered that these are to the effect that individual errors of observation, or individual differences in objects belonging to the same generic group, are entirely due to the aggregate action of variable influences in different combinations, and that these influences must be.

1. all independent in their effects,
2. all equal,
3. all admitting of being treated as simple alternatives "above average" or "below average,"
4. the usual tables are calculated on the further supposition that the variable influences are infinitely numerous.

This is, of course, an informal restatement of Laplace's Central Limit Theorem. The same argument had been advanced by Herschel. Galton was fully aware that conditions (1–4) never actually occur in nature and tried to show that they were unnecessary. His argument, however, was vague and inadequate. Over the past two centuries the Central Limit Theorem has been greatly generalized and a newer version, known as Lindeberg's Theorem, exists which makes it possible to dispense with requirement (2). Thus, De Moivre's curve (1) emerges as the limiting, observable distribution even when the aggregated "atoms" possess a variety of nonidentical distributions. In general it seems to be commonplace for statisticians to attribute the great success of the normal distribution to these generalized versions of the Central Limit Theorem. Quetelet's belief that all deviations from the mean were to be regarded as errors in a process that seeks to replicate a certain ideal has been relegated to the same dustbin that contains the phlogiston and aether theories.

8.5 *Why* Normal?

A word must be said about the origin of the term *normal.* Its aptness is attested by the fact that three scientists independently initiated its use to describe the error curve

$$\phi(x) = \frac{1}{\sqrt{2\pi}} e^{-x^2/2}.$$

These were the American C. S. Peirce in 1873, the Englishman Sir Francis Galton in 1879, and the German Wilhelm Lexis, also in 1879. Its widespread use is probably due to the influence of the great statistician Karl E. Pearson, who had this to say in 1920:

> *Many years ago [in 1893] I called the Laplace-Gaussian curve the* normal *curve, which name, while it avoids the international question of priority, has the disadvantage of leading people to believe that all other distributions of frequency are in one sense or another abnormal.*

At first it was customary to refer to Gauss's error curve as the *Gaussian curve* and Pearson, unaware of De Moivre's work, was trying to reserve some of the credit of discovery to Laplace. By the time he realized his error the *normal curve* had become one of the most widely applied of all mathematical tools and nothing could have changed its name. It is quite appropriate that the name of the error curve should be based on a misconception.

Acknowledgements

The author thanks his colleagues James Church and Ben Cobb for their helpfulness, Martha Siegel for her encouragement, the editor Frank A. Farris and the anonymous referees for their constructive critiques.

Chapter 9

Map Colorings

Until its resolution in 1976, the Four Color Problem was arguably the most notorious mathematical problem of all time. This problem asks whether four colors suffice to color every planar map so that adjacent countries, i.e., countries that share a border of positive length, receive different colors. This question's deceptive simplicity attracted many would-be solvers who oft times spent years on their search for a solution. Most returned empty-handed, or worse, with a false proof. Some were fortunate enough to have devised a new twist on the original problem that was sufficiently interesting to attract the attention of other aficionados. *The Heawood Conjecture*, one of the earliest of these offshoots, proved to be also one of the most fascinating and difficult. This other coloring conjecture guessed at the number of colors required by maps on other, more complicated, surfaces. Surprisingly enough, even though this later problem seems more difficult than its planar progenitor, Heawood's conjecture was actually verified a decade earlier. In my opinion this verification marks a milestone in the development of the modern combinatorial approach to geometry. It is my intention here to formulate this problem, recount its history, and discuss its relationship to the original Four Color Problem, as well as to other branches of mathematics.

9.1 Heawood's Conjecture

It is generally agreed that the Four Color Conjecture was first formulated by Francis Guthrie, a graduate student at University College, London, in 1852. Appel and

Haken's proof of this conjecture is described by them in [1] and has also received wide publicity elsewhere, so I will not discuss it here and only refer to it when it provides an interesting parallel or contrast with the other map coloring theorem. The first false proof of the Four Color Conjecture to be published was given in 1879 by A. B. Kempe [14], barrister and part-time mathematician. The validity of this proof was not challenged until 1890 when P. J. Heawood published a paper [10] in which he accomplished several things. First, he pointed out the error in Kempe's reasoning. Next, he salvaged the remains of Kempe's fallacious proof by using its techniques to show that every planar map can be colored with five colors. Finally, he went on to state and solve, so he thought, the same problem in a new context. This came to be known as the Heawood Conjecture.

Before describing this new context, a minor clarification is in order. Whenever a map is mentioned above and in the sequel, it is to be understood that each country forms a single contiguous geographical unit, unlike pre-Bangladesh Pakistan. Also, in order to minimize redundancy it will be assumed that whenever a map is colored, the coloring satisfies this map-coloring constraint: adjacent countries receive different colors.

The plane is not the only surface on which maps can be drawn; they can of course also be drawn on the surface of a sphere. However, coloring maps on spheres is really not different from coloring them on the plane. The well-known stereographic projection of Figure 1, often employed by map makers, can be used to convert any spherical map to a planar one, and the coloring pattern of the planar projection of a spherical map can also be used to color the original map on the sphere. Nothing is therefore to be gained by reformulating the Four Color Problem for the sphere. Maps on the surface of a cone or a pyramid do not provide any new challenges either, for these surfaces can be easily deformed into spherical ones in a manner that preserves the adjacency pattern of any maps

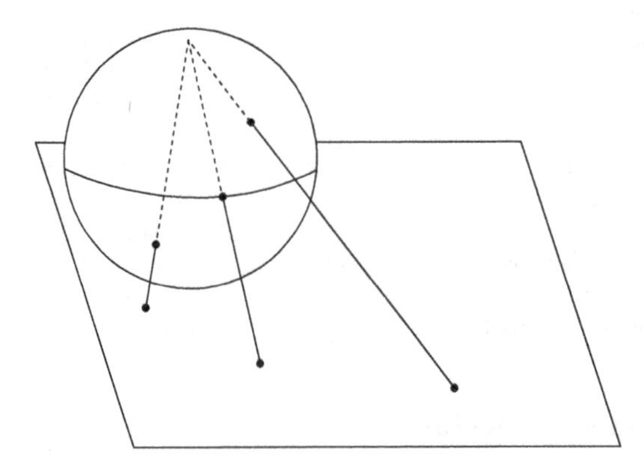

Figure 9.1: Stereographic projection. A technique used to represent spherical maps on flat paper. The particular version portrayed here will represent regions near the south pole fairly faithfully, but will greatly distort northern countries. Nevertheless, it does preserve adjacencies.

drawn on them. So, if one is to find a surface that will yield a genuinely new coloring problem, then this surface cannot be deformable into a sphere in any "nice" way. One such surface is a torus—the surface of the doughnut. In M_5 (Figure 2) we have a toroidal map consisting of five countries, every two of which are adjacent. Such a map clearly requires five colors. In fact, M_6 and M_7 (Figures 3, 4) are toroidal maps with six and seven countries in which every two countries are adjacent to each other. Consequently, these maps require six and seven colors, respectively.

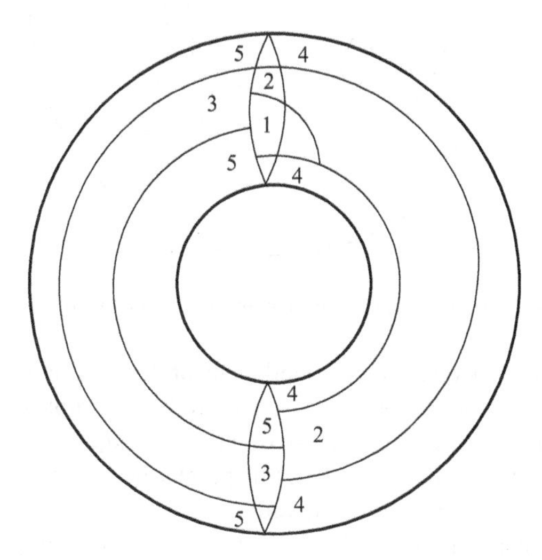

Figure 9.2: M_5, a toroidal map of five countries, every two of which touch each other.

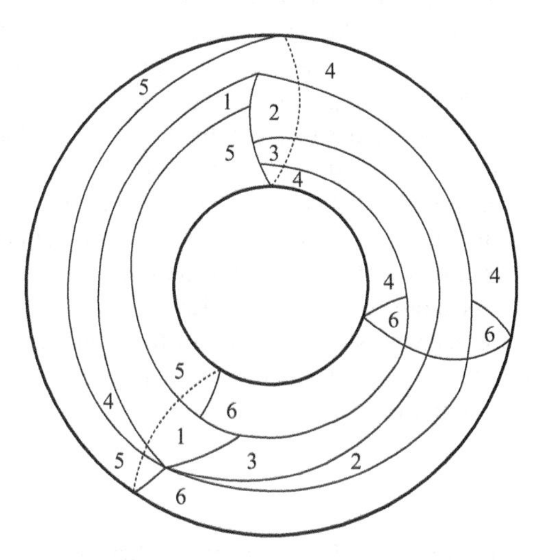

Figure 9.3: M_6, a toroidal map of six countries, every two of which touch each other.

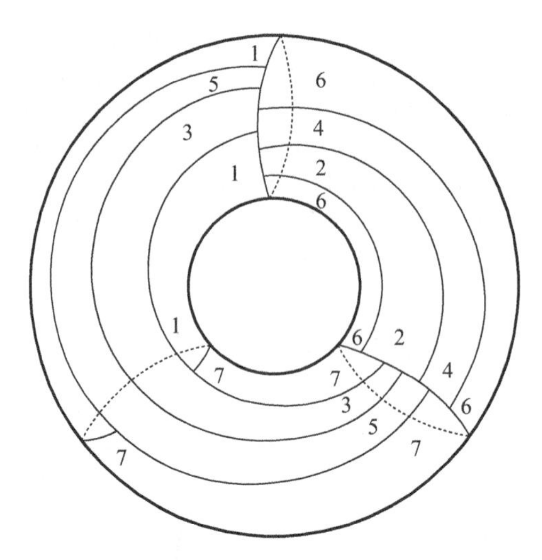

Figure 9.4: M_7, a toroidal map of seven countries, every two of which are adjacent.

The subsequent discussion centers around maps of this type and it will therefore be convenient to have a name for them. Accordingly, a complete m-map is a map which consists of m countries, every two of which are adjacent. Such a map clearly requires m colors. It is natural to ask at this point whether there exists a *complete* 8-map. The answer is negative for the torus.

The mathematician A. F. Möbius (see [2]) had already pointed out in the 1840s that it is not possible to draw a complete 5-map in the plane. In later years this observation was sometimes misconstrued as a solution of the Four Color Problem. It is, of course, no such thing. The nonexistence of a complete 5-map in the plane does not preclude the existence of an incomplete planar map with a large number of countries, whose complexity actually requires five or more colors. We now have an analogous situation on the torus, for it supports a complete 7-map and no complete 8-map can be drawn on it. Does this mean that every toroidal map can be colored with seven colors? Such indeed is the case. What is surprising is the relative ease with which this answer was obtained. So simple is this proof that its main points deserve to be brought out here.

In any map, let n denote the number of its nodes, namely, the number of points at which three or more countries meet. Let b denote the number of borderlines in the map, in other words, curves that go from one node to another. The Euler-Poincaré formula [15], one of the central facts of geometry, implies that if a toroidal map has m countries, then

$$n - b + m = 0.$$

For the maps M_5, M_6, and M_7, this formula takes the forms $9 - 14 + 5 = 0$, $9 - 15 + 6 = 0$, and $14 - 21 + 7 = 0$, respectively. For the sake of accuracy let me point out that the Euler-Poincaré formula does not apply to maps in which one country has the shape of a ring and completely surrounds one or several other countries. Such maps, however, can be easily disposed of by certain standard reduction techniques and will be ignored in the sequel.

Let A denote the average number of borderlines that occur on the boundaries of the m countries of a given map. Then mA is the count of all the occurrences of borderlines on the boundaries of these countries. Since each of the b borderlines occurs on two of these boundaries, it follows that

$$b = \frac{mA}{2}.$$

Turning our attention to the nodes, observe that A must also be the average number of nodes on the boundaries of these countries, since each country has an equal number of nodes and borderlines on its boundary. However, in contrast with the borderlines, we cannot specify the number of countries on whose boundary a given node will occur. Still we can say that each node appears on the boundaries of at least three countries. This gives us the following weak analog of equation (2), which is nevertheless sufficient for our purpose:

$$n \leq \frac{mA}{3}$$

Combining the three, we obtain

$$\frac{mA}{3} - \frac{mA}{2} + m \geq n - b + m = 0,$$

$$m\left(1 - \frac{A}{6}\right) \geq 0.$$

Since m is a positive quantity, we may conclude that $A < 6$. In other words, the average number of borderlines on the boundary of a typical country in any toroidal map does not exceed six. Consequently we obtain the following surprising fact: in every toroidal map there must be a country that is adjacent to no more than six other countries. This fact immediately points out the impossibility of drawing a complete 8-map on the torus, since in such a map every country is necessarily adjacent to seven other countries. This fact can also be used to define an algorithm for coloring any toroidal map with $6+1=7$ colors. Thus, every map on the torus can be colored with seven colors and some such map (the complete 7-map) actually requires seven colors. In other words, the answer to the toroidal analog of the Four Color Problem is that seven colors suffice. Heawood was the first to formulate this analog clearly and to answer it. He did much more, however.

Why does the torus allow for more complicated maps than the sphere? (The complexity of a map is here equated with the number of colors it requires.) One might say that this additional complexity is made possible by the hole in the torus—the doughnut hole. The adjacency pattern of a spherical map is constrained by the fact that a country that lies completely in the northern hemisphere cannot possibly be adjacent to one that lies entirely in the southern hemisphere. However, if a tunnel is bored from the north pole of the sphere to its south pole, thus converting it essentially into a doughnut, it now becomes possible for a northern country to reach out and touch a southern one without crossing the equator, namely, along the walls of the tunnel. From this point of view, a tunnel through a surface may be thought of as a bridge that connects two parts of the surface and allows for a higher degree of complexity in the adjacency pattern of its maps. So if we wish to find a surface that supports a complete 8-map (which the torus does not), it is clear what must be done. A tunnel should be bored through the torus or, equivalently, two nonintersecting tunnels should be bored through the sphere. The resulting surface, the double torus, does indeed support the complete 8-map M_8 of Figure 5.

We now have a technique for creating surfaces that would seem to allow for maps requiring arbitrarily many colors. All that needs to be done is to bore enough tunnels through the sphere. We will call the surface that results from boring g nonintersecting tunnels through the sphere *the surface of genus* g and denote it by S_g. Thus S_2 is the double torus, the torus itself is S_1, and the surface of the unperforated sphere is S_0. The great mathematician Bernard Riemann brought these surfaces into the foreground of mathematics in 1851 [18] when he showed that they play a focal role in the calculus of complex variables. His theory was so central to nineteenth century mathematics that it has been said [4, p. 121] that at one time all research mathematicians had to be familiar with it. It is therefore not surprising that shortly after the Four Color Problem was formulated for planar maps, mathematicians would ask the same question in the context of Riemann's surfaces.

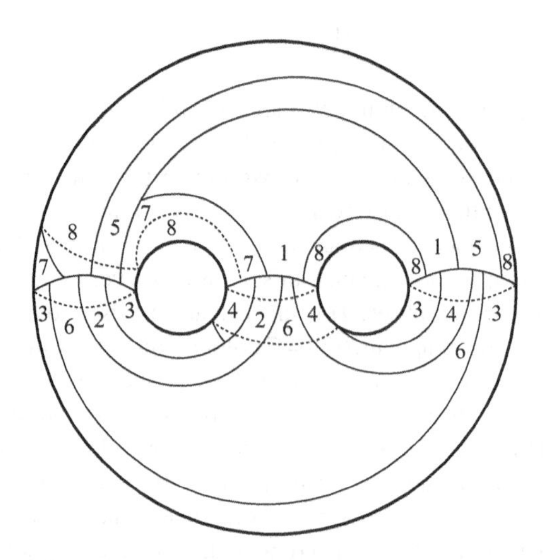

Figure 9.5: M_8, a complete 8-map on the double torus S_2.

The Euler-Poincaré formula states that whenever a map with m countries, b borderlines, and n nodes is drawn on the surface S_g, then these parameters are linked by the equation

$$n - b + m = 2 - 2g.$$

Heawood used this fact to show that if $g \geq 1$, *then any such map can be colored with no more than*

$$H_g = \left[\frac{1}{2}(7 + \sqrt{1 + 48g})\right]$$

colors, where the brackets denote the integer part of the enclosed number. This guarantees that any map on the torus can be colored with $H_1 = \left[\frac{1}{2}(7 + \sqrt{1 + 48})\right] = 7$ colors, that every map on the double torus can be colored with $H_2 = \left[\frac{1}{2}(7 + \sqrt{1 + 96})\right] = [8.42...] = 8$ colors, and that every map on S_3 can be colored with 9 colors.

Heawood's proof that the number of colors specified above is actually sufficient is paraphrased as follows. Call those borderlines along which a country touches its neighbors its contacts (clearly every country possesses at least as many contacts as neighbors, and sometimes more). Suppose that there is an integer x such that every map on S_g has a country with fewer than x contacts. Then we claim that x colors suffice to color every map on S_g. For in any such map we can annihilate the country with less than x contacts, making those around it close up in the space which it occupied (Figure 6), and obviously the original map can be done in x colors if the reduced map can (for whatever the coloring of the countries around, there would be a color to spare for the annihilated country). Having thus described an induction process, Heawood now needed only to find the smallest x that satisfied this condition, and he used an averaging argument that is essentially the same as the one used above to argue that the torus does not support a complete 8-map. First, however, Heawood observed that every map on any surface can be converted by the operations described in Figure 7 to a map in which every node appears on the boundary of exactly three countries, this new map requiring no more colors than the

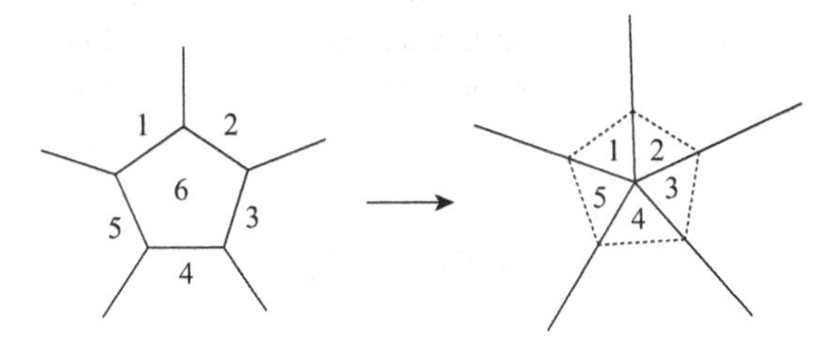

Figure 9.6: A reduction process for maps that diminishes the number of countries without increasing the number of required colors.

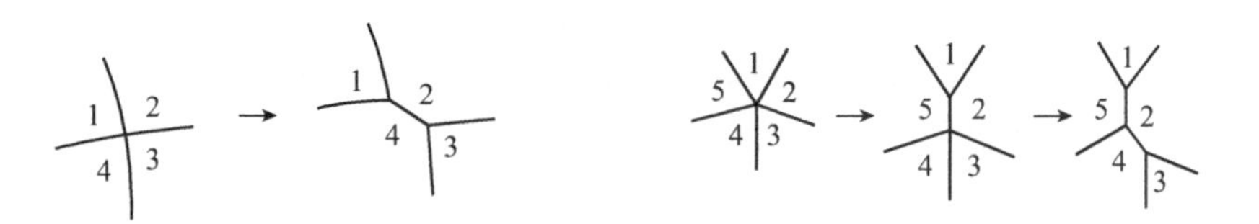

Figure 9.7: How to transform any map to one requiring no more colors, but in which every node is on the boundary of exactly three countries.

original one. Consequently, with n, b, m, A denoting the same quantities as before, equation 2 (page 313) still holds, whereas the inequality can now be replaced by the equation

$$n = \frac{mA}{3}.$$

If we now substitute the above equations, we obtain

$$A = 6\left(1 + \frac{2g-2}{m}\right).$$

As was argued for the torus, every map on S_g clearly has a country with no more than A adjacencies, so any x satisfying

$$x > A$$

should do. The expression for A, however, still contains an m which depends on the map rather than the surface. To eliminate this m Heawood observed that A diminishes as m increases, and since every map with no more than x countries clearly must have a country with fewer than x contacts, it follows that in order for the inequality to hold we need only guarantee that

$$x > 6\left(1 + \frac{2g-2}{x+1}\right).$$

Thus, if we let H_g denote the smallest of all the integers x that satisfy the above inequality, then the maps on the surface S_g can be colored with H_g colors. Since for all smaller values of x the reverse inequality must hold, this can be rephrased as saying that H_g is the largest integer x that satisfies the inequality

$$x - 1 \leq 6\left(1 + \frac{2g-2}{x}\right).$$

This is a straightforward quadratic inequality in x, whose solution yields the value specified on page 9.8.

The above argument is essentially the same as the one presented by Heawood himself. I have even gone so far as to leave intact some of the rougher edges and omissions in his proof, partly for the sake of brevity, and partly to prepare the reader for Heawood's real mistake, which is to be discussed shortly. Cleaner arguments can be found in [2] and [20].

To demonstrate that H_g colors are in fact required by some map on S_8, Heawood also drew a complete 7-map on the torus—essentially the same as M_7—but only *asserted* the existence of a complete H_g-map on the surface S_g for all $g \geq 2$. "For more highly connected surfaces," Heawood stated, "it will be observed that there are generally contacts enough and to spare for the above number of divisions [countries] each to touch each." In other words, Heawood thought it was easy to see that a complete 9-map could be drawn on S_3. Anyone who tries to draw such a map will quickly discover that not only is this quite a difficult task, but it also sheds no light on the question of how to draw a complete 10-map on S_4. Thus, what Heawood accomplished was to show that no map on the surface of genus $g \geq 1$ requires more than H_g colors. What he failed to show was that a map requiring this number of colors could actually be drawn on S_g. So a question he believed he had both raised and answered, in fact, remained open. The number H_g defined above acquired the name *Heawood number* and the assertion that *the Heawood number* H_g *is the largest number of colors required to color any map on the surface of genus* $g \geq 1$ came to be known as the *Heawood Conjecture*.

The observant reader will have noted the qualification $g \geq 1$ that occurs in the statement of Heawood's conjecture. Were this qualification absent, the conjecture would apply to the surface S_0 (the sphere), and it would assert that any spherical map could be colored with no more than $H_0 = \left[\frac{1}{2}(7 + \sqrt{1 + 0})\right] = 4$ colors! In other words, the removal of the inequality $g \geq 1$ would make the Heawood Conjecture contain the Four Color Conjecture as a special case. Unfortunately, the condition $g \geq 1$ cannot be disregarded. In Heawood's proof, the quantity

$$A = 6\left(1 + \frac{2g - 2}{m}\right)$$

was observed to be a *decreasing* function of m. This of course fails to be the case when $g = 0$, and so the proof breaks down for this value of g. This is a good example of the many near misses that proliferate in the history of the Four Color Problem.

9.2 Heffter's Contributions

The deficiency in Heawood's paper was pointed out one year after its publication by Lothar Heffter [11]. Heffter also realized that the problem needed a new approach. Trying to draw complicated maps on two-dimensional drawings of perforated spheres was a very cumbersome way to attack this problem. As an alternative he suggested that the adjacency pattern of a map (which, after all, is all that matters here) be recorded in the following manner. Suppose an inhabitant of some country were to inspect its borders by traveling along them in, say, the counterclockwise direction (counterclockwise from the point of view of an observer stationed right above his country). This inhabitant might then record, in order, the neighboring countries along whose border he is traveling. For instance, in the map M_4 of Figure 8, the inspector of country 1, if he starts his tour from a location on the border his country shares with country 2, would write 2, 4, 3 in his logbook. Had he started from a location on the border his country shares with country 3, his entry would have been 3, 2, 4. These two records are considered to be the same since it is only the cyclic ordering of the neighbors that matters to us. Now, country 2's inspector, in touring his borders, would write 1, 3, 4. The information obtained from these tours and those of the inspectors of countries 3 and 4, can be tabulated as the array A_4 in Figure 9. When applied to the toroidal complete 7-map M_7, this process yields the array A_7 in Figure 10. The array A_7 has a very exciting pattern. Every row can be derived from the previous one by the addition of 1 to each entry of the latter. This arithmetic is modulo 7; that is, we stipulate that $7 + 1 = 1$ and that the first row follows the seventh one.

Before we go on to discuss these arrays in general, it should be pointed out that two miracles occurred in the passage from the complete 7~map M_7 to its array A_7. In the first place, the array displays a pattern, or a symmetry, that is totally

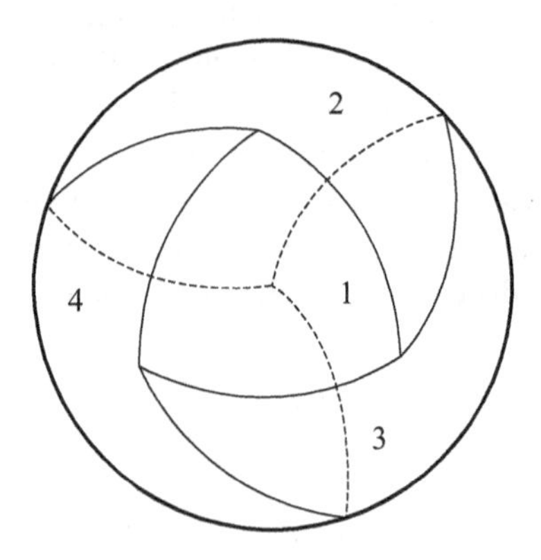

Figure 9.8: M_4, a complete 4-map on the sphere.

Country	Adjacency record		
1)	2	4	3
2)	1	3	4
3)	4	2	1
4)	3	1	2

Figure 9.9: A_4, the adjacency pattern for the map A_4.

Country	Adjacency record					
1)	2	4	3	7	5	6
2)	3	5	4	1	6	7
3)	4	6	5	2	7	1
4)	5	7	6	3	1	2
5)	6	1	7	4	2	3
6)	7	2	1	5	3	4
7)	1	3	2	6	4	5

Figure 9.10: A_7, the adjacency pattern for the map M_7 in Figure 4.

Country	Adjacency record		
1)	2	3	4
2)	3	4	1
3)	4	1	2
4)	1	2	3

Figure 9.11: B_4, an array that is not the adjacency pattern of any map.

obscured in the original map. Symmetry, of course, is one of the strongest tools of mathematics and of science. Hence, finding it in such an unexpected place is indeed an undisguised blessing. The second observation is even more surprising. The numbers 1,2,3,... were used as a matter of convenience only. We could, and perhaps should, have used labels such as "Spain" or "Union of the Free Toroidal Republics." Nevertheless, we find that these countries, when symbolized by numbers, follow a very rigid arithmetical pattern. This phenomenon has been observed and exploited in other coloring problems, but not in the proof of the Four Color Theorem. Both of these miracles were crucial to the eventual resolution of Heawood's conjecture.

Let us now return to the arrays themselves. The array A_4 obtained from the complete 4-map M_4 does not quite conform to the nice pattern that was discovered in A_7. It seems that in certain columns one might need to subtract rather than add 1. Still, there is enough regularity in the rows of this array to justify some hope for a general pattern.

An obvious question at this point is: *which arrays correspond to some complete m-map?* Such an array must clearly have m rows, and the ith row must be some permutation of the numbers $1,2,\ldots,i+1,\ldots,m$. Let us call this requirement the shape constraint. An example of another array that satisfies the shape constraint is B_4, shown in Figure 11. A little experimentation, however, will quickly convince the reader that the array B_4 cannot correspond to any complete 4-map. This can be reasoned out by observing that according to the first row of this array, country 1's inspector, traveling counterclockwise, encounters country 2 just before country 3.

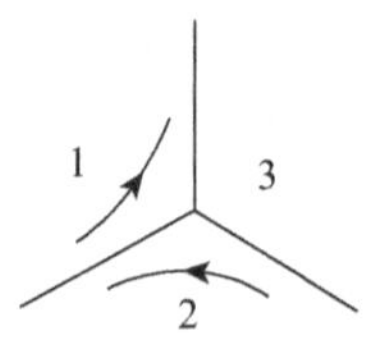

Figure 9.12

Hence, the node formed by countries 1, 2, and 3 must form the configuration of Figure 12. However, according to this portion of the map, country 2's inspector must encounter country 3 just before he encounters country 1, which contradicts the information in the second row of B_4. This observation generalizes to the following *consistency constraint* to which all arrays that describe complete m-maps must conform: *If in the ith row of the array, j is followed by k, then in the k th row, j must be preceded by i.*

In other words,

if the ith row is

$i)$ ... j k ...

then the kth row is

$k)$ ... i j ...

Surprisingly, there are no other constraints. *Any array that satisfies both the shape and consistency constraints describes a complete m-map.* Moreover, as Heffter demonstrated, this complete m-map verifies the Heawood Conjecture for the surface S_g for $g = (m-3)(m-4)/12$. Since the genus g (number of tunnels) of a surface is necessarily an integer, it follows that the product $(m-3)(m-4)$ must be divisible by 12, from which it follows that m itself, when divided by 12, must yield 0, 3, 4, or 7 as remainder. Since, in general, division by 12 yields one of 12 possible remainders, this approach can yield the appropriate complete m-map for at most one-third of the possible values of m. As it happens, even this estimate is overly optimistic. Heffter was aware that an array possessing the additional cyclic structure displayed by A_7 could only be constructed when the remainder of m divided by 12 was 7. All of these limitations notwithstanding, he felt that this was an approach well worth pursuing. He translated the problem into a purely number theoretic one and then showed that the required cyclic arrays A exist for those values of m that satisfy conditions i)—iii) below:

i) m leaves remainder 7 when divided by 12,
ii) $(m + 2)/3$ is a prime number,
iii) $2^k - 1$ is not divisible by $(m + 2)/3$ for $k = 1,2,3,\ldots, (m-7)/6$.

Among the values that satisfy these conditions we find the numbers m = 19, 31, 55, 67, 139, 175, 199. Consequently, Heffter verified the Heawood Conjecture for the surfaces S_g where g = 20, 63, 221, 336, 1530, 2451, 3185. Actually, he accomplished a little more than that. Since $H_{21}=\left[\frac{1}{2}(7+\sqrt{1+48\times 21})\right]=[19.38...]=19=H_{20}$, and since every map that can be drawn on S_{20} can also be drawn on S_{21}, it follows that Heffter's construction of A_{19} also verifies the Heawood Conjecture for S_{21}, as it does for S_{22} as well. Similarly, the existence of the array A_{199} verifies the Heawood Conjecture for all surfaces with at least 3185 but no more than 3217 tunnels. The algebra Heffter used in his work was deep and difficult enough that he could not decide whether the class of surfaces to which his proof applies was finite or infinite. Even today it is still not known whether Heffter's solution applies to infinitely many surfaces.

Heffter's contributions can be summarized as follows. He pointed out the error in Heawood's paper and thereby attracted the attention of the mathematical world to Heawood's beautiful conjecture. He showed how this geometrical problem could be translated first to a combinatorial one of arrays, and then to an algebraic problem in the theory of numbers. He went on to solve this number theoretic problem in some special cases. Finally, a fact that was not mentioned above, he also confirmed the Heawood conjecture for all surfaces of genus at most 7 by constructing the appropriate arrays.

9.3 The Resolution of Heawood's Conjecture

Nothing was contributed toward the resolution of Heawood's conjecture in the sixty years that followed the publication of Heffter's work. That mathematicians were aware of it is attested to by the fact that it is mentioned in Hilbert and Cohn-Vossen's 1932 book *Geometry and the Imagination* [12], where it is called the problem of contiguous regions. The lack of progress on this problem during the first half of this century can be attributed to three factors. First comes its inherent difficulty. Next, some mathematicians believed that the problem had indeed been solved by Heawood. Courant and Robbins, in their 1941 book *What is Mathematics?* [5, p. 248], wrote: "A remarkable fact connected with the four color problem is that for surfaces more complicated than the plane or the sphere the corresponding theorems have actually been proved." (As late as 1980, I met a topologist who believed the same.) Finally, and perhaps most significantly, to many mathematicians this problem seemed to represent a blind alley. It had been shown toward the end of the 19th century that the combinatorial approach could be very fruitful in analyzing geometrical problems in all dimensions. However, there were no indications that a solution to Heawood's problem would lead anywhere else. Even the Four Color Problem, which many had discounted for the same reason, has corollaries which mathematicians consider to be interesting, although perhaps not "important."

The first breakthrough to follow Heffter's pioneering work was produced by G. A. Dirac [7], a relative of the famed physicist, in 1952. To understand his accomplishment, let us recall the issue. It was known that every map on the surface of genus g could be colored with no more than H_g colors. To show that this number of colors might actually be required, it would suffice to draw a complete H_g-map on this surface. But is the existence of such a complete map on the surface really necessary? Conceivably the surface S_3 might support a map that requires nine colors even though it might not admit a complete 9-map. That, after all, was the whole point of the Four Color Problem. It was known that no complete 5-map could be drawn in the plane, but that did not preclude the possibility of some other planar map actually requiring five colors. Dirac showed that this situation could not arise on the other surfaces. Specifically, he demonstrated that if the surface of genus g supported a map that required H_g colors, then it would also support a complete H_g-map. Actually, he only proved this for the cases $g = 3$ and $g \geq 5$. His arithmetic got in the way of the cases $g = 0,1,2,4$. Had his proof applied to the case $g = 0$, he would have produced a proof of the Four Color Conjecture—yet another near miss. However, the other values he missed, $g = 1,2,4$, were already covered by Heffter's work.

Next, in what has been called "a tour de force of combinatorial brilliance" ([22, p. 317]), Gerhard Ringel [19] constructed in 1954 a set of arrays which confirmed the existence of a complete H_g-map on the surface S_g for all those values of H_g that leave a remainder of 5 upon division by 12. At last the Heawood conjecture had been verified for an infinite number of surfaces.

The special significance that the number 12 has for this problem was pointed out earlier in the discussion of Heffter's work. It was noted there that arrays that satisfied both the shape and the consistency constraints could only be obtained when m left remainders of 0, 3, 4, or 7 upon division by 12. Since the remainder 5 is not one of these, Ringel had to relax the constraints somewhat. He chose to relax the shape constraint as is evident in the arrays $A_5^{(-1)}$ in Figure 13 and $A_{17}^{(-1)}$ in Figure 14. Specifically, in $A_5^{(-1)}$, row 4 does not contain the entry 5 and, vice versa, row 5 does not contain the entry 4. This means that in the associated map $M_5^{(-1)}$ of Figure 15 the two countries 4 and 5 are not adjacent to each other. The superscript (-1) in the notation for the array and the map records the fact that both the array and the map lack one adjacency to being complete. Similarly, in the map represented by $A_{17}^{(-1)}$ countries 16 and 17 are not adjacent to each other. Now it so happens that the array $A_{17}^{(-1)}$ actually represents an "almost complete" 17-map on the surface S_{15}. Connect the nonadjacent countries 16 and 17 by boring

Country	Adjacency record			
1)	2	4	3	5
2)	3	4	1	5
3)	1	4	2	5
4)	1	2	3	
5)	3	2	1	

Figure 9.13: $A_5^{(-1)}$, the adjacency pattern of the map $M_5^{(-1)}$.

1)	6	4	11	8	13	3	9	17	2	16	15	12	7	5	14	10
4)	9	7	14	11	1	6	12	17	5	16	3	15	10	8	2	13
7)	12	10	2	14	4	9	15	17	8	16	6	3	13	11	5	1
10)	15	13	5	2	7	12	3	17	11	16	9	6	1	14	8	4
13)	3	1	8	5	10	15	6	17	14	16	12	9	4	2	11	7
2)	6	14	7	10	5	9	3	16	1	17	12	15	11	12	4	8
5)	9	2	10	13	8	12	6	16	4	17	15	3	11	1	7	11
8)	12	5	13	1	11	15	9	16	7	17	3	6	2	4	10	14
11)	15	8	1	4	14	3	12	16	10	17	6	9	5	7	13	2
14)	3	11	4	7	2	6	15	16	13	17	9	12	8	10	1	5
3)	16	2	9	1	13	7	6	8	17	10	12	11	11	5	15	4
6)	16	5	12	4	1	10	9	11	17	13	15	14	2	8	3	7
9)	16	8	15	7	4	13	12	14	17	1	3	2	5	11	6	10
12)	16	11	3	10	7	1	15	2	17	4	6	5	8	14	9	13
15)	16	14	6	13	10	4	3	5	17	7	9	8	11	2	12	1
16)	1	2	3	4	5	6	7	8	9	10	11	12	13	14	15	
17)	14	13	6	11	10	3	8	7	15	5	4	12	2	1	9	

$A_{17}^{(-1)}$

Figure 9.14: An array that describes the adjacency pattern of a nearly complete map on S_{15}. This map has seventeen countries of which every two, except 16 and 17, are adjacent to each other.

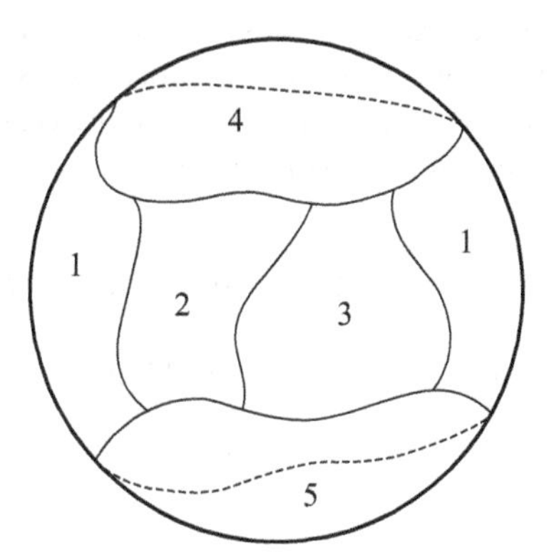

Figure 9.15: $M_5^{(-1)}$, an almost complete 5-map on the sphere. Every two countries, except 4 and 5, touch each other.

an additional tunnel through S_{15}. This gives us a complete 17-map on the surface S_{16}. Since $H_{16} = 17$ this array implies the validity of the Heawood Conjecture for the surface of genus 16.

Due to the special role played by the number 12, the problem was now recognized as possessing 12 cases, depending on H_g's remainder when divided by 12. By 1961 Ringel had also resolved the cases of remainder 7, 10, and 3, making use of the same technique as in the case of 5. The year 1963 saw the emergence of some new tools. William Gustin [9] discovered a way of encoding arrays in a form that greatly resembles electrical networks. In these, the numbers originally used to label countries are interpreted as denoting the intensity of a (fictitious) current flowing along the branches of the network in the direction indicated by the arrowheads. Kirchhoff's Current Law, which states that the total current entering a node equals the total current leaving it, is also satisfied by these networks. Because of this resemblance, Gustin's networks have been dubbed "Current Graphs." To indicate the manner in which an array may be stored in a current graph, we decode CG_{31} (Figure 16). Choosing our departure point arbitrarily, we start out with the network branch labeled 5 and proceed to node A. From here we could choose to go to either B or C. The correct choice is indicated by the way in which the node is drawn. A solid dot indicates that at this node the traveler should choose the left fork, whereas a hollow dot dictates a choice of the right fork. Hence we go on to B and record a 13 in our logbook, to follow the previous entry 5. Node B is also a solid dot, and so we go to D from here and record a 14 in the logbook. By the time node J is reached, the logbook's entries will be 5, 13, 14, 12, 15, 11. Since the node J is represented by a hollow dot, we choose the right fork and go on to I, adding the entry 6 to the log. From I we again bear right, toward H, but this time, since we are progressing against the arrowhead, instead of recording 4 in the log, we enter $31 - 4 = 27$. From the solid dot of H, we choose the left fork to F and record $31 - 15 = 16$ in the log. The solid and hollow nodes are so placed that the initial current 5 will not be encountered again before all the possible currents from 1 to 30 have been recorded in the logbook. In other words, this procedure is set up so that each of the branches of the network will be traversed exactly twice, once in each direction. If we consider the above tour as terminating when 5 is reencountered, then the final log is:

5, 13, 14, 12, 15, 11, 6, 27, 16, 28, 7, 29, 17, 30, 8,

26, 20, 4, 10, 3, 19, 2, 9, 1, 18, 23, 22, 24, 21, 25.

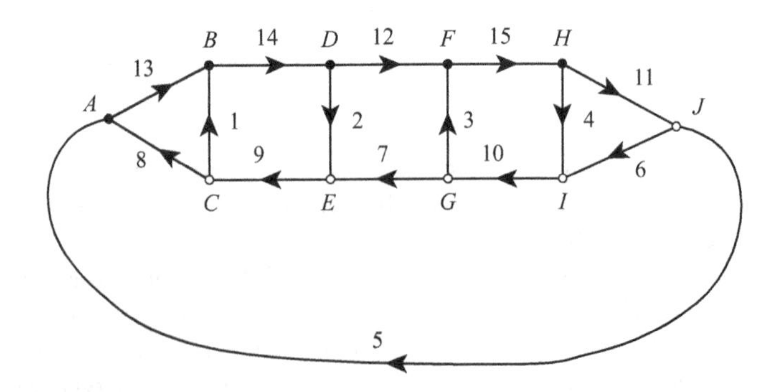

Figure 9.16: CG_{31}, a current graph which encodes a cómplete 31-map on S_{63}.

The ith row of the array A_{31} is now obtained by adding i to each entry of the above log, with the stipulation that $31 + i$ is to be replaced by i. This array satisfies both the shape and consistency constraints and so it attests to the existence of a complete 31-map on the surface S_{31}. Since $H_{63} = [\frac{1}{2}(7 + \sqrt{1 + 48 \times 63})] = 31$, the Heawood Conjecture for that surface is verified.

Mysterious as the above procedure may seem, we know why it works. The distribution of the currents among the links of the network and the relative placements of the solid dots versus the hollow dots among the nodes are such as to guarantee that the shape constraint is fulfilled. It can also be shown that Kirchhoff's Current Law is transformed in the array into the consistency constraint. On the other hand, the heuristics underlying the idea that a map can be encoded as an electrical network are still not well understood. It should be pointed out that complex function theory, within which Riemann's surfaces were first recognized, has very strong connections to potential theory. Indeed, some very deep mathematical theorems become "obvious" when translated into statements about electrons and electrical fields. Be that as it may, these current graphs are of course much more tractable than the arrays they represent, and these arrays are in turn much more tractable than the actual maps they represent. Nevertheless, even these graphs are far from easy to find, especially as for some of the cases they have to be slightly modified in order to work. It was not until 1968 that the combined efforts of W. Gustin, R. K. Guy, C. M. Terry, L. R. Welch, and, most of all, G. Ringel and J. W. T. Youngs (see [20] for details) resolved all of the remaining eight cases of remainders 0, 1, 2, 4, 6, 8, 9, 11. Their work left the cases $H_g = 18, 20$, and 23 unresolved, but these gaps were filled within one year by J. Mayer [16] (professor of French literature at the University of Montpellier). It took three quarters of a century to verify completely the statement that Heawood felt was too obvious to require justification.

9.4 Coloring Maps on Other Surfaces

The work described above completely resolved the coloring problem on Riemann's surfaces, except for the sphere. An issue that has so far been sidestepped is the question of whether there are any other surfaces for which interesting coloring problems could be posed. The reader is reminded that maps of higher complexity were made possible by boring tunnels through the sphere, thus motivating the definition of the Riemann surfaces. Is there anything else that could be done to a sphere to allow for more complex maps? The answer is yes.

The map $\tilde{M}_5$ (Figure 17) displays a complete 5-map on the Möbius strip. This strip is obtained by making a 180° twist in a long ribbon and then gluing its two ends to each other. Now this map can be transferred to the sphere in the following manner. Observe that the edge of the Möbius strip consists of a single closed loop, as opposed to, say, the two loops that would have formed the edge had the ribbon not been twisted before its ends were glued. If a disk is cut out of the surface of the

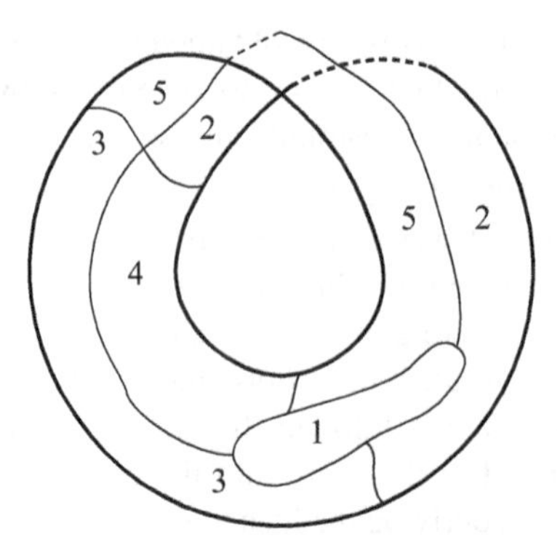

Figure 9.17: $\tilde{M}_5$, a complete 5-map on the Möbius strip.

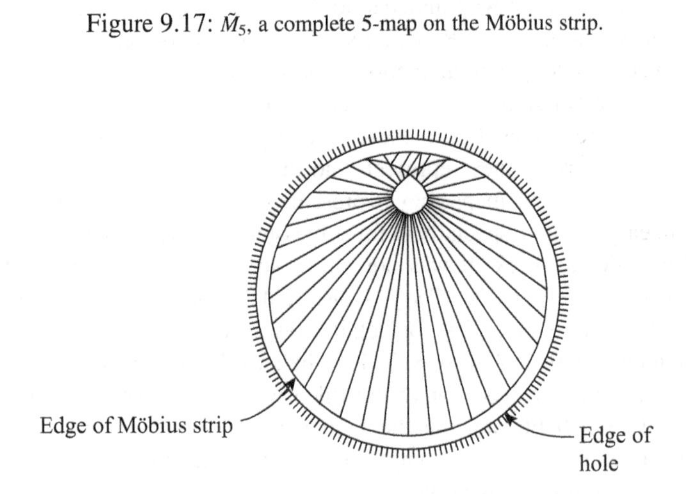

Figure 9.18: Trying to patch a hole with a Möbius strip.

sphere, a closed loop is left as the edge of the perforated sphere. Since all closed loops are essentially identical, it should be possible to patch the hole in the sphere by sewing its edge to the edge of the Möbius strip. Most of the readers who try to carry out this patching process will soon discover that the twist in the strip has a nasty tendency to get in the way (see Figure 18). However, my four-dimensional readers should experience no such difficulties. That extra dimension will provide them with all the room they need to maneuver. For this reason mathematicians have accepted the resulting hybrid as a genuine surface, even though its internal structure prevents it from being realized in our three-dimensional space. It meets the criteria that mathematicians have set for surfaces, namely, they can have no edges and at each point they must be "reasonably" flat.

Of course, once this twisted surface is accepted, one must be prepared to allow for the possibility of adding more than one twist, just as we allowed for the possibility of boring several tunnels in the sphere. The surface obtained by adding g twists to the surface of the sphere is called the *one-sided surface of genus* g, and is denoted by $\tilde{S}_g$. It is one-sided because the twisted patches would allow an ant living on the outside of the sphere to move to its inside simply by walking along a twist in the manner depicted by Escher's famous woodcut. The one-sided surface $\tilde{S}_1$ is the projective plane and $\tilde{S}_2$ is the well-known Klein bottle. By way of contrast, Riemann's surfaces are all two-sided. An ant on the outside has no way of getting into the inside. And what would happen if twists were added to the other two-sided surfaces? Would we then obtain a whole new bewildering collection of surfaces with mixed numbers of tunnels and twists? Fortunately, the answer is no. It has been known since the turn of the century that when a twist is added to the two-sided surface of genus g, one simply obtains the one-sided surface of genus $2g + 1$. Similarly, when a tunnel is bored into the one-sided surface of genus g, the result is the one-sided surface of genus $g + 2$. Moreover, there are no other surfaces. The two-sided and the one-sided surfaces comprise the totality of all surfaces (see, for example, [15]).

One-sided surfaces were still very new when Heawood formulated his problem. At the time they were probably regarded as a curiosity rather than a significant phenomenon, which may explain why Heawood, as well as others who should have known better, ignored them. They were mentioned by Heffter, but he felt, for good reasons, that once the coloring problem was resolved for the two-sided surfaces, a certain theoretical connection between them and their one-sided siblings could be utilized to solve the same problem for the latter as well. In the event, Heffter's good reasons notwithstanding, history did not bear him out.

The Euler-Poincaré formula for the one-sided surfaces states that

$$n - b + m = 2 - g$$

holds for maps on $\tilde{S}_g$. From this it follows that every map on this surface can be colored with

$$\tilde{H}_g = \left[\frac{1}{2}(7 + \sqrt{1 + 24g})\right]$$

colors, and it is of course natural to conjecture that every $\tilde{S}_g$ supports a map that actually requires that many colors. This was indeed done in 1910 by the topologist H. Tietze [21], who also pointed out that the complete map $\tilde{M}_g$ of Figure 19 was implicit in some work by Möbius [17]. Since $\tilde{H}_1 = 6$, this map solves the coloring problem for $\tilde{S}_1$. Every map on $\tilde{S}_1$ can be colored with six colors, and some such map in fact requires as many as six colors.

In 1934, P. Franklin [8] uncovered a surprising fact. He showed that while $\tilde{H}_2 = 7$, every map on the Klein bottle $\tilde{S}_2$ could be colored with no more than

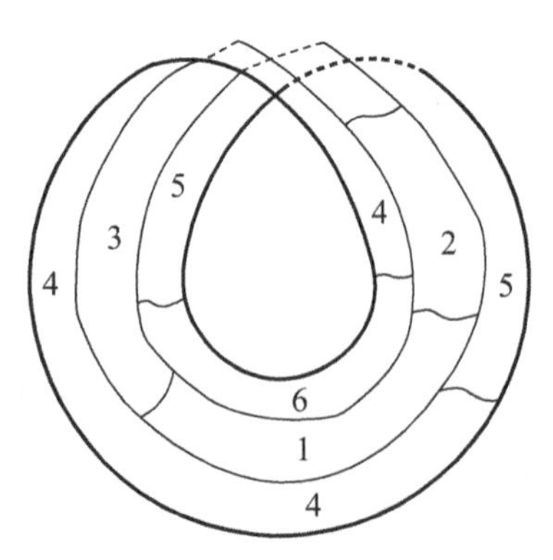

Figure 9.19: $\mathcal{M}_6$, a complete 6-map on the Möbius strip.

six colors! This was the first known failure of the conjecture that each of the numbers H_g and $\tilde{H}_g$ is indeed required by some map on the corresponding surface. Once such an exception is found, of course, it becomes reasonable to expect others to occur. By 1943, the work of I. N. Kagno [13], H. S. M. Coxeter [6], and R. C. Bose [3] showed that no such failures could occur on $\tilde{S}_3$, $\tilde{S}_4$, $\tilde{S}_5$, $\tilde{S}_6$, or $\tilde{S}_7$. They did this by constructing arrays for the appropriate complete m-maps. These arrays are very similar to the ones used for the description of maps on the two-sided surfaces. The main difference is that the consistency rule is replaced by: if the ith row is

$$i) \; \ldots \; j \; k \; \ldots$$

then the kth row is

either $k) \ldots i \, j \ldots$

or $k) \ldots j \, i \ldots$

and the jth row is

either $j) \ldots k \, i \ldots$

or $j) \ldots i \, k \ldots$

In 1954, in the same paper that contained his first major contribution toward the resolution of the Heawood Conjecture, Ringel completely solved Tietze's coloring problem for one-sided surfaces. He accomplished this by producing Heffter-style arrays for the requisite complete maps on all of these surfaces. This was a truly formidable achievement. His proof, by the way, showed that the Klein bottle provided the only exception to the rule that, in general, the number $\tilde{H}_g$ is the solution of the coloring problem on $\tilde{S}_g$. The reason for the earlier resolution of the problem for one-sided surfaces is that the consistency constraint for these is considerably less restrictive than the one for the two-sided surfaces. This pattern recurs frequently in the theoretical study of surfaces. Many questions are formulated first in the context of two-sided surfaces because they are so easily visualized. They are first resolved, however, for one-sided surfaces, despite the fact that the natural habitat of these surfaces is in four-dimensional space.

In 1967 Youngs simplified Ringel's solution by replacing his arrays with current graphs. The complete solution to the problems posed by Heawood and Tietze is now known as the *Ringel-Youngs Theorem.* In 1974 Ringel published his book *Map Color Theorem* [20], which contains both the complete details of the solution and an explanation of the underlying theory of current graphs.

I had several reasons in mind when I decided to recount the history of the Ringel-Youngs Theorem. It is one of my favorite theorems, both because it deals with surfaces and because its proof is so rich. Moreover, I felt that its history makes for a good story. Finally, I saw this as a way to bring the reader closer to the way mathematicians actually operate. Problems, be they solved or unsolved, give rise to more problems. We saw the Four Color Problem motivate the Heawood conjecture, and the latter eventually gave rise to Tietze's one-sided analog. The path leading to a problem's solution is often littered with mistakes, such as those made by Kempe and Heawood. It is the good mathematician who can extract useful information from his own and other people's mistakes, and use them as a basis for new investigations. We saw mere notational conventions transformed into crucial breakthroughs, while other promising technical approaches dwindled into blind alleys. The final solution came as a result of fusion of disparate mathematical disciplines and the cooperative efforts of several mathematicians. These are only some of the elements that go into the production of a mathematical proof, but they are also ones that do not appear in the final product.

The author thanks all those readers of earlier drafts of this article for their helpful critical comments.

Appendix A

Sopra Le Scoperte Dei Dadi

(Galileo, *Opere*, Firenze, Barbera, 8 (1898), pp. 591–594)
Translated by E. H. Thorne
The fact that in a dice-game certain numbers are more advantageous than others has a very obvious reason, i.e., that some are more easily and more frequently obtained than others, which depends on their being able to be made up with more variety of numbers. Thus, a 3 and an 18, which are throws which can only be made in one way with three numbers (that is the latter with 6.6.6 and the former with 1.1.1, and in no other way), are more difficult to make than, e.g., 6 or 7, which can be made up in several ways, that is, a 6 with 1.2.3 and with 2.2.2 and with 1.1.4, and a 7 with 1.1.5, 1.2.4, 1.3.3, and 2.2.3. Nevertheless, although 9 and 12 can be made up in as many ways as 10 and 11, and therefore they should be considered as being of equal utility to these, yet it is known that long observation has made dice-players consider 10 and 11 as being more advantageous than 9 and 12. And it is clear that 9 and 10 can be made up by an equal diversity of numbers (and this is also true of 12 and 11): since 9 is made up of 1.2.6, 1.3.5, 1.4.4., 2.2.5, 2.3.4, 3.3.3 which are six triple numbers, and 10 of 1.3.6, 1.4.5, 2.2.6, 2.3.5, 2.4.4, 3.3.4, and in no other ways, and these also are six combinations. Now I, to oblige him who has ordered me to produce whatever occurs to me about such a problem, will expound my ideas, in the hope not only of solving this problem but of opening the way to a precise understanding of the reasons for which all the details of the game have been with great care and judgment arranged and adjusted.

And to achieve my end with the greatest clarity of which I am capable, I will begin by considering how, since a die has six faces, and when thrown it can equally well fall on any one of these, only six throws can be made with it, each different from all the others. But if together with the first die we throw a second, which also has six faces, we can make 36 throws each different from all the others, since each face of the first can be combined with each face of the second, and in consequence can make six different throws, whence it is clear that such combinations are 6 times 6, i.e., 36. And if we add a third die, since each of its six faces can be combined with each one of the 36 combinations of the other two dice, we shall find that the combinations of three dice are 6 times 36, i.e., 216, each different from the others. But because the numbers in the combinations in three-dice throws are only 16, that is, 3.4.5, etc. up to 18, among which one must divide the said 216 throws, it is necessary that to some of these numbers many throws must belong; and if we can find how many belong to each, we shall have prepared the way to find out what we want to know, and it will be enough to make such an investigation from 3 to 10, because what pertains to one of these numbers will also pertain to that which is immediately greater.

Three special points must be noted for a clear understanding of what follows. The first is that the sum of the points of three dice which is composed of three equal numbers can only be produced by one single throw of the dice: and thus a 3 can only be produced by the three ace-faces, and a 6, if it is to be made up of 3 twos, can only be made by a single throw. Secondly: the sum which is made up of three numbers, of which two are the same and the third different, can be produced by three throws; as e.g., a 4 which is made up of a 2 and two aces, can be produced by three different throws; that is, when the first die shows 2 and the second and the third show the ace; or the second die a 2 and the first and third the ace; or the third a 2 and the first and the second the ace. And so e.g., an 8, when it is made up of 3.3.2, can be produced also three ways: i.e., when the first die shows 2 and the others 3 each, or when the second die shows 2 and the first and third 3, or finally when the third shows 2 and the first and second 3. Thirdly: the sum of points which is made up of three different numbers can be produced in six ways. As for example, an 8 which is made up of 1.3.4, can be made with six different throws: first, when the first die shows 1, the second 3 and the third 4; second, when the first die still shows 1, but the second 4 and the third 3; third, when the second die shows 1, and the first 3 and the second 4; fourth, when the second still shows 1, and the first 4 and the third 3; fifth, when the third die shows 1, the first 3 and the third 4; sixth, when the third shows 1, the first 4 and the second 3.

Therefore, we have so far declared these three fundamental points; first, that the triples, that is the sum of three-dice throws, which are made up of three equal numbers, can only be produced in one way; second, that the triples, which are made up of two equal numbers and the third different, are produced in three ways; third, that those triples, which are made up of three different numbers, are produced in six ways. From these fundamental points we can easily deduce in how many

1																
3																
6	10		9		8		7		6		5		4		3	
10	631	6	621	6	611	3	511	3	411	3	311	3	211	3	111	1
15	622	3	531	6	521	6	421	6	321	6	221	3				
21	541	6	522	3	431	6	331	3	222	1						
25	532	6	441	3	422	3	322	3								
27	442	3	333	1												
108	433	3	333	1												
108		27		25		21		15		10		6		3		1
216																

different ways, or rather in how many different throws, all the numbers of the three dice may be formed, which will be easily understood from the above table:
on top of which are noted the points of the throws from 10 down to 3, and beneath these the different triples from which each of these can result; next to which are placed the number of ways in which each triple can be produced, and under these is finally shown the sum of all the possible ways of producing these throws. So, e.g., in the first column we have the sum of points 10, and beneath it six triples of numbers with which it can be made up, which are 6.3.1, 6.2.2, 5.4.1, 5.3.2, 4.4.2, 4.3.3. And since the first triple 6.3.1 is made up of three different numbers, it can (as is declared above) be made by six different dice-throws, therefore next to this triple a 6 is noted: and since the second triple 6.2.2 is made up of two equal numbers and a third which is different, it can only be produced by three different throws, and therefore a 3 is noted next to it: the third triple 5.4.1, being made up of three different numbers, can be produced by six throws, therefore a 6 is noted next to it and so on with all the other triples. And finally at the bottom of the little column of numbers these throws are all added up: there one can see that the sum of points 10 can be made up by 27 different dice-throws, but the sum of points 9 by 25 only, the 8 by 21, the 7 by 15, the 6 by 10, the 5 by 6, the 4 by 3, and finally the 3 by 1: which all added together amount to 108. And there being a similar number of throws for the higher sums of points, that is for the points 11, 12, 13, 14, 15, 16, 17, 18, one arrives at the sum of all possible throws which can be made up with the faces of three dice, which is 216. And from this table any who understands the game can very accurately measure all the advantages, however small they may be, of the zare, the incontri and any other special rule and term observed in this game.

Table A

Cumulative Normal Probabilities

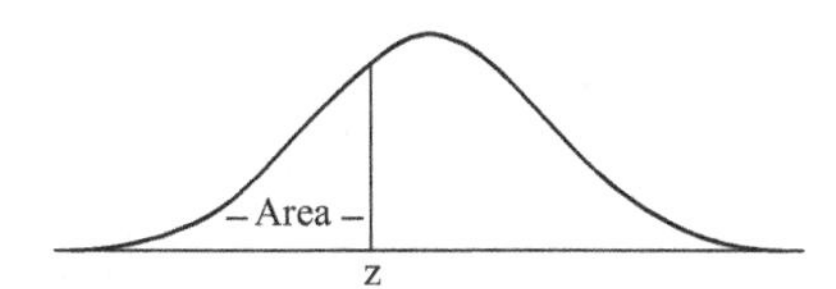

TABLE A (*Cumulative Normal Probabilities*)

Areas under the normal curve.

z	.00	.01	.02	.03	.04	.05	.06	.07	.08	.09
-0.0	0.5000	0.4960	0.4920	0.4880	0.4840	0.4801	0.4761	0.4721	0.4681	0.4641
-0.1	0.4602	0.4562	0.4522	0.4483	0.4443	0.4404	0.4364	0.4325	0.4286	0.4247
-0.2	0.4207	0.4168	0.4129	0.4090	0.4052	0.4013	0.3974	0.3936	0.3897	0.3859
-0.3	0.3821	0.3783	0.3745	0.3707	0.3669	0.3632	0.3594	0.3557	0.3520	0.3483
-0.4	0.3446	0.3409	0.3372	0.3336	0.3300	0.3264	0.3228	0.3192	0.3156	0.3121
-0.5	0.3085	0.3050	0.3015	0.2981	0.2946	0.2912	0.2877	0.2843	0.2810	0.2776
-0.6	0.2743	0.2709	0.2676	0.2643	0.2611	0.2578	0.2546	0.2514	0.2483	0.2451
-0.7	0.2420	0.2389	0.2358	0.2327	0.2296	0.2266	0.2236	0.2206	0.2177	0.2148
-0.8	0.2119	0.2090	0.2061	0.2033	0.2005	0.1977	0.1949	0.1922	0.1894	0.1867
-0.9	0.1841	0.1814	0.1788	0.1762	0.1736	0.1711	0.1685	0.1660	0.1635	0.1611
-1.0	0.1587	0.1562	0.1539	0.1515	0.1492	0.1469	0.1446	0.1423	0.1401	0.1379
-1.1	0.1357	0.1335	0.1314	0.1292	0.1271	0.1251	0.1230	0.1210	0.1190	0.1170
-1.2	0.1151	0.1131	0.1112	0.1093	0.1075	0.1056	0.1038	0.1020	0.1003	0.0985
-1.3	0.0968	0.0951	0.0934	0.0918	0.0901	0.0885	0.0869	0.0853	0.0838	0.0823
-1.4	0.0808	0.0793	0.0778	0.0764	0.0749	0.0735	0.0721	0.0708	0.0694	0.0681
-1.5	0.0668	0.0655	0.0643	0.0630	0.0618	0.0606	0.0594	0.0582	0.0571	0.0559
-1.6	0.0548	0.0537	0.0526	0.0516	0.0505	0.0495	0.0485	0.0475	0.0465	0.0455
-1.7	0.0446	0.0436	0.0427	0.0418	0.0409	0.0401	0.0392	0.0384	0.0375	0.0367
-1.8	0.0359	0.0351	0.0344	0.0336	0.0329	0.0322	0.0314	0.0307	0.0301	0.0294
-1.9	0.0287	0.0281	0.0274	0.0268	0.0262	0.0256	0.0250	0.0244	0.0239	0.0233
-2.0	0.0228	0.0222	0.0217	0.0212	0.0207	0.0202	0.0197	0.0192	0.0188	0.0183
-2.1	0.0179	0.0174	0.0170	0.0166	0.0162	0.0158	0.0154	0.0150	0.0146	0.0143
-2.2	0.0139	0.0136	0.0132	0.0129	0.0125	0.0122	0.0119	0.0116	0.0113	0.0110
-2.3	0.0107	0.0104	0.0102	0.0099	0.0096	0.0094	0.0091	0.0089	0.0087	0.0084
-2.4	0.0082	0.0080	0.0078	0.0075	0.0073	0.0071	0.0069	0.0068	0.0066	0.0064
-2.5	0.0062	0.0060	0.0059	0.0057	0.0055	0.0054	0.0052	0.0051	0.0049	0.0048
-2.6	0.0047	0.0045	0.0044	0.0043	0.0041	0.0040	0.0039	0.0038	0.0037	0.0036
-2.7	0.0035	0.0034	0.0033	0.0032	0.0031	0.0030	0.0029	0.0028	0.0027	0.0026
-2.8	0.0026	0.0025	0.0024	0.0023	0.0023	0.0022	0.0021	0.0021	0.0020	0.0019
-2.9	0.0019	0.0018	0.0018	0.0017	0.0016	0.0016	0.0015	0.0015	0.0014	0.0014
-3.0	0.0013	0.0013	0.0013	0.0012	0.0012	0.0011	0.0011	0.0011	0.0010	0.0010

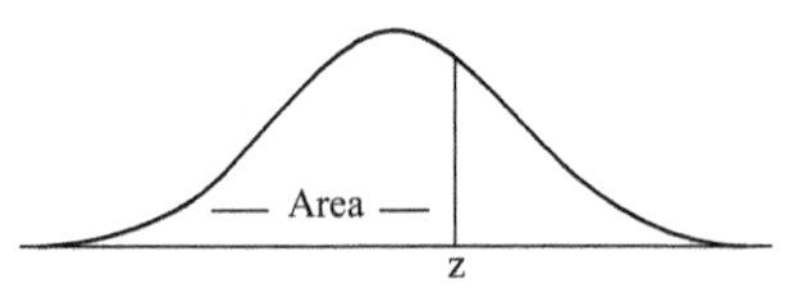

TABLE A (*Continued*)

Areas under the normal curve.

z	.00	.01	.02	.03	.04	.05	.06	.07	.08	.09
0.0	0.5000	0.5040	0.5080	0.5120	0.5160	0.5199	0.5239	0.5279	0.5319	0.5359
0.1	0.5398	0.5438	0.5478	0.5517	0.5557	0.5596	0.5636	0.5675	0.5714	0.5753
0.2	0.5793	0.5832	0.5871	0.5910	0.5948	0.5987	0.6026	0.6064	0.6103	0.6141
0.3	0.6179	0.6217	0.6255	0.6293	0.6331	0.6368	0.6406	0.6443	0.6480	0.6517
0.4	0.6554	0.6591	0.6628	0.6664	0.6700	0.6736	0.6772	0.6808	0.6844	0.6879
0.5	0.6915	0.6950	0.6985	0.7019	0.7054	0.7088	0.7123	0.7157	0.7190	0.7224
0.6	0.7257	0.7291	0.7324	0.7357	0.7389	0.7422	0.7454	0.7486	0.7517	0.7549
0.7	0.7580	0.7611	0.7642	0.7673	0.7704	0.7734	0.7764	0.7794	0.7823	0.7852
0.8	0.7881	0.7910	0.7939	0.7967	0.7995	0.8023	0.8051	0.8078	0.8106	0.8133
0.9	0.8159	0.8186	0.8212	0.8238	0.8264	0.8289	0.8315	0.8340	0.8365	0.8389
1.0	0.8413	0.8438	0.8461	0.8485	0.8508	0.8531	0.8554	0.8577	0.8599	0.8621
1.1	0.8643	0.8665	0.8686	0.8708	0.8729	0.8749	0.8770	0.8790	0.8810	0.8830
1.2	0.8849	0.8869	0.8888	0.8907	0.8925	0.8944	0.8962	0.8980	0.8997	0.9015
1.3	0.9032	0.9049	0.9066	0.9082	0.9099	0.9115	0.9131	0.9147	0.9162	0.9177
1.4	0.9192	0.9207	0.9222	0.9236	0.9251	0.9265	0.9279	0.9292	0.9306	0.9319
1.5	0.9332	0.9345	0.9357	0.9370	0.9382	0.9394	0.9406	0.9418	0.9429	0.9441
1.6	0.9452	0.9463	0.9474	0.9484	0.9495	0.9505	0.9515	0.9525	0.9535	0.9545
1.7	0.9554	0.9564	0.9573	0.9582	0.9591	0.9599	0.9608	0.9616	0.9625	0.9633
1.8	0.9641	0.9649	0.9656	0.9664	0.9671	0.9678	0.9686	0.9693	0.9699	0.9706
1.9	0.9713	0.9719	0.9726	0.9732	0.9738	0.9744	0.9750	0.9756	0.9761	0.9767
2.0	0.9772	0.9778	0.9783	0.9788	0.9793	0.9798	0.9803	0.9808	0.9812	0.9817
2.1	0.9821	0.9826	0.9830	0.9834	0.9838	0.9842	0.9846	0.9850	0.9854	0.9857
2.2	0.9861	0.9864	0.9868	0.9871	0.9875	0.9878	0.9881	0.9884	0.9887	0.9890
2.3	0.9893	0.9896	0.9898	0.9901	0.9904	0.9906	0.9909	0.9911	0.9913	0.9916
2.4	0.9918	0.9920	0.9922	0.9925	0.9927	0.9929	0.9931	0.9932	0.9934	0.9936
2.5	0.9938	0.9940	0.9941	0.9943	0.9945	0.9946	0.9948	0.9949	0.9951	0.9952
2.6	0.9953	0.9955	0.9956	0.9957	0.9959	0.9960	0.9961	0.9962	0.9963	0.9964
2.7	0.9965	0.9966	0.9967	0.9968	0.9969	0.9970	0.9971	0.9972	0.9973	0.9974
2.8	0.9974	0.9975	0.9976	0.9977	0.9977	0.9978	0.9979	0.9979	0.9980	0.9981
2.9	0.9981	0.9982	0.9982	0.9983	0.9984	0.9984	0.9985	0.9985	0.9986	0.9986
3.0	0.9987	0.9987	0.9987	0.9988	0.9988	0.9989	0.9989	0.9989	0.9990	0.9990

Table B

Cumulative Binomial Probabilities

TABLE B (*Cumulative Binomial Probabilities*)

$n = 2$

	p								
k	0.1	0.2	03	0.4	0.5	0.6	0.7	0.8	0.9
0	0.8100	0.6400	0.4900	0.3600	0.2500	0.1600	0.0900	0.0400	0.0100
1	0.9900	0.9600	0.9100	0.8400	0.7500	0.6400	0.5100	0.3600	0.1900
2	1.0000	1.0000	1.0000	1.0000	1.0000	1.0000	1.0000	1.0000	1.0000

$n = 3$

	p								
k	0.1	0.2	03	0.4	0.5	0.6	0.7	0.8	0.9
0	0.7290	0.5120	0.3430	0.2160	0.1250	0.0640	0.0270	0.0080	0.0010
1	0.9720	0.8960	0.7840	0.6480	0.5000	0.3520	0.2160	0.1040	0.0280
2	0.9990	0.9920	0.9730	0.9360	0.8750	0.7840	0.6570	0.4880	0.2710
3	1.0000	1.0000	1.0000	1.0000	1.0000	1.0000	1.0000	1.0000	1.0000

$n = 4$

	p								
k	0.1	0.2	03	0.4	0.5	0.6	0.7	0.8	0.9
0	0.6561	0.4096	0.2401	0.1296	0.0625	0.0256	0.0081	0.0016	0.0000
1	0.9477	0.8192	0.6517	0.4752	0.3125	0.1792	0.0837	0.0272	0.0037
2	0.9963	0.9728	0.9163	0.8208	0.6875	0.5248	0.3483	0.1808	0.0523
3	0.9999	0.9984	0.9919	0.9744	0.9375	0.8704	0.7599	0.5904	0.3439
4	1.0000	1.0000	1.0000	1.0000	1.0000	1.0000	1.0000	1.0000	1.0000

$n = 5$

	p								
k	0.1	0.2	03	0.4	0.5	0.6	0.7	0.8	0.9
0	0.5905	0.3277	0.1681	0.0778	0.0312	0.0102	0.0024	0.0003	0.0000
1	0.9185	0.7373	0.5282	0.3370	0.1875	0.0870	0.0308	0.0067	0.0005
2	0.9914	0.9421	0.8369	0.6826	0.5000	0.3174	0.1631	0.0579	0.0086
3	0.9995	0.9933	0.9692	0.9130	0.8125	0.6630	0.4718	0.2627	0.0815
4	1.0000	0.9997	0.9976	0.9898	0.9688	0.9222	0.8319	0.6723	0.4095
5	1.0000	1.0000	1.0000	1.0000	1.0000	1.0000	1.0000	1.0000	1.0000

TABLE B (*Continued*)

$n = 5$

k	p = 0.1	0.2	03	0.4	0.5	0.6	0.7	0.8	0.9
0	0.5314	0.2621	0.1176	0.0467	0.0156	0.0041	0.0007	0.0000	0.0000
1	0.8857	0.6554	0.4202	0.2333	0.1094	0.0410	0.0109	0.0016	0.0000
2	0.9841	0.9011	0.7443	0.5443	0.3438	0.1792	0.0705	0.0170	0.0013
3	0.9987	0.9830	0.9295	0.8208	0.6562	0.4557	0.2557	0.0989	0.0159
4	0.9999	0.9984	0.9891	0.9590	0.8906	0.7667	0.5798	0.3446	0.1143
5	1.0000	0.9999	0.9993	0.9959	0.9844	0.9533	0.8824	0.7379	0.4686
6	1.0000	1.0000	1.0000	1.0000	1.0000	1.0000	1.0000	1.0000	1.0000

$n = 6$

k	p = 0.1	0.2	03	0.4	0.5	0.6	0.7	0.8	0.9
0	0.5314	0.2621	0.1176	0.0467	0.0156	0.0041	0.0007	0.0000	0.0000
1	0.8857	0.6554	0.4202	0.2333	0.1094	0.0410	0.0109	0.0016	0.0000
2	0.9841	0.9011	0.7443	0.5443	0.3438	0.1792	0.0705	0.0170	0.0013
3	0.9987	0.9830	0.9295	0.8208	0.6562	0.4557	0.2557	0.0989	0.0159
4	0.9999	0.9984	0.9891	0.9590	0.8906	0.7667	0.5798	0.3446	0.1143
5	1.0000	0.9999	0.9993	0.9959	0.9844	0.9533	0.8824	0.7379	0.4686
6	1.0000	1.0000	1.0000	1.0000	1.0000	1.0000	1.0000	1.0000	1.0000

$n = 7$

k	p = 0.1	0.2	03	0.4	0.5	0.6	0.7	0.8	0.9
0	0.4783	0.2097	0.0824	0.0280	0.0078	0.0016	0.0002	0.0000	0.0000
1	0.8503	0.5767	0.3294	0.1586	0.0625	0.0188	0.0038	0.0004	0.0000
2	0.9743	0.8520	0.6471	0.4199	0.2266	0.0963	0.0288	0.0047	0.0002
3	0.9973	0.9667	0.8740	0.7102	0.5000	0.2898	0.1260	0.0333	0.0027
4	0.9998	0.9953	0.9712	0.9037	0.7734	0.5801	0.3529	0.1480	0.0257
5	1.0000	0.9996	0.9962	0.9812	0.9375	0.8414	0.6706	0.4233	0.1497
6	1.0000	1.0000	0.9998	0.9984	0.9922	0.9720	0.9176	0.7903	0.5217
7	1.0000	1.0000	1.0000	1.0000	1.0000	1.0000	1.0000	1.0000	1.0000

TABLE B (*Continued*)

$n=8$									
	p								
k	0.1	0.2	03	0.4	0.5	0.6	0.7	0.8	0.9
0	0.4305	0.1678	0.0576	0.0168	0.0039	0.0007	0.0000	0.0000	0.0000
1	0.8131	0.5033	0.2553	0.1064	0.0352	0.0085	0.0013	0.0000	0.0000
2	0.9619	0.7969	0.5518	0.3154	0.1445	0.0498	0.0113	0.0012	0.0000
3	0.9950	0.9437	0.8059	0.5941	0.3633	0.1737	0.0580	0.0104	0.0004
4	0.9996	0.9896	0.9420	0.8263	0.6367	0.4059	0.1941	0.0563	0.0050
5	1.0000	0.9988	0.9887	0.9502	0.8555	0.6846	0.4482	0.2031	0.0381
6	1.0000	0.9999	0.9987	0.9915	0.9648	0.8936	0.7447	0.4967	0.1869
7	1.0000	1.0000	0.9999	0.9993	0.9961	0.9832	0.9424	0.8322	0.5695
8	1.0000	1.0000	1.0000	1.0000	1.0000	1.0000	1.0000	1.0000	1.0000

$n=9$									
	p								
k	0.1	0.2	03	0.4	0.5	0.6	0.7	0.8	0.9
0	0.3874	0.1342	0.0404	0.0101	0.0020	0.0003	0.0000	0.0000	0.0000
1	0.7748	0.4362	0.1960	0.0705	0.0195	0.0038	0.0004	0.0000	0.0000
2	0.9470	0.7382	0.4628	0.2318	0.0898	0.0250	0.0043	0.0003	0.0000
3	0.9917	0.9144	0.7297	0.4826	0.2539	0.0994	0.0253	0.0031	0.0000
4	0.9991	0.9804	0.9012	0.7334	0.5000	0.2666	0.0988	0.0196	0.0009
5	0.9999	0.9969	0.9747	0.9006	0.7461	0.5174	0.2703	0.0856	0.0083
6	1.0000	0.9997	0.9957	0.9750	0.9102	0.7682	0.5372	0.2618	0.0530
7	1.0000	1.0000	0.9996	0.9962	0.9805	0.9295	0.8040	0.5638	0.2252
8	1.0000	1.0000	1.0000	0.9997	0.9980	0.9899	0.9596	0.8658	0.6126
9	1.0000	1.0000	1.0000	1.0000	1.0000	1.0000	1.0000	1.0000	1.0000

TABLE B (*Continued*)

$n = 10$

	p								
k	0.1	0.2	03	0.4	0.5	0.6	0.7	0.8	0.9
0	0.3487	0.1074	0.0282	0.0060	0.0010	0.0001	0.0000	0.0000	0.0000
1	0.7361	0.3758	0.1493	0.0464	0.0107	0.0017	0.0000	0.0000	0.0000
2	0.9298	0.6778	0.3828	0.1673	0.0547	0.0123	0.0016	0.0000	0.0000
3	0.9872	0.8791	0.6496	0.3823	0.1719	0.0548	0.0106	0.0009	0.0000
4	0.9984	0.9672	0.8497	0.6331	0.3770	0.1662	0.0473	0.0064	0.0001
5	0.9999	0.9936	0.9527	0.8338	0.6230	0.3669	0.1503	0.0328	0.0016
6	1.0000	0.9991	0.9894	0.9452	0.8281	0.6177	0.3504	0.1209	0.0128
7	1.0000	0.9999	0.9984	0.9877	0.9453	0.8327	0.6172	0.3222	0.0702
8	1.0000	1.0000	0.9999	0.9983	0.9893	0.9536	0.8507	0.6242	0.2639
9	1.0000	1.0000	1.0000	0.9999	0.9990	0.9940	0.9718	0.8926	0.6513
10	1.0000	1.0000	1.0000	1.0000	1.0000	1.0000	1.0000	1.0000	1.0000

$n = 11$

	p								
k	0.1	0.2	03	0.4	0.5	0.6	0.7	0.8	0.9
0	0.3140	0.0859	0.0198	0.0036	0.0005	0.0000	0.0000	0.0000	0.0000
1	0.6970	0.3220	0.1130	0.0302	0.0059	0.0007	0.0000	0.0000	0.0000
2	0.9100	0.6170	0.3130	0.1190	0.0327	0.0059	0.0006	0.0000	0.0000
3	0.9810	0.8390	0.5700	0.2960	0.1130	0.0293	0.0043	0.0002	0.0000
4	0.9970	0.9500	0.7900	0.5330	0.2740	0.0994	0.0216	0.0020	0.0000
5	1.0000	0.9880	0.9220	0.7530	0.5000	0.2470	0.0782	0.0117	0.0003
6	1.0000	0.9980	0.9780	0.9010	0.7260	0.4670	0.2100	0.0504	0.0028
7	1.0000	1.0000	0.9960	0.9710	0.8870	0.7040	0.4300	0.1610	0.0185
8	1.0000	1.0000	0.9990	0.9940	0.9670	0.8810	0.6870	0.3830	0.0896
9	1.0000	1.0000	1.0000	0.9990	0.9940	0.9700	0.8870	0.6780	0.3030
10	1.0000	1.0000	1.0000	1.0000	1.0000	0.9960	0.9800	0.9140	0.6860
11	1.0000	1.0000	1.0000	1.0000	1.0000	1.0000	1.0000	1.0000	1.0000

TABLE B (*Continued*)

$n = 12$									
	p								
k	0.1	0.2	03	0.4	0.5	0.6	0.7	0.8	0.9
0	0.2824	0.0687	0.0138	0.0022	0.0002	0.0000	0.0000	0.0000	0.0000
1	0.6590	0.2749	0.0850	0.0196	0.0032	0.0003	0.0000	0.0000	0.0000
2	0.8891	0.5583	0.2528	0.0834	0.0193	0.0028	0.0002	0.0000	0.0000
3	0.9744	0.7946	0.4925	0.2253	0.0730	0.0153	0.0017	0.0000	0.0000
4	0.9957	0.9274	0.7237	0.4382	0.1938	0.0573	0.0095	0.0006	0.0000
5	0.9995	0.9806	0.8822	0.6652	0.3872	0.1582	0.0386	0.0039	0.0000
6	0.9999	0.9961	0.9614	0.8418	0.6128	0.3348	0.1178	0.0194	0.0005
7	1.0000	0.9994	0.9905	0.9427	0.8062	0.5618	0.2763	0.0726	0.0043
8	1.0000	0.9999	0.9983	0.9847	0.9270	0.7747	0.5075	0.2054	0.0256
9	1.0000	1.0000	0.9998	0.9972	0.9807	0.9166	0.7472	0.4417	0.1109
10	1.0000	1.0000	1.0000	0.9997	0.9968	0.9804	0.9150	0.7251	0.3410
11	1.0000	1.0000	1.0000	1.0000	0.9998	0.9978	0.9862	0.9313	0.7176
12	1.0000	1.0000	1.0000	1.0000	1.0000	1.0000	1.0000	1.0000	1.0000

$n = 15$									
	p								
k	0.1	0.2	03	0.4	0.5	0.6	0.7	0.8	0.9
0	0.2059	0.0352	0.0047	0.0005	0.0000	0.0000	0.0000	0.0000	0.0000
1	0.5490	0.1671	0.0353	0.0052	0.0005	0.0000	0.0000	0.0000	0.0000
2	0.8159	0.3980	0.1268	0.0271	0.0037	0.0003	0.0000	0.0000	0.0000
3	0.9444	0.6482	0.2969	0.0905	0.0176	0.0019	0.0000	0.0000	0.0000
4	0.9873	0.8358	0.5155	0.2173	0.0592	0.0093	0.0007	0.0000	0.0000
5	0.9978	0.9389	0.7216	0.4032	0.1509	0.0338	0.0037	0.0001	0.0000
6	0.9997	0.9819	0.8689	0.6098	0.3036	0.0950	0.0152	0.0008	0.0000
7	1.0000	0.9958	0.9500	0.7869	0.5000	0.2131	0.0500	0.0042	0.0000
8	1.0000	0.9992	0.9848	0.9050	0.6964	0.3902	0.1311	0.0181	0.0003
9	1.0000	0.9999	0.9963	0.9662	0.8491	0.5968	0.2784	0.0611	0.0022
10	1.0000	1.0000	0.9993	0.9907	0.9408	0.7827	0.4845	0.1642	0.0127
11	1.0000	1.0000	0.9999	0.9981	0.9824	0.9095	0.7031	0.3518	0.0556
12	1.0000	1.0000	1.0000	0.9997	0.9963	0.9729	0.8732	0.6020	0.1841
13	1.0000	1.0000	1.0000	1.0000	0.9995	0.9948	0.9647	0.8329	0.4510
14	1.0000	1.0000	1.0000	1.0000	1.0000	0.9995	0.9953	0.9648	0.7941
15	1.0000	1.0000	1.0000	1.0000	1.0000	1.0000	1.0000	1.0000	1.0000

TABLE B (*Continued*)

$n = 20$									
	p								
k	0.1	0.2	03	0.4	0.5	0.6	0.7	0.8	0.9
0	0.1216	0.0115	0.0008	0.0000	0.0000	0.0000	0.0000	0.0000	0.0000
1	0.3917	0.0692	0.0076	0.0005	0.0000	0.0000	0.0000	0.0000	0.0000
2	0.6769	0.2061	0.0355	0.0036	0.0000	0.0000	0.0000	0.0000	0.0000
3	0.8670	0.4114	0.1071	0.0160	0.0013	0.0000	0.0000	0.0000	0.0000
4	0.9568	0.6296	0.2375	0.0510	0.0059	0.0003	0.0000	0.0000	0.0000
5	0.9887	0.8042	0.4164	0.1256	0.0207	0.0016	0.0000	0.0000	0.0000
6	0.9976	0.9133	0.6080	0.2500	0.0577	0.0065	0.0003	10.0000	0.0000
7	0.9996	0.9679	0.7723	0.4159	0.1316	0.0210	0.0013	0.0000	0.0000
8	0.9999	0.9900	0.8867	0.5956	0.2517	0.0565	0.0051	0.0001	0.0000
9	1.0000	0.9974	0.9520	0.7553	0.4119	0.1275	0.0171	0.0006	0.0000
10	1.0000	0.9994	0.9829	0.8725	0.5881	0.2447	0.0480	0.0026	0.0000
11	1.0000	0.9999	0.9949	0.9435	0.7483	0.4044	0.1133	0.0100	0.0000
12	1.0000	1.0000	0.9987	0.9790	0.8684	0.5841	0.2277	0.0321	0.0004
13	1.0000	1.0000	0.9997	0.9935	0.9423	0.7500	0.3920	0.0867	0.0024
14	1.0000	1.0000	1.0000	0.9984	0.9793	0.8744	0.5836	0.1958	0.0113
15	1.0000	1.0000	1.0000	0.9997	0.9941	0.9490	0.7625	0.3704	0.0432
16	1.0000	1.0000	1.0000	1.0000	0.9987	0.9840	0.8929	0.5886	0.1330
17	1.0000	1.0000	1.0000	1.0000	0.9998	0.9964	0.9645	0.7939	0.3231
18	1.0000	1.0000	1.0000	1.0000	1.0000	0.9995	0.9924	0.9308	0.6083
19	1.0000	1.0000	1.0000	1.0000	1.0000	1.0000	0.9992	0.9885	0.8784
20	1.0000	1.0000	1.0000	1.0000	1.0000	1.0000	1.0000	1.0000	1.0000

Appendix B

Chapter 1: References and Further Readings

Bennet, Deborah, *Randomness*, Cambridge, MA: Harvard University Press, 1998.
Bernstein, Peter, *Against the Gods: The Remarkable Story of Risk*, New York: John Wiley & Sons, 1996.
David, Florence Nightingale, *Games, Gods, and Gambling*, New York: Dover Publications, 1998.
Everitt, Brian S., *Chance Rules: An Informal Guide to Probability, Risk, and Statistics*, New York: Springer-Verlag, 1999.
Feller, William, *An Introduction to Probability Theory and its Applications*, 2 vols, New York: John Wiley & Sons, 1968.
Galilei, Galileo, Sopra le Scoperte dei Dadi, *Opere, Firenze, Barbera,* 8 (1898), pp. 591–594.
Hald, Anders, *A History of Probability and Statistics and Their Applications before 1750*, New York: John Wiley & Sons, 1990.
_____, *A History of Mathematical Statistics From 1750 to 1930*, New York: John Wiley & Sons, 1998.
Keynes, John M., *A Treatise on Probability*, New York: Harper and Row, 1992.
McGervey, John d., *Probabilities in Everyday Life*. New York: Ivy Books, 1992.
Packel, Edward, *The Mathematics of Games and Gambling*, Washington, DC: Mathematical Association of America, 1981.
Ross, Ken, *A Mathematician at the Ballpark: Odds and Probabilities for Baseball Fans*, New York: Pearson Education, 2004.

Chapter 2: References and Further Readings

Albert, James, *Teaching Statistics Using Baseball*, Washington, DC: Mathematical Association of America, 2003.
Gould, Stephen J., *The Mismeasure of Man*, New York: W. W. Norton, 1996.

Hald, Anders, *A History of Probability and Statistics and Their Applications before 1750*, New York: John Wiley & Sons, 1990.

_____, *A History of Mathematical Statistics From 1750 to 1930*, New York: John Wiley & Sons, 1998.

Jacoby, Russell, and Naomi Glauberman (eds), *The Bell Curve Debate: History, Documents, Opinions*, New York: Times Books, 1995.

Porter, T.M., *The Rise of Statistical Thinking 1820–1900*, Princeton: Princeton University Press, 1986.

Quetelet, Adolphe, *Letters addressed to H. R. H. the grand duke of Saxe Coburg and Gotha, on the Theory of Probabilities as Applied to the Moral and Political Sciences*, trans. O. G. Downes, London: Layton, 1849.

Stigler, Stephen M., *The History of Statistics: The Measurement of Uncertainty before 1900*, Cambridge: The Belknap Press of Harvard University Press, 1986.

_______, *Statistics on the Table,* Cambridge: Harvard University Press, 1999.

Walpole Ronald E., Myers, Raymond H., Myers, Sharon L., Ye, Keying, and Yee, Keying, *Probability and Statistics for Engineers and Scientists*, (7th Edition), New Jersey: Prentice Hall, 2002.

Chapter 3: References and Further Readings

Arrow, Kenneth, *Social Choice and Individual Values,* 2nd ed. New Haven: Yale University Press, 1963.

Barry, Brian W. and Russell Hardin, *Rational Man and Irrational Society? An Introduction and Sourcebook.* Beverly Hills: Sage Publications, 1982.

Bauer, Alissa, "Rank System Strips Teams of Postseason Opportunities." *University Daily Kansan,* p. 2, December 8, 2004.

Black, Duncan, *The Theory of Committees and Elections.* Cambridge: Cambridge University Press, 1958.

Borda, Jean-Charles de, *Memoire sur les Elections au Scrutin'*. Paris: His l'Academie Royale des Sciences, 1781.

Condorcet, Marquis de, *Essai sur* L' *application* L' *analyse a la probabilite dés Décisions Rendues a la Pluralite des* Voix (trans. *Essay* on *the Application of Mathematics to the Theory of Decision-Making*). Paris: L'Imprimerie Royale, 1785.

Devlin, Keith, "Election Math." *Devlin's Angle,* MAA, 2004. *Online. devlin/devlin_114_04.html.*

Fishburn, Peter C., "Paradoxes of Voting." *American Political Science Review* **68**(1974), No. 2, 537–556.

Harsanyi, John C., "Cardinal Welfare, Individualistic Ethics, and lnterpersonal Comparisons of Utility." *Journal of Political Economy* **61**(1955), 309–321.

Hilburn, Robert, "It's Time to Nominate a New System." *Los Angeles Times*, February 28, 2002.

Johnson, Paul E., *Social Choice: Theory and Research.* Thousand Oaks, CA: Sage, 1998.

Jones, Bradford, Benjamin Radcliffe, Charles Taber, and Richard Timpone, "Condorcet Winners and the Paradox of Voting." *American Political Science Review* **89**(1995), No. 1, 137–144.

Mackenzie, Dana, "Making Sense Out of Consensus: How Did That Guy Win?" *SIAM News 33*, 2000a.

Mackenzie, Dana, "May the Best Man Lose." *Discover* **21**(2000b), No. 11, 85–91.

May, Kenneth O, "A Set of Independent Necessary and Sufficient Conditions for Simple Majority Decision." *Econometrica* **20**(1952), 680–784.

Nash, John F, "The Bargaining Problem." *Econometrica* **18**(1950), 155–162.

Nelson, Randy A., Michael R. Donihue, Donald M. Waldman, and Calbraith Wheaton, "What's An Oscar Worth?" *Economic Inquiry* **39**(2001), No. 1, 1–16.

Novaselic, Krist, "Instant Runoff Voting Used for First Time in San Francisco Elections" 2004. http://www.fixour.us/editorials/11.08.04.html.

Novoselic, Krist, *Of Grunge & Government: Let's Fix This Broken Democracy!* New York: Akashic Books, 2004.

Reilly, Benjamin, "Social Choice in the South Seas: Electoral Innovation and the Borda Count in the Pacific Island Countries." *International Political Science Review* **23** (2002), No. 4, 355–372.

Riker, William H, *Liberalism Against Populism:* A *Confrontation Between the Theory of Democracy and the Theory of Social Choice.* San Francisco, CA: W. H. Freeman, 1982.

Riker, William H, *The Art of Political Manipulation.* New Haven, CT: Yale University Press, 1986.

Saari, Donald G, *Geometry of Voting.* Berlin: Springer-Verlag, 1994.

Schulze, Markus, "A New Monotonic and Clone-Independent Single-Winner Election Method." Voting Matters **17**(2003), 9–19.

Woodling, Chuck, "Maybe It's Time To Do Away with AP Poll." *Lawrence Journal World,* p. 3C, December 24, 2004.

Chapter 4: References and Further Readings

Coleman, A., *Game Theory and Experimental Games,* Pergamon Press, Oxford, 1982.

Davenport, W. C., Jamaican Fishing Village, *Papers in Carribean Anthropology,* **59**, (I. Rouse ed.) Yale University Press, 1960.

Davis, M. D., *Game Theory*, Basic Books, New York, 1983.

Dixit, A. and Nalebuff B., *Thinking Strategically*, W. W. Norton & Company, New York, 1991.

Dresher, M., *The Mathematics of Games and Strategy, Theory and Applications,* Dover, New York, 1981.

Gale, D., *The Theory of Linear Economic Models,* McGraw Hill, New York, 1960.

Haywood O. G. Junior, Military Decision and Game Theory. *J. of the Operations Research Society of America,* 2(1954), 365–385.

Heckathorn, D. D., The Dynamics and Dilemmas of Collective action, *American Sociological Review*, **61**(1996) No. 2, 250–277.

Jones, A. J., *Game Theory: Mathematical Models of Conflict,* Halstead Press, Chichester, England.

Luce, R. D., and Raiffa, H., *Games and Decision: Introduction and Critical Survey,* John Wiley and Sons, London 1957.

Maynard Smith, J. and Parker, G. A., The Logic of Animal Conflict, *Nature*, **246**(1973), 15–18.

Maynard Smith, J., *Evolution and the Theory of Games*, Cambridge University Press, London 1982.

Montgomery J. D., Equilibrium Wage Dispersion and Interindustry Wage Differentials, *Quarterly Journal of Economics* 1991, 163–179.

Morris, P., *Introduction to Game Theory*, Springer-Verlag, New York, 1994.

Nash, J., Non-cooperative games, *Annals of Mathematics*, **54**(1950), 286–295.

Packel E., *The Mathematics of Games and Gambling,* The Mathematical Association of America, Washington D. C., 1981.

Poundstone, W., *Prisoner's Dilemma*, Doubleday: New York, 1992.
Rapoport A., Guyer, M. J., and Gordon, D. G., *The* 2 x 2 *Game*, The University of Michigan Press: Ann Arbor, 1976.
Sinervo, B., and Lively, C. M., The Rock-Paper-Scissors game and the evolution of alternative male strategies, *Nature*, **380**(1996), 240–243.
Straffin P. D., *Game Theory and Strategy*, The Mathematical Association of America: Washington, D. C., 1993.
Thomas L. C., *Games, Theory and Applications*, Ellis Horwood Limited: Chichester, England, 1984.
Ventsel, Y. S., *Elements of Game Theory,* Mir Publishers: Moscow, 1980.
von Neumann, J., Zur Theorie der Geselltschaftsspiele, *Mathematische Annalen*, **100**(1928), 295–300.
von Neumann, J. and Morgenstern O., *Theory of Games and Economic Behavior,* Princeton University Press: Princeton 1947.
Williams, J. D., *The Compleat Strategyst*, Dover, New York, 1986.
Zagare, F. C., *Game Theory: Concepts and Applications,* Sage Publications: Beverly Hills, 1984.

Chapter 5: References and Further Readings

Dorfman, Robert, Samuelson, Paul A., and Solow, Robert M., *Linear Programming and Economic Analysis,* New York: Dover Publications, 1987.
Gass, Saul I., *An Illustrated Guide to Linear Programming*, New York: McGraw-Hill, 1970.
______, *Decision Making, Models, and Algorithms*, Florida: Krieger, 1991.
Kolman, Bernard and Beck, Robert E., *Elementary Linear Programming with Applications*, (2nd Edition), New York: Academic Press, 1995.

Chapter 6: References and Further Readings

Ausdley, W. and G., *Designs and Patterns from Historic Ornaments,* Dover Publications: New York, 1989.
Budden, F. J., *Fascination of Groups,* Cambridge University Press, Cambridge, 1972.
Farmer, D. W., *Groups and Symmetry: A Guide to Discovering Mathematics*, American Mathematical Society: Washington DC, 1996.
Grünbaum, B., and Shepard, G. C., *Tilings and Patterns*, W. H. Freeman and Co., 1987.
Jones, O., *The Grammar of Chinese Ornament,* Studio Editions: London, 1987.
Martin, G. E., *Transformation Geometry: An Introduction to Symmetry*, Springer-Veralg: New York, 1982.
Schattschneider, D., The Plane Symmetry Groups, Their Recognition and Notation, *Amer. Math. Monthly*, **85**(1978) no. 6, 439–450.
Weyl, H., *Symmetry*, Princeton University Press: Princeton, 1952.

Chapter 7: References and Further Readings

Coxeter, H. S. M., *Regular Polytopes*, Dover Publications: New York, 1973.
Hilton, P., and Pedersen, Jean, *Build Your Own Polyhedra,* Addison Wesley: Reading, MA, 1994.
Weyl, H., *Symmetry*, Princeton University Press: Princeton, 1952.

Chapter 8: References and Further Readings

[1] R. Adrain, Research concerning the probabilities of the errors which happen in making observations, *Analyst* **1**(1808), 93–109. Reprinted in Stigler 1980.

[2] D. Bernoulli, Dijudicatio maxime probabilis plurium observationam discrepantium atque versimillima inductio inde formanda. *Acta Acad. Sci. Imp. Petrop.*, **1**(1777) 3–23. Reprinted in *Werke* **2**(1982), 361–375. Translated into English by C. G. Allen as "The most probable choice between several discrepant observations and the formation therefrom of the most likely induction," *Biometrica*, **48**(1961) 1–18; reprinted in Pearson and Kendall (1970).

[3] ______, Specimen philosophicum de compensationibus horologicis, et veriorimensura temporis. *Acta Acad. Sci. Imp. Petrop.* **2**(1777), 109–128. Reprinted in *Werke* **2**(1982), 376–390.

[4] F. W. Bessel, Untersuchungen über die Wahrscheinlichkeit der Beobachtungsfehler, *Astron. Nachr.* **15**(1838), 369–404. Reprinted in *Abhandlungen* **2**.

[5] R. Cotes, Aestimatio Errorum in Mixta Mathesi, per Variationes Partium Trianguli Plani et Sphaerici. In *Opera Miscellanea* Cambridge, 1722.

[6] F. N. David, *Games, Gods, and Gambling*, New York: Hafner Pub. Co., 1962.

[7] C. Eisenhart, The background and evolution of the method of least squares. Unpublished. Distributed to participants of the ISI meeting, 1963. Revised 1974.

[8] W. Feller, *An Introduction to Probability Theory and its Applications*, 2 vols, John Wiley & Sons: New York,1968.

[9] G. Galilei, *Dialogue Concerning the Two Chief World Systems - Ptolemaic & Copernican* (S. Drake translator), 2nd ed., Berkeley, Univ. California Press, 1967.

[10] F. Galton, *Hereditary Genius: An Inquiry into its Laws and Consequences*. London: Macmillan, first ed. 1863, 2nd ed. 1892.

[11] ______, Statistics by intercomparison, with remarks on the law of frequency of error. *Philosophical Magazine*, 4th series, **49**(1875), 33–46.

[12] C. F. Gauss, *Theoria Motus Corporum Celestium*. Hamburg: Perthes et Besser, 1809. Translated as *Theory of Motion of the Heavenly Bodies Moving about the Sun in Conic Sections* (trans. C. H. Davis), Boston, Little, Brown 1857. Reprinted: New York: Dover 1963.

[13] C. C. Gillispie, Mémoires inédits ou anonymes de Laplace sur la théorie des erreurs, les polyômes de Legendre, et la philosophie des probabilités. *Révue d'histoire des sciences* **32**(1979), 223–279.

[14] G. J. 'sGravesande, Démonstration mathématique de la direction de la providence divine. *Oeuvres* **2**(1774), 221–236.

[15] G. Hagen, *Grundzüge der Wahrscheinlichskeit-Rechnung*, Dümmler: Berlin, 1837.

[16] A. Hald, *A History of Probability and Statistics and Their Applications before 1750.* New York: John Wiley & Sons, 1990.

[17] ______, *A History of Mathematical Statistics From 1750 to 1930*. New York: John Wiley & Sons, 1998.

[18] J. F. W. Herschel, Quetelet on probabilities, *Edinburgh Rev.* **92**(1850) 1–57.

[19] J. Kepler, *New Astronomy* (W. H. Donahue, translator) Cambrdige: Cambridge University Press, 1992.

[20] J. Lancaster, private communication, 2004.

[21] P. S. Laplace, Mémoire sur la probabilité des causes par les évènements. *Mémoires de l'Academie royale des sciences presentés par divers savan* **6**(1774), 621–656. reprinted in Laplace, 1878–1912, Vol. 8, pp 27–65. Translated in [38].

[22] _______, Mémoire sur les approximations des formules qui sont fonctions de très grand nombres et sur leur applications aux probabilités. *Mémoires de l'Academie des sciences de Paris*, 1809, pp. 353–415, 559–565. Reprinted in *Oeuvres complètes de Laplace*, Paris: Gauthier-Villars, vol. 12, pp. 301–353.

[23] _______, Recherches sur le milieu qu'il faut choisir entre les résultat de plusieurs obervations. In [13] pp. 228–256.

[24] J. C. Maxwell, Illustrations of the dynamical theory of gases. *Phil. Mag.* **19**(1860), 19–32. Reprinted in *The Scientific papers of James Clerk Maxwell*. Cambridge, UK: Cambridge University Press, 1890, and New York, Dover, 1952.

[25] A. de Moivre, *Approximatio ad Summam Terminorum Binomii in Seriem Expansi*. Printed for private circulation, 1733.

[26] A. de Morgan, Least Squares, *Penny Cyclopaedia*. London: Charles Knight, 1833–1843.

[27] E. S. Pearson, and M. Kendall (eds.), *Studies in the history of statistics and probability*. Vol. 1, London: Griffin, 1970.

[28] K. Pearson, Notes on the History of Correlation. *Biometrika* **13**(1920) 25–45. Reprinted in [27].

[29] R. L. Placket, The principle of the arithmetic mean, *Biometrika* **45**(1958), 130–135. Reprinted in [27].

[30] _______, Data Analysis Before 1750, *International Statistical Review*, **56**(1988), No. 2, 181–195.

[31] T. M. Porter, *The Rise of statistical Thinking 1820–1900*, Princeton: Princeton University Press, 1986.

[32] A. Quetelet, *Lettres à S. A. R. Le Duc Régnant de Saxe Cobourg et Gotha, sur la théorie des probabilités, appliquée aux sciences morales et politique*. Brussels: Hayez, 1846.

[33] _______, *Letters addressed to H. R. H. the grand duke of Saxe Coburg and Gotha, on the Theory of Probabilities as Applied to the Moral and Political Sciences*, trans. O. G. Downes. London: Layton, 1849. Translation of [32].

[34] _______, *A Treatise on Man and the Development of his Faculties*, New York: Burt Franklin, 1968.

[35] O. B. Sheynin, *The history of the theory of errors*. Deutsche Hochschulschriften [German University Publications], 1118. Hänsel-Hohenhausen, Egelsbach, 1996, 180 pp.

[36] T. Simpson, A Letter to the Right Honourable George Macclesfield, President of the Royal Society, on the Advantage of taking the Mean, of a Number of Observations, in practical Astronomy, *Phil. Trans.* **49**(1756), 82–93.

[37] _______, *Miscellaneous Tracts on Some Curious, and Very Interesting Subjects in Mechanics, Physical-Astronomy, and Speculative Mathematics*. London: Nourse, 1757.

[38] S. M. Stigler, Memoir on the probability of the causes of events. *Statistical Science* **1**(1986) 364–378.

[39] _______, *The History of Statistics: The Measurement of Uncertainty before 1900*. Cambridge: The Belknap Press of the Harvard University Press, 1986.

[40] _______, *American Contributions to Mathematical Statistics in the Nineteenth Century*. 2 vols., New York: Arno Press, 1980.

[41] _______, *Statistics on the Table*, Cambridge: Harvard University Press, 1999.

Chapter 9: References and Further Readings

[1] K. Appel and W. Haken, *The four color problem*, Mathematics Today, ed. L. A. Steen, Springer-Verlag. 1978, pp. 153–180.

[2] N. L. Biggs, E. K. Lloyd, and R. J. Wilson, *Coloring maps on surfaces,* Graph Theory, 1736–1936, Oxford University Press, 1976, pp. 109–130.

[3] R. C. Bose, *On the construction of balanced incomplete block designs*, Ann. of Eugenics, 9 (1939) 353–399.

[4] H. Cohn, *Conformal Mapping on Riemann Surfaces*, Dover, 1967.

[5] R. Courant and H. Robbins, *What is Mathematics?*, Oxford, 1941.

[6] H. S. M. Coxeter, *The map-coloring of unorientable surfaces*, Duke Math. J., 10 (1943) 293–304.

[7] G. A. Dirac, *Map colour theorems*, Can. J. Math., 4 (1952) 480–490.

[8] P. Franklin, *A six color problem*, J. Math. Physics, 13 (1934) 363–369.

[9] W. Gustin, *Orientable embeddings of Cayley graphs*, Bull. Amer. Math. Soc., 69 (1963) 272–275.

[10] P. J. Heawood. *Map color theorem*, Quart. J. Math., 24 (1890) 332–338.

[11] L. Heffter, *Uber das Problem der Nachbargebiete*, Math. Ann. 38 (1891) 477–508. An English translation of much of this is in reference [2].

[12] D. Hilbert and S. Cohn-Vossen, *Geometry and the Imagination*, Chelsea, 1952.

[13] I. N. Kagno, *A note on the Heawood color formula*, J. Math. Physics, 14 (1935) 228–231.

[14] A. B. Kempe, *On the geographical problem of the four colors*, Amer. J. Math., 2 (1879) 193–200.

[15] W. S. Massey, *Algebraic Topology: An Introduction*, Harcourt, Brace & World, 1967.

[16] J. Mayer, *Le Probléme des Regions Voisines sur les Surfaces Closes Orientables*. J. Comb. Th, 6 (1969) 177–195.

[17] A. F. Möbius, *Uber die Bestimmung des lnhaltes eines Polyeders*, Ber. K. Sachs. Ges. Wiss. Leipzig Math-Phys C1., 17 (1865) 31–68, also Werke, vol. 2, pp. 473–512.

[18] B. Riemann. *Grundlagen fur eine allgemeine Theorie des Funktionen einer verandliche Grosse*, Collected Works, Dover. 1953.

[19] G. Ringel, *Bestimmung der Maximalzahl der Nachbargebiete auf nichtorientierbaren Flachen*, Math. Ann., 127 (1954) 181–214.

[20] _______. *Map Color Theorem*, Springer-Verlag, 1974.

[21] H. Tietze, *Eine Bemerkungen uber das Problem der Kartenfarbens auf cinseitigen Flachen*, Jahrsber. Deutsch. Math. Vereinigung, 19 (1910), pp. 155–159. An English translation of much of this is in reference [2].

[22] J. W. T. Youngs, *The Heawood map-coloring conjecture*. Graph Theory and Theoretical Physics, Academic Press, 1967, pp. 313–354.

Appendix C

Solutions

Exercises 1.1

1. a) 10/36 = 27.8% b) 16/36 = 44.4% c) 8/36 = 22.2%
   d) 24/36 = 66.7% e) 9/36 = 25% f) 9/36 = 25%
   g) 10/36 = 27.8% h) 11/36 = 30.6% i) 35/36 = 97.2%

2. a) 14/64 = 21.9% b) 11/32 = 34.4% c) 12/64 = 18.8%
   d) 34/64 = 53.1% e) 1/4 = 25% f) 1/4 = 25%
   g) 14/64 = 21.9% h) 15/64 = 23.4% i) 63/64 = 98.4%

3. (1, 1) (1, 2) (1, 3) (1, 4)
   (2, 1) (2, 2) (2, 3) (2, 4)
   (3, 1) (3, 2) (3, 3) (3, 4)
   (4, 1) (4, 2) (4, 3) (4, 4)

4. a) 6/16 = 37.5% b) 10/16 = 62.5% c) 1/4 = 25%
   d) 14/16 = 87.5% e) 1/4 = 25% f) 1/4 = 25%
   g) 6/16 = 37.5% h) 7/16 = 43.8% i) 15/16 = 93.4%

5. a) 7/24 = 29.2% b) 11/24 = 45.8% c) 6/24 = 25%
   d) 17/24 = 70.8% e) 6/24 = 25% f) 6/24 = 25%
   g) 8/24 = 33.3% h) 9/24 = 37.5% i) 23/24 = 95.8%

6. (1,1), (1,2), (1,3), (1,4), (2,1), (2,2), (2,3), (2,4), (3,1), (3, 2), (3,3), (3,4), (4,1), (4,2), (4,3), (4,4). 1/16 = 6.2%.

7. a) 22/144 = 15.5% b) 34/144 = 23.6% c) 20/144 = 13.9%
d) 54/144 = 37.5% e) 36/144 = 25% f) 36/144 = 25%
g) 22/144 = 15.5% h) 23/144 = 16% i) 143/144 = 99.3%

8. a) 2/64 = 3.1% b) 3/64 = 4.7% c) 4/64 = 6.2%
d) 5/64 =7.8% e) 6/64 = 9.4% f) 7/64 = 10.9%
g) 8/64 = 12.5%

9. a) 2/144 = 1.4% b) 3/144 = 2.1% c) 4/144 = 2.8%
d) 5/144 = 3.5% e) 6/144 = 4.2% f) 7/144 = 4.9%
g) 8/144 = 5.6% h) 9/144 = 6.2% i) 10/144 = 6.9%
j) 11/144 = 7.6% k) 12/144 = 8.3% l) 11/144 = 7.6%
m) 10/144 = 6.9% n) 9/144 = 6.2%

10. a) 13/52 = 25% b) 26/52 = 50% c) 4/52 = 7.7%
d) 12/52 = 23.1% e) 12/52 = 23.1% f) 2/52 = 6.24%

Exercises 1.2

1. a) 1/216 = 0.5% b) 3/216 = 1.4% c) 6/216 = 2.8%
d) 10/216 = 4.6% e) 15/216 = 6.9% f) 21/216 = 9.7%
g) 27/216 = 12.5% h) 25/216 = 11.6% i) 21/216 = 9.7%
j) 15/216 = 6.9% k) 10/216 = 4.6% l) 6/216 = 2.8%
m) 3/216 = 1.4% n) 1/216 = 0.5%

2. a) 1/512 = 0.2% b) 3/52 = 0.6% c) 6/512 = 1.2%
d) 10/512 = 2.0% e) 15/512 = 2.9% f) 21/512 = 4.1%
g) 28/512 = 5.5% h) 36/512 = 7% i) 42/512 = 8.24%
j) 46/512 = 9.0% k) 48/512 = 9.4% l) 48/512 = 9.4%
m) 46/512 = 9.0% n) 42/512 = 8.2%

3. a) 1/64 = 1.5% b) 3/64 = 4.7% c) 6/64 = 9.4%
d) 10/64 = 15.6% e) 12/64 = 18.8% f) 12/64 = 18.8%
g) 10/64 = 15.6%

4. a) 1/1728 = 0.1% b) 3/1728 = 0.2% c) 6/1728 = 0.3%
d) 10/1728 = 0.6% e) 15/1728 = 0.9% f) 21/1728 = 1.2%
g) 28/1728 = 1.6% h) 36/1728 = 2.1% i) 45/1728 = 2.6%
j) 55/1728 = 3.2% k) 66/1728 = 3.8% l) 78/1728 = 4.5%
m) 88/1728 = 5.1% n) 96/1728 = 5.6%

5. a) 1+1+1+1 b) 2+1+1+1 c) 3+1+1+1, 2+2+1+1
   d) 4+1+1+1, 3+2+1+1, 2+2+2+1
   e) 5+1+1+1, 4+2+1+1, 3+3+1+1, 3+2+2+1, 2+2+2+2
   f) 6+1+1+1, 5+2+1+1, 4+3+1+1, 4+2+2+1, 3+3+2+1, 3+2+2+2
   g) 7+1+1+1, 6+2+1+1, 5+3+1+1, 5+2+2+1, 4+4+1+1, 4+3+2+1, 4+2+2+2, 3+3+3+1, 3+3+2+2

6. 4321, 4312, 4231, 4213, 4132, 4123, 3421, 3412, 3241, 3214, 3142, 3124, 2431, 2413, 2341, 2314, 2143, 2134, 1432, 1423, 1342, 1324, 1243, 1234.

7. 3321, 3312, 3231, 3213, 3123, 3132, 2313, 2331, 2133, 1332, 1323, 1233

8. 2211, 2121, 2112, 1221, 1212, 1122.

9. 2111, 1211, 1121, 1112

10. Every permutation of xxxx looks the same.

11. a) 1/256 = 0.4% b) 4/256 = 1.6% c) 10/256 = 3.9%
    d) 20/256 = 7.8% e) 31/256 = 12.11% f) 40/256 = 15.6%
    g) 44/256 = 17.2%

12. a) 1/1296 = 0.08% b) 4/1296 = 0.3% c) 10/1296 = 0.8%
    d) 20/1296 = 1.5% e) 35/1296 = 2.7% f) 56/1296 = 4.3%
    g) 80/1296 = 6.2%

13. a) 1/4096 = 0.02% b) 4/4096 = 0.1% c) 10/4096 = 0.2%
    d) 20/4096 = 0.5% e) 35/4096 = 0.8% f) 56/4096 = 1.1%
    g) 84/4096 = 2.1%

14. a) 1/20736 b) 4/20736 c) 10/20736
    d) 20/20736 e) 35/20736 f) 56/20736
    g) 84/20736

15. Join two pentagon-based pyramids at their base.

16. Make use of the answer to Exercise 15 and the hint.

17. Join two heptagon-based pyramids at their base and label diametrically opposite facets with the same number.

Exercises 1.3

	a)	b)	c)	d)
1.	a) 40.2%	b) 40.2%	c) 16.1%	d) 0%
2.	a) 33.5%	b) 40.2%	c) 20.1%	d) 0.002%
3.	a) 0%	b) 0.1%	c) 31.1%	d) 13.3%
4.	a) 0.02%	b) 27.3%	c) 3.9%	
5.	a) 13.2%	b) 32.9%	c) 0.4%	
6.	a) 0%	b) 12.1%		
7.	a) 37.7%	b) 39.9%	c) 17.6%	
8.	a) 84.9%	b) 30.4%	c) 15.3%	d) 49.6%
9.	a) 86.8%	b) 25.2%	c) 12%	d) 73.6%
10.	a) 3.9%	b) 99.1%	c) 3%	d) 90.6%
11.	a) 49.7%	b) 79.7%	c) 29.4%	d) 48.6%
12.	a) 83.7%	b) 47.2%	c) 30.9%	d) 80.1%
13.	a) 34.1%	b) 88.9%	c) 23%	d) 71.4%
14.	a) 97.3%	b) 8.2%	c) 5.5%	d) 40.1%

Exercises 1.4

	a)	b)	c)	d)
1.	a) 8.9%	b) 93.1%	c) 5.6%	d) 86.2%
2.	a) 7.1%	b) 94.8%	c) 6.4%	d) 90%
3.	a) 69.8%	b) 38.3%	c) 73.8%	d) 8.1%
4.	a) 70.4%	b) 37.4%	c) 76.2%	d) 7.8%
5.	a) 77.8%	b) 24.4%	c) 2.2%	d) 76.5%
6.	a) 0.9%	b) 99.4%	c) 6.6%	d) 81%

7. a) 18% b) 89.9% c) 14.5% d) 17.8%

8. a) 1.5% b) 99.6% c) 9.7% d) 66.2%

9. a) 4.6% b) 96.8% c) 7% d) 93.7%

10. a) 92.4% b) 14.2% c) 13.2% d) 29.8%

11. a) 99.7% b) 0.3% c) 0.08% d) 97.1%

Exercises 2.1

1. 1 2. 1 3. 3.53 4. 5.83 5. 0.47

6.

Data	Freq.
0	8
1	11
2	8
3	5
4	2
5	1

mean = 1.57, st dev = 1.29

7.

Index	Data	Frequency	Relative F.
1	60	5	12.5%
2	61	3	7.5%
3	62	3	7.5%
4	63	4	10%
5	64	5	12.5%
6	65	3	7.5%
7	66	2	5%
8	67	6	15%
9	68	4	10%
10	69	5	12.5%
		40	100

Mean = 64.7, standard deviation = 3.0

8.

Groups	Freq.
0.5–0.9	4
1.0–1.4	10
1.5–1.9	6
2.0–2.4	8

2.5–2.9	4
3.0–3.4	3

Mean = 1.82, st dev = 0.75

9.

Index	Group	Frequency	Relative F.	Mark
1	5–9	1	2%	7
2	10–14	5	10%	12
3	15–19	8	16%	17
4	20–24	12	24%	22
5	25–29	13	26%	27
6	30–34	6	12%	32
7	35–39	5	10%	37
		50	100	

Mean = 23.9, standard deviation = 7.5

10. Relative frequencies: 13.5%, 37.2%, 26.7%, 10.7%, 9.1%, 2.7%
Mean = 21.8, standard deviation = 12.5

11.

Index	Group	Frequency	Relative F.	Mark
1	30–36	9	4.5%	33
2	37–43	73	36.5%	40
3	44–50	96	48%	47
4	51–57	9	4.5%	54
5	58–64	7	3.5%	61
6	65–71	6	3%	68
		200	100%	

Mean = 45.25, standard deviation = 6.8

12. Relative frequencies: 13.2%, 39.7%, 27.9%, 10.3%, 8.8%
Mean = 15.88, standard deviation = 0.33

13.

Index	Group	Frequency	Relative F.	Mark
1	16–18	111	1.9%	17
2	19–21	1380	23.4%	20
3	22–24	1158	19.7%	23
4	25–27	928	15.7%	26
5	28–30	847	14.4%	29
6	31–33	591	10.0%	32
7	34–36	878	14.9%	35

Mean = 26.2, standard deviation = 5.35

15. Mean = 2; (cc) = 4(1–2) + 4(3–2) = –4 + 4 = 0

16. (1–4.4) + (2–4.4) + (3–4.4) + (6–4.4) + (10–4.4) = 0

Exercises 2.2

1. a) Median = 3, average = 4
   b) $|1-3| + |2-3| + |3-3| + |4-3| + |10-3| = 11$
   $|1-4| + |2-4| + |3-4| + |4-4| + |10-4| = 12$
   c) $|1-3|^2 + |2-3|^2 + |3-3|^2 + |4-3|^2 + |10-3|^2 = 55$
   $|1-4|^2 + |2-4|^2 + |3-4|^2 + |4-4|^2 + |10-4|^2 = 50$

2. a) 6.5, 7 b) 14, 14 c) 26.5, 24

3. a) Median = 5.3, average = 5.4 b) 1.1, 1.2 c) 0.55, 0.50

4. a) 10.3, 10.3 b) 0.4, 0.4 c) 0.06, 0.06

5. a) 9.1, 9.14 b) 0.2, 0.24 c) 0.02, 0.012

6. a) 1.85, 1.8 b) 1, 1 c) 0.235, 0.22

7. a) 8.6, 8.6 b) 0.6, 0.6 c) 0.06, 0.06

8. a) 8.7, 8.667 b) 0.2, 0.333 c) 0.04, 0.033

Exercises 2.3

1. a) $y = -x + 2$ b) (0, 0) c) 0 e) $y = -x + 2$, (0, 0), $0 = 0$.

2. a) $y = 2x$ b) 0, 0 c) 4 e) $y = 2x$; 0, 0; $0 \geq 0$

3. a) $y = x$ b) (0, 0, 0) c) 3 e) $y = x$, (0, 0, 0), $0 = 0$

4. a) $y = 3x/2 - 1/6$ b) 1/6, −1/3, 1/6 c) 22/3 = 7.333
   e) $y = 3x/2$; 0, −0.5, 0; $1/4 > 1/6$

5. a) $y = 7/3$ b) (−1/3, 2/3, −1/3) c) 7/3 e) $y = 2$, (0, 1, 0), $2/3 < 1$

6. a) $y = 5.05 - 0.9x$ b) −0.05, −0.15, 0.65, −0.45 c) 3.25
   e) $y = 5 - x$; 0, 0, 1, 0; $1 > 0.65$

7. a) $y = -1.06x + 36.29$ b) (0.31, −0.39, 0.21) c) 15.09
   e) $y = -1.06x + 36.7$, (0, −2/3, 0), $0.292 < 0.444$

8. a) $y = 100.77 - 3.84x$ b) 0.385, −1.538, 1.154 c) 62.31
   e) $y = 100 - 3.75x$; 0, −2.5, 0; $3.85 < 6.25$

9. $y = -316.3x + 6437.3$, 3274.4

10. $y = 36.05 + 0.51x$, 66.49

11. $y = 1.58x + 133.51$, 212.42

Exercises 2.4

1. a) 70.3% b) 29.6% c) 88.4% d) 89.4% e) 72.1%
   f) 15.3%

2. a) 30,852 b) 32,084 c) 25,602
   d) 26,610 e) 29,200 f) 27,800

3. a) 78 b) 81 c) 65 d) 67 e) 74
   f) 70

4. a) 48.6% b) 3.0% c) 34.4%
   d) 14.2% e) 83.1% f) 47.2%

5. a) 11,949 b) 2,536 c) 3084 d)11,575 e) 3423
   f) Either 0 or 1271, depending on the interpretation of "exactly.
   g) 6847

6. a) 72.0 inches b) 74.0 inches
   c) 63.1 inches d) 64.8 inches
   e) 69.2 inches f) 66.8 inches

7. a) 4.8% b) 90.4% c) 4.8% d) 70.65% e) 9.6%
   f) 1.2%

8. a) 0.3% b) 67.3% c) 20.9%
   d) 88.5% e) 11.5% f) 0.1%
   g) 4.6% h) 0.0%

9. a) 62.5% b) 99.4% c) 30.2% d) 30.9% e) 30.2%
   f) 93.3% g) 30.2% h) 18.4%

10. a) 38.1% b) 72.6% c) 22.0%
    d) 50% e) 68.3% f) 27.4%
    g) 54.9% h) 16.2%

11. a) 15.9% b) 84% c) 0.1% d) 49.5% e) 68.3%
    f) 89%

12. a) 0.4% b) 59.0% c) 15.49%
    d) 74.9% e) 25.1% f) 1%

13. a) 44% b) 84.1% c) 14.6% d) 59.9% e) 77.5%
    f) 15.9% g) 45.3% h) 26.1%

14. a) 4.10 b) 4.37 c) 2.98
    d) 3.20 e) 3.75 f) 3.45

15. a) [25, 29.8], [22.6, 32.2], [20.2, 34.6] b) 68.3%, 95.4%, 99.7%

16. a) [12, 18.2], [8.9, 21.3], [5.8, 24.4] b) 68.3%, 95.4%, 99.7%

Exercises 2.5

1. a) (0.61, 0.82), 0.12, 4145 b) (0.63, 0.80), 0.09, 2401
   c) (0.65, 0.78), 0.075, 1692

2. a) C.I. = (0.54, 0.70) MOE = 0.08 $n = 7368$
   b) C.I. = (0.56, 0.68) MOE = 0.06 $n = 4269$
   c) C.I. = (0.57, 0.67) MOE = 0.05 $n = 3007$

3. a) (0,07, 0.17), 0.07, 16577 b) (0.09, 0.16), 0.06, 9604
   c) (0.09, 0.15), 0.05, 6766

4. a) C.I. = (0.81, 0.90) MOE = 0.06 $n = 2653$
   b) C.I. = (0.82, 0.89) MOE = 0.05 $n = 1537$
   c) C.I. = (0.83, 0.89) MOE = 0.04 $n = 1083$

5. a) (0.01, 0.12), 0.11, 16577 b) (0.03, 0.11), 0.08, 9604
   c) (0.03, 0.10), 0.07, 6766

6. a) (14.72, 14.88) b) (14.74, 14.86)
   c) (14.75, 14.85)

7. a) (3.02, 4.18) b) (3.16, 4.04) c) (3.23, 3.98)

8. a) (15.68, 15.72) b) (15,68, 15.72) c) (15.68, 15.72)

9. a) (239, 245) b) (239, 245) c) (240, 244)

10. a) (28,881, 29,802) b) (28,991, 29,693)
    c) (29,048, 29,636)

11. p-value = 0.4%. Reject the 60% belief at all levels.

12. p-value = 0. Reject the null hypothesis in a, b, and c.

13. p-value = 0.1%. Reject the 90% belief at all levels.

14. p-value = 33.9%. Retain the null hypothesis in a, b, and c.

15. p-value = 9.9%. Reject the 75% at level 10% only.

16. p-value = 1.8%. Retain the null hypothesis in a and reject it in b and c.

17. p-value = 1.5%. Conclude they are different at levels 5% and 10% only.

18. p-value = 1.2%. Retain the null hypothesis in a and reject it in b and c.

19. p-value = 4.6%. Conclude there was a change at the 5% and 10% levels only.

20. p-value = 1.2%. Retain the null hypothesis in a and reject it in b and c.

Exercises 2.6

1. b) 2.5, 1.12 d) 2.5, 0.65 f) 2.5, 0.37

2. b) 3.5, 1.18 c) 5/2, 3, 7/2, 7/2, 4, 9/2 d) 3.5, 0.66
   e) 3, 10/3, 11/3, 4 f) 3.5, 0.37

3. b) 3.5, 2.06 d) 3.5, 1.19 f) 3.5, 0.69

4. b) 2, .71 c) 3/2, 3/2, 2, 2, 3/2, 5/2 d) 2, 0.41
   e) 5/3, 2, 2, 7/3 f) 2, 0.24

5. b) 2.25, 0.83 d) 2.25, 0.48 f) 2.25, 0.28

6. b) 1.75, 0.43 c) 1.5, 1.5, 1.5, 2, 2, 2 d) 1.75, 0.25
   e) 2, 5/3, 5/3, 5/3 f) 1.75, 0.14

7. b) 2, 0 d) 2, 0 f) 2, 0

8. b) 3, 1.41 c) 1.5, 2, 2.5, 3, 2.5, 3, 3.5, 3.5, 4, 4.5 d) 3, .87
   e) 2, 7/3, 8/3, 8/3, 3, 10/3, 3, 10/3, 11/3, 4
   f) 3, 0.58

9. b) 2.4, 1.02 d) 2.4, 0.62 f) 2.4, 0.42

10. b) 2.2, .75 c) 1.5, 1.5, 2, 2, 2, 2.5, 2.5, 2.5, 2.5, 3
 d) 2.2, 0.46 e) 5/3, 2, 2, 2, 2, 7/3, 7/3, 7/3, 8/3, 8/3
 f) 2.2, 0.31

11. b) 2, 0.63 d) 2, 0.39 f) 2, 0.26

12. b) 1.8, 0.4 c) 1.5, 1.5, 1.5, 1.5, 2, 2, 2, 2, 2, 2
 d) 1.8, 0.24 e) 5/3, 5/3, 5/3, 5/3, 5/3, 5/3, 2, 2, 2, 2
 f) 1.8, 0.16

13. b) 3, 0 d) 3, 0 f) 3, 0

Exercises 3.2

1. $steak \succ_{Joe} eggplant$

2. No.

3. Mary's preference violates the principle of transitivity.

Exercises 3.3

1. No.

2. The system does not violate anonymity. We can't say if it is neutral, however, because the question does not give enough information.

3. This is a well-behaved example because candidate x will win with any procedure.
 (a) There is no majority rule winner.
 (b) Plurality vote totals: $x=34$, $y=16$, $z=0$. x wins.
 (c) No.
 (d) Eliminating z does not affect any plurality votes.
 (e) $x \succ_P y$ and $x \succ_p z$.

4. (a) x.
 (b) No.
 (c) y.
 (d) No. x loses head-to-head against y and z.

5. Note how the winner changes according to the different methods used.
   (a) w.
   (b) No.
   (c) $x \succ_P w$.
   (d) Eliminate y. Plurality totals are $x=24$, $z=40$, and $w=36$. z wins the plurality, but there is no majority winner. Eliminate x, and the final runoff has $w \succ_P z$.

6. (a) No.
   (b) No particular percentage is required; the replacement is selected by plurality.
   (c) Fear that a bad candidate might win stage 2 could cause them to vote strategically to keep Davis, while the hope that one's favorite might win in stage two could cause some to vote to remove Davis, even if he is preferred to the actual winner of stage two.

7. Strong monotonicity requires that, if there is a tie, then the change of a single voter's opinion must break the tie. A "hung jury" is the legal equivalent of a tie vote in an election. A jury that is divided 9-3 is "hung," unable to decide. If someone changes, so the total is 10-2, the jury is still unable to decide.

8. (a) Let the sum of the votes be Σ. Majority rule requires that if $\Sigma > 0$, then $x \succ_M y$, while $\Sigma < 0$ implies $y \succ_M x$, and $\Sigma = 0$ means a tie.
   (b) The sum will change to $\Sigma + 1$, and so if $\Sigma > 0$, then x will win.
   (c) x.
   (d) Yes.

Exercises 3.4

1. Make a table to show that, with 2 candidates, the Borda count and the majority election give the same results.

		Borda Votes	Borda Count
Group:	1	2	
Number of Members	81	54	
Alternatives			
Fred	1	0	81
Barney	0	1	54

2. (a) z.
   (b) x.
   (c) Group 1: (2,1,0)
   Group 2: (1,2,0)
   Group 3: (0,1,2)

(d) y.
(e) y is the Condorcet winner because $y \succ_P z$ and $y \succ_P x$.

3. (a) Compare each candidate against each of the others:
$x \succ_P y, z \succ_P x, x \succ_P w$
$y \succ_P w, z \succ_P y$
$z \succ_P w$
z is the Condorcet winner.
(b) x wins the Borda count. The overall ranking is $x \succ_{BC} z \succ_{BC} y \succ_{BC} w$.

	Borda votes						Borda count
Groups	1	2	3	4	5	6	
Number of Members	200	150	150	300	100	100	
Alternatives							
w	0	0	0	3	1	0	1000
x	2	1	3	1	2	3	1800
y	1	3	2	0	3	2	1450
z	3	2	1	2	0	1	1750

(c) After eliminating w, the Borda ranking is $z \succ_{BC} x \succ_{BC} y$.

	Borda votes						Borda count
Groups	1	2	3	4	5	6	
Number of Members	200	150	150	300	100	100	
Alternatives							
x	1	0	2	1	1	2	1100
y	0	2	1	0	2	1	750
z	2	1	0	2	0	0	1150

(d) No. Deleting a candidate never affects the Condorcet winner.

4. The Borda counts are (8, 7, 6), so w wins.
To change that, insert a losing alternative z as follows:

	Borda votes							Borda count
Groups	1	2	3	4	5	6	7	
Alternatives								
w	3	3	0	0	2	2	3	13
x	2	2	3	3	1	1	2	14
y	1	1	2	3	3	3	1	14
z	0	0	1	1	0	0	0	2

5. (a) *Pierre*. Note that *Pierre* $\succ p$ Paul with a vote of 41–40.
   (b) *Paul*. The Borda totals for (*Pierre, Paul, Jacques*) are (101, 109, 33).
   (c) If you change 1 vote, you can make *Paul* the Condorcet winner. On the other hand, if you try to make *Pierre* win the Borda count, you must change 3 ballots. Consider voter group 5, the ones who vote (0, 2, 1). Change 3 of those voters into group 1, so their votes become (2, 0, 1). The Borda totals change to (107, 103, 33). (Note, if you only change 2 voters, the result is a tie).

Exercises 3.5

1. 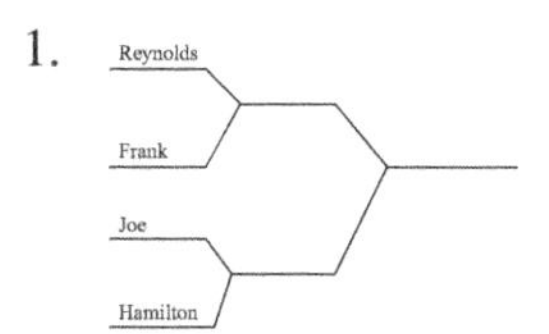

2. (a) *Reynolds* and *Hamilton* win in stage one, and in stage 2, the winner is Hamilton.
   (b) No.
   (c) Yes, *Joe* defeats *Hamilton*.
   (d) *Frank* is the plurality winner. *Hamilton* wins a majority/runoff election.
   (e) *Frank* wins the Borda count.

3. (a) $x \succ_P y$ and $x \succ_P z$.
   (b) Yes.
   (c) No, it is impossible.

4. (a) Compare x against y, then pair the winner off against z. y will win.
   (b) No, y is a Condorcet winner.

5. (a) Note that x beats y (6-5), y beats z (6-5), and z beats x (8-3).

	Voters					
Groups	1	2	3	4	5	6
Number of Members	3	3	2	1	1	1
Ranking						
First	z	y	x	y	y	x
Second	x	z	w	z	z	y
Third	y	w	z	w	w	z
Fourth	w	x	y	x	x	w

   (b) Our favorite is x. So our tournament would pit y against z in the first round, and y would advance to the final vote, and x would win.

Exercises 3.6

1. The round robin would end up in a tie, with each candidate winning one contest.
2. Voter 1 might change preferences so that $x \succ_1 z \succ_1 y$ or more voters of type 3 could be inserted.
3. The Borda totals for (x, y, z) are $(56, 77, 62)$. The aggregated pairwise votes are as follows. Cells represent votes earned for the row candidate against the column candidate.

	x	y	z	Aggregated Pairwise Vote (row sum)
x	.	22	34	56
y	33	.	34	77
z	31	31	.	62

4. 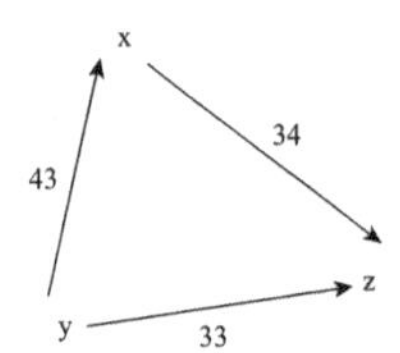

   y wins

5. 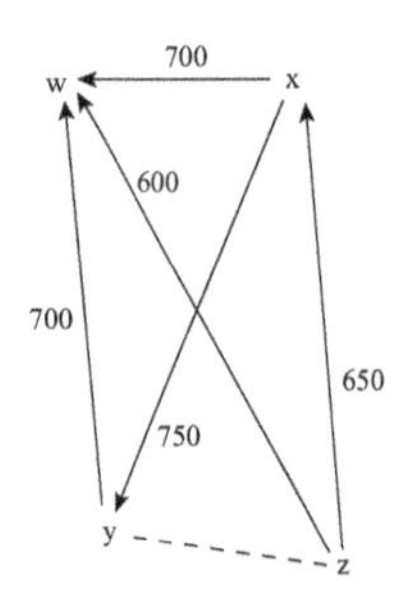

6. Schwartz set is $\{z\}$, the winner.
7. Show that if $x \succ_{BC} y \approx_{BC} z$, then the valid pairwise outcomes are:

$$\begin{aligned}
&1.\ x \succ_P y,\ x \succ_P z,\ y \approx_P z \\
&2.\ x \succ_P y,\ x \succ_P z,\ y \succ_P z \\
&3.\ x \succ_P y,\ z \succ_P x,\ y \succ_P z \\
&4.\ x \succ_P y,\ x \approx_P z,\ y \succ_P z \\
&5.\ x \approx_P y,\ x \succ_P z, y \succ_P z
\end{aligned} \tag{1}$$

   Let $q_{x,y}$ to represent the number of voters who prefer x to y.

$$q_{x,y} = |\{i \in N : x \succ i\ y\}|.$$

Because there are no indifferent voters, a tie means $q_{x,y} = q_{y,x} = N/2$. Note that

$$q_{x,y} > q_{y,x} \textit{ if and only if } q_{x,y} - q_{x,y} > 0 \textit{ if and only if } x \succ_P y.$$

The result of the aggregated pairwise vote, which is equivalent to the Borda count, can be expressed as

$$\begin{aligned} &\textit{aggregated pairwise vote for } x : q_{x,y} + q_{x,z} \\ &\textit{aggregated pairwise vote for } y : q_{y,x} + qy,z \\ &\textit{aggregated pairwise vote for } z : q_{z,x} + q_{z,y} \end{aligned}$$

If the Borda outcome is $x \succ_{BC} y \approx_{BC} z$, then, by definition:

$$\begin{aligned} q_{x,y} + q_{x,z} &> q_{y,x} + q_{y,z} \\ q_{x,y} + q_{x,z} &> q_{z,x} + q_{z,y} \\ q_{y,z} + q_{y,x} &= q_{z,x} + q_{z,y} \end{aligned} \quad (2)$$

Keep in mind that $q_{y,x} = N - q_{x,y}, q_{z,x} = N - q_{x,z}$, and $q_{z,y} = N - q_{y,z}$.

Re-arrange the inequalities so that they have the pairwise majority rule conditions on the left-hand side:

$$\begin{aligned} q_{x,y} - q_{y,x} &> q_{y,z} - q_{x,z} \\ q_{x,z} - q_{z,x} &> q_{z,y} - q_{x,y} \\ q_{y,z} - q_{z,y} &= q_{z,x} - q_{y,x} \end{aligned} \quad (3)$$

Consider 3 cases, $y \approx_P z$, $y \succ_P z$, and $z \succ_P y$.
Case 1: $y \approx_P z$. We now claim that if this (equivalently, $q_{y,z} = q_{z,y}$) is true, then the previous inequalities imply both $q_{x,y} > q_{y,x}$ and $q_{x,z} > q_{z,x}$. Proceed as follows. Since $q_{z,y} = N - qy,z$, then $qy,z = qz,y = N/2$. The third condition in 3, along with $qy,z = qz,y$, implies $qz,x = qy,x$. That is to say, either x defeats both y and z, and by the same margin, or y and z defeat x (by the same margin). It is obvious that x must defeat both, since it is the winner of the Borda count (and the aggregated pairwise vote).

This means that $x \succ_P z$ and $x \succ_P y$ are implied by the condition $y \approx_P z$, so one word has to be the first one listed in 1. That is to say, in order to be consistent with the Borda outcome $x \succ_{BC} y \approx_{BC} z$, then if y ties z in a plurality election, then x must defeat both y and z (and by the same margin).

Case 2: $y \succ_P z$. Beginning with the information supplied by the Borda outcomes, we can use the identities $q_{x,y} = N - q_{y,x}$, $q_{y,z} = N - q_{z,y}$, and $q_{z,x} = N - q_{x,z}$ to arrive at the following:

$$\begin{aligned} q_{x,y} - \frac{N}{2} &> \frac{1}{2}q_{y,z} - \frac{1}{2}q_{x,z} \\ q_{x,z} - \frac{N}{2} &> \frac{N}{2} - \frac{1}{2}q_{y,z} - \frac{1}{2}q_{x,y} \\ q_{y,z} &= \frac{N}{2} - \frac{1}{2}q_{x,z} + \frac{1}{2}q_{x,y} \end{aligned} \quad (4)$$

Insert the value of $q_{y,z}$ implied by the third expression in the second one to obtain

$$q_{x,z} - \frac{N}{2} > \frac{N}{4} + \frac{1}{4}q_{x,z} - \frac{3}{4}q_{x,y}$$

which can be rearranged as:

$$\frac{3}{4}q_{x,z} - \frac{3N}{4} > -\frac{3}{4}q_{x,y},$$

or

$$q_{x,z} + q_{x,y} > N \tag{5}$$

We do not know for sure if x defeats y or x defeats z in pairwise votes, but we can say for sure that x will defeat at least one (possibly both) of them.

This narrows down the possibilities quite a bit. This case assumes $y \succ_P z$, so it implies that the only majority rule outcomes consistent with the Borda result $x \succ_{BC} y \approx_{BC} z$ are:

$$\begin{array}{l} 1\ x \succ_P y, x \succ_P z, y \succ_P z \\ 2\ y \succ_P x, x \succ_P z, y \succ_P z \\ 3\ x \succ_P y, z \succ_P x, y \succ_P z \\ 4\ x \approx_P y, x \succ_P z, y \succ_P z \\ 5\ x \succ_P y, x \approx_P z, y \succ_P z \end{array} \tag{6}$$

Possibilities 2 and 4 can be ruled out. Here's why. If $y \succ_P z$, then $z \succ_P x$. To see why, recall $x \succ_{BC} y \approx_{BC} z$, so it must be that

$$qy,z + qy,x = qz,x + qz,y$$

Equivalently,

$$qy,z - qy,x = qz,x - qz,y$$

If $y \succ_P z$, then $qy,z - qz,y > 0$ and so $qz,x - qy,x > 0$ and using the identities like $qx,y + qy,x = N$, we find $qx,y - qx,z > 0$. If $x\ P\ z$, then $qx,z > N/2$, and it must be that $qx,y > N/2$ (equivalently, $x\ P\ y$). The possibilities 2 and 4 violate that restriction. That leaves possibilities 1, 3, and 5, which correspond to three of the words in the solution.

Case 3: $z \succ_P y$. If we repeat the same exercise as the previous case, we arrive at the same set of possible words, except that the last one "swaps" the positions of y and z.

$$\begin{array}{l} 1\ x \succ_P y, x \succ_P z, y \succ_P z \\ 3\ x \succ_P y, z \succ_P x, y \succ_P z \\ 5\ x \approx_P y, x \succ_P z, y \succ_P z \end{array} \tag{7}$$

Notice that two of these solutions are identical to the ones found in the previous case, only the third one is unique. Collecting up the unique solutions, we have finished our work. All five of the words that are compatible with the Borda outcome $x \succ_{BC} y \approx_{BC} z$ have been found.

Exercises 3.7

1. (a) The plurality totals for (x, y, z) are (22, 12, 31), so y is eliminated first. The 12 votes for y are transferred to x.
   (b) x wins the instant runoff, 34 – 31.
   (c) Yes, the results agree. x would win a majority/runoff contest.

2. (a) The plurality totals for (w, x, y, z) are (300, 250, 250, 200), so z is eliminated first. 200 votes are transferred to x.
   (b) After eliminating z, for candidates (w, x, y) the totals are (300, 450, 250). Candidate y is eliminated and 250 votes are transferred to x, the winner.
   (c) There is a tie for second place, so the runoff includes (w, x, y), and the vote totals would be (300, 550, 250), so x wins.

3. (a) The plurality totals for (*Hamilton, Joe, Frank, Reynolds*) are (22, 12, 31, 0), so *Reynolds* is eliminated first. Since *Reynolds* had no first-place votes, deleting him does not change the outcome. Next, *Joe* is deleted and 12 votes are transferred to *Hamilton*.
   (b) *Hamilton* wins the instant runoff.
   (c) A majority/runoff procedure picks *Hamilton* and *Frank* to compete in the runoff election, and *Hamilton* wins that runoff election.

4. The winner of the instant runoff is x, but the winner of the majority/runoff is y.

5. Note that x is a Condorcet winner, but it is eliminated in the first stage of the instant runoff process. Transfer 10 votes to y and then y is the winner.

	Voters		
Groups:	1	2	3
Number of Members	10	35	33
Ranking			
First	x	y	z
Second	y	x	x
Third	z	z	y

6. One approach would be to introduce some books that would be preferred to *Grapes of Wrath* by some of the voters and then call for a plurality election

or an instant runoff. As long as *Grapes* is the first choice of only a small group, it will be defeated. Another approach would be to introduce just one new book that is preferred to *Grapes* by the Superman fans as well as about one-half of the people who prefer *Grapes* to *Superman*, but for other *Grapes* supporters, this new book is absolutely the worst possible choice. Propose a sequence of comparisons. Begin by holding a vote betwen *Grapes* and the new book. With the help of the Superman supporters, the new book defeats *Grapes* and then Superman will defeat this new entry at the final stage.

Exercises 4.1

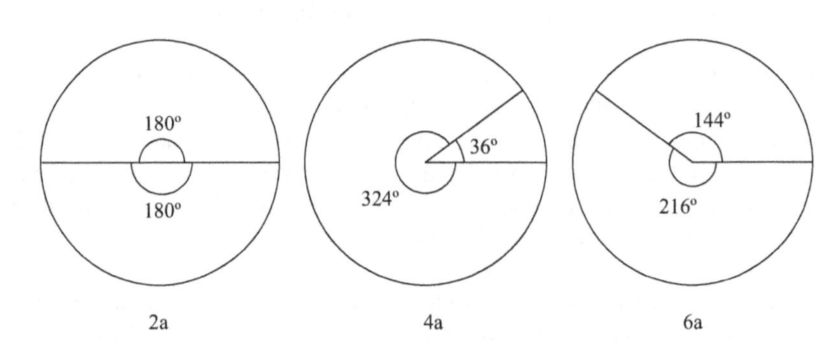

2a 4a 6a

1.	a) 144°, 108°, 72°, 36°	b) 0	c) 0.5	d) −0.8	e) 0.1	f) (2, 3)
2.	a) See figure above	b) 0.5	c) 0	d) 0	e) −0.5	f) e
3.	a) 36°, 72°, 108°, 144°	b) 0.5	c) −0.1	d) −1.3	e) −0.6,	f) (1, 3)
4.	a) See figure above	b) 81%	c) 95%	d) 88%	e) 86.6%	
5.	a) 108°, 252°	b) 83%	c) 85%	d) 84%	e) 83.8%	
6.	a) See figure above	b) 84%	c) 80%	d) 82%	e) 82.4%	

Exercises 4.2

1) 2.94	2) 2.2	3) 1.8	4) 1.0
5) 0.37	6) −0.12	7) −3.5	8) 1.24
9) 0.12	10) −3	11) −4	

12. See figure below

13.

	B-day	No
Flowers	1.5	1
No	−10	0

14. See figure below.

15.

	Rain	Shine
Umbrellas	250	−150
Glasses	−150	350

16. See figure below.

Stove \ Stranger	H	T
H	–20	30
T	10	–20

Gunning & Kappler \ Nature	Defeat	No Defeat
(1) Cheap	–10	–1
(2) Guarantee	–6	–6
(3) Insure	0	–10

Stockpile \ Winter	Mild	Normal	Severe
10	–100	–175	–300
15	–150	–150	–250
20	–100	–200	–200

Exercises 4.3

1. a) [1, 0], [1, 0] b) 1, 3 c) (2, 3) d) All outcomes
   e) None f) [1, 0], [0, 1], (1, 4) g) [1, 0], [1, 0], (2, 3)
2. a) [0, 1], [0, 1] b) 1, 3 c) (1, 4)
   d) All four e) None f) [0, 1], [0, 1], (1, 4)
   g) [1, 0], [0, 1], (2, 3)

3. a) [0, 1], [1, 0] b) 2, 2 c) (2, 2) d) (3, 1), (2, 5)
   e) None f) [0, 1], [0, 1], (2, 5) g) [0, 1], [1, 0], (2, 2)

4. a) [1, 0], [0, 1] b) 2, 2 c) (2, 5)
   d) (3, 1), (2, 5) e) (2, 5), (2, 2) f) [0, 1], [0, 1], (2, 2)
   g) [0, 1], [0, 1], (2, 2)

5. a) [1, 0, 0], [1, 0] or [0, 1] b) 4, 1 c) (5, 4) or (4, 1) d) (5, 5)
   e) (5, 4), (5,5) f) [1, 0, 0], [1, 0], (5, 4) or [0, 0, 1], [0, 1], (5, 5)

6. a) [0, 1, 0] or [0, 0, 1], [0, 1] b) 5, 4 c) (7, 8) or (5, 5) d) (7, 8)
   e) (7, 6), (7, 8) f) [0, 0, 1], [1, 0], (7, 6) or [0, 1, 0], [0, 1], (7, 8)
   g) [0, 1, 0], [0, 1], (7, 8)

7. a) [1, 0, 0], [0, 0, 1] b) 7, 2 c) (7, 2) d) All outcomes
   e) (7, 2) f) [1, 0, 0], [0, 0, 1], (7, 2) g) [1, 0, 0], [0, 0, 1], (7, 2)

8. a) [1, 0, 0], [1, 0, 0] b) 7, −6 c) (9, 0)
   d) (9, 0) e) (9, 0) f) [1, 0, 0], [1, 0, 0], (9, 0)
   g) [1, 0, 0], [1, 0, 0], (9, 0)

9. a) [1, 0, 0, 0], [1, 0, 0, 0] or [0, 0, 1, 0] b) 0, −1 c) (1, 2) or (1, 1)
   d) (0, 3), (3, −1), (2, 2) e) (2, 2), (2, 2) f) [0, 1, 0, 0], [1, 0, 0, 0], (2, 1) or [0, 0, 1, 0], [0, 0, 0, 1], (2, 2) or [0, 0, 0, 1], [0, 1, 0, 0] (2, 2)
   g) [0, 0, 0, 1], [0, 1, 0, 0], (2, 2) or [0, 0, 1, 0], [0, 0, 0, 1], (2, 2)

10. a) [0, 0, 1, 0], [0, 0, 0, 1] b) 0, 1 c) (0, 3)
    d) (3, 0), (0, 3), (2, 2) e) (1, 2), (1, 1)
    f) [0, 0, 0, 1], [0, 1, 0, 0], (2, 2) g) [1, 0, 0, 0], [0, 0, 0, 1], (1, 2)

11–18. See next page.

19. Yes. First maximize Ruth's payoff and then maximize Charlie's payoff. Or vice versa.

20. See next page.

	Enter	Wait
Enter	(4, 4)	(1, 2)
Wait	(2, 1)	(3, 3)

(11)

	Concert	TV
Concert	(1, 2)	(3, 4)
TV	(4, 3)	(2, 1)

(12)

		NFL	
		Fall	Spring
USFL	Fall	(7, 3)	(10, 5)
	Spring	(5, 10)	(3.5, 1.5)

(13)

	1	2
1	(64, 16)	(48, 24)
2	(60, 12)	(40, 16)

(14)

	Gun	No gun
Gun	(3, 3)	(1, 4)
No gun	(4, 1)	(2, 2)

(15)

	Lots	Less
Lots	(4, 4)	(2, 3)
Less	(3, 2)	(1, 1)

(16)

	Keep mum	Demand raise
Keep mum	(4, 4)	(3, 2)
Demand raise	(2, 3)	(1, 1)

(17)

	Keep mum	Demand raise
Keep mum	(4, 4)	(1, 3)
Demand raise	(3, 1)	(2, 2)

(18)

(8, 1)	(6, 3)
(4, 5)	(2, 7)

(20)

Exercises 4.4

1. (0.86, 1.36) 2. 1.2, 0.7 3. (1.66, 1.88)

4. 0.3, 1.7 5. (0, 1) 6. 1, 0.7

7. a) Any strategy b) [1, 0]

8. a) [0, 1] b) [1, 0] or [0, 1]

9. a) Any strategy b) Any strategy

10. a) [0, 0, 1] b) [0, 1, 0]

11. a) [0, 0, 1] b) [0, 1, 0, 0]

12. a) [0, 0, 0, 1] b) [0, 1, 0] 13. No

14. No 15. Yes 16. No 17. Yes

18. No 19. Yes 20. Yes 21. No

22. No 23. Yes

24. a) 0 b) 1

25. a) 63.6% b) 36.4% 26. a) 4/11 b) 7/11

27. a) 1 b) 0 28. a) 0.2 b) 0.8

29. a) 0 b) 1 30. a) 2/3 b) 1/3

31. a) 100% b) 0% 32. a) 100% b) None

33. a) 100% b) 0%

34. $(a-b, a-b)$

Exercises 4.5

1. [1/3, 2/3], [2/3, 1/3], 5/3
2. [1, 0], [0,1], 2
3. Any strategy, [0, 1], 1
4. [2/3, 1/3], [2/3, 1/3], 5/3
5. [0, 1], [1, 0], 4
6. [3/7, 4/7], [4/7, 3/7], −5/9
7. [1, 0], [1, 0], 1
8. [0, 1], [1, 0], 0
9. [1,0], [1,0], 2
10. [1, 0], [0,1], 1
11. [1, 0], any strategy, 1
12. [0, 1], [1, 0], 0
13. [7/13, 6/13], [4/13, 9/13], 11/13
14. [5/7, 2/7], [1/2, 1/2], 1
15. [1, 0], [0, 1], 2
16. [1/3, 2/3], [1/3, 2/3], 5/3
17. [1, 0], [1, 0], −3
18. [7/13, 6/13], [4/13, 9/13], c−15/13
19. [1, 0], [1, 0], 0
20. [0, 1], [1, 0], 1

21. [1, 0], [0, 1], –2

22. [3/8, 5/8], [5/8, 3/8], 17/16

23. [1, 0], [1, 0], 2

24. [0, 1], [1, 0], 2

25. [3/7, 4/7], [4/7, 3/7], 9/7

26. [1, 0], [0, 1], 0

27. [0, 1], [0, 1], 0

28. [0, 1], [1, 0], 2

29. [2/3, 1/3], [2/3, 1/3], –1/3

30. [5/7, 2/7], [1/2, 1/2], 0

31. [1, 0], [0, 1, 0], 1

32. [0, 1], [0, 1, 0, 0], –1

33. [0, 0, 1, 0], [1, 0], 2

34. [0, 0, 1], [0, 0, 1], 2

35. [0, 1, 0], [0, 1, 0, 0], –2

36. [1, 0, 0, 0], [1, 0, 0], 1

37. [0, 0, 1, 0], [0, 0, 1, 0], 0

38. [0, 1, 0, 0] or [0, 0, 0, 1], [0, 0, 0, 1, 0], 2

39. [1, 0, 0, 0, 0], [0, 1, 0, 0], 1

40. [0,0,0,0,1], [0,0,0.5,0,0.5], –0.5

41. This game has a saddle point and Frank had better buy the flowers.

42. [9/19, 0, 10/19], [9,19, 10/19], –100/19 (See Exercise 4.2.14)

43. See Exercise 4.2.15. The oddments are [5:4]. Merrill should invest $161.11 in umbrellas and $88.89 in sunglasses. The value of the game is $72.22.

44. See Exercise 4.2.14 for the game. This game has a saddle point and its solution is [0, 0, 1], [0, 0, 1], –200.

45. a) Yes b) Yes c) No d) [0.2, 0.8] e) No f) No
g) [0.6, 0.4], but he probably won't make it.

46. a) No b) [0, 1] c) No d) [3, 1] e) No f) No
g) [0.5, 0.5]

47. a) Yes b) Yes c) No d) [1,0] e) No f) Yes g) [0, 1]

48. a) No b) No c) [0, 1] d) [5/12, 7/12] e) No f) [0, 1]
g) [0.5, 0.5]

Exercises 5.1

1, 3, 5, 7, 9. See figure below

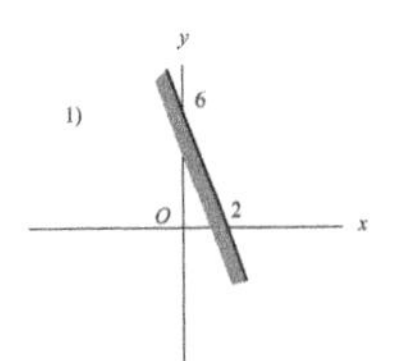

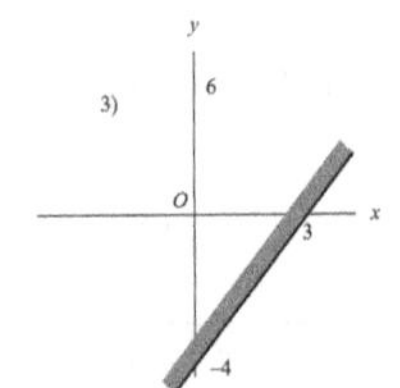

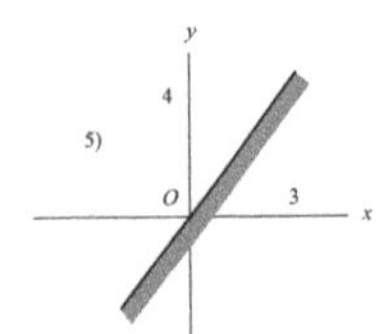

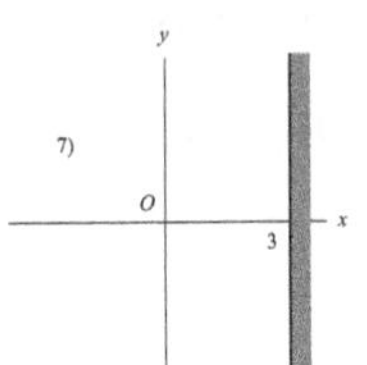

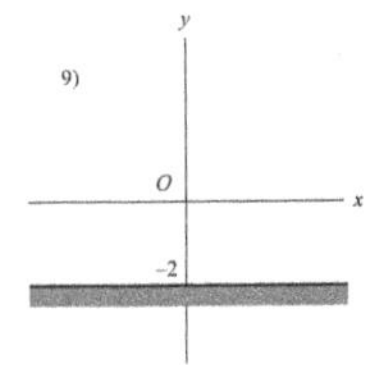

2, 4, 6, 8, 10. See figure below

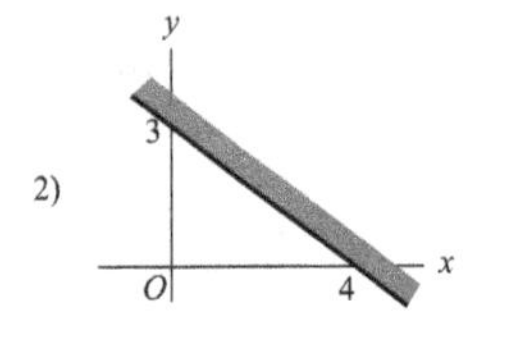

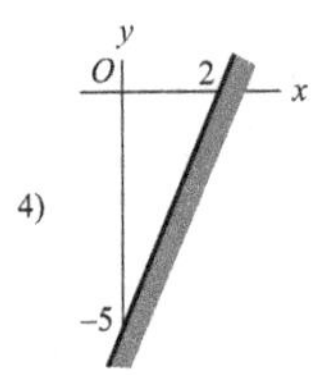

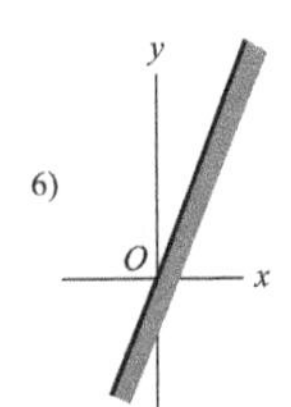

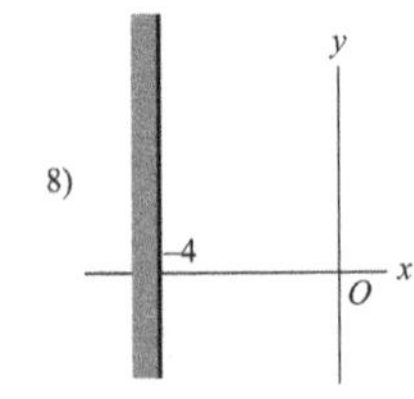

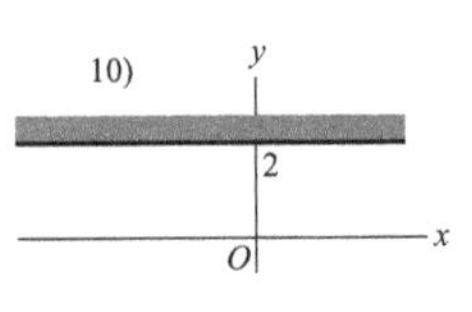

11, 13, 15, 17, 19. See figure below

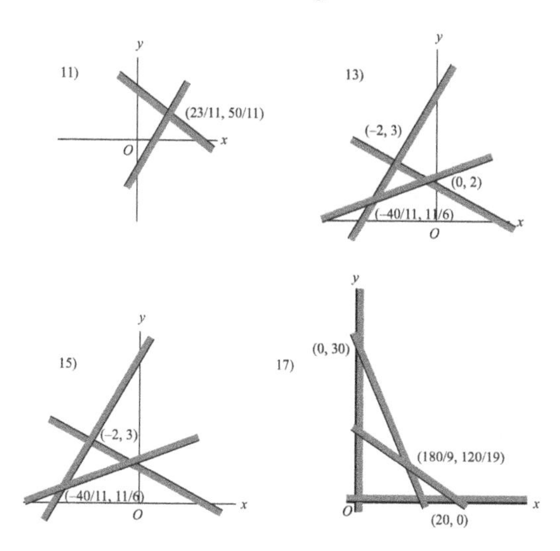

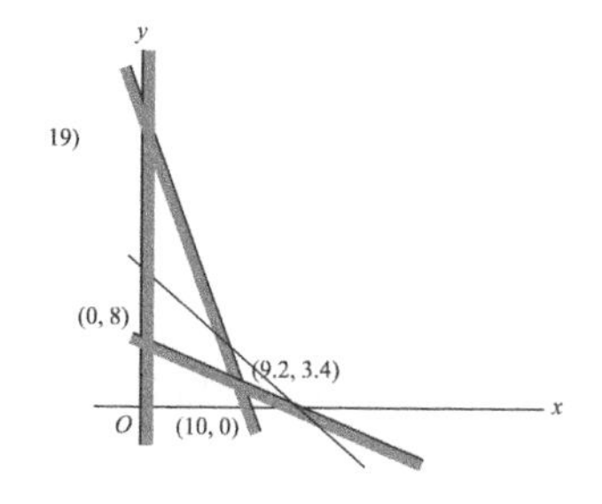

12. See figure below, (5/2, 3/2), (4/9, −14/3)

14. The empty set.

16. See figure below, (5/22, 3/2), (4/9, −14/3)

18. See figure below, (25.2, 21.7), (0, 23), (0, 400)

20. See figure below, (30, 50), (0, 60), (0, 95)

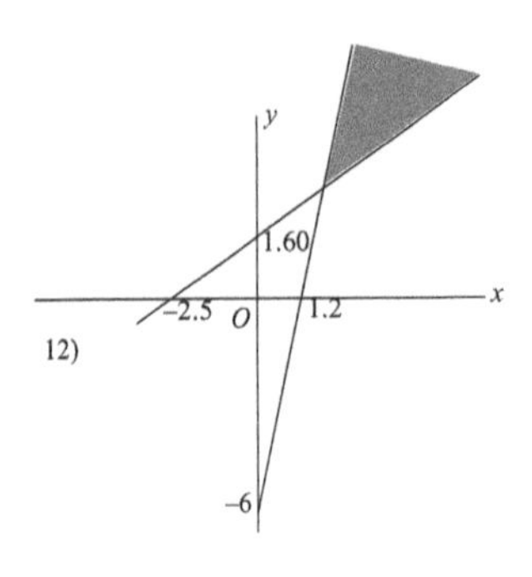

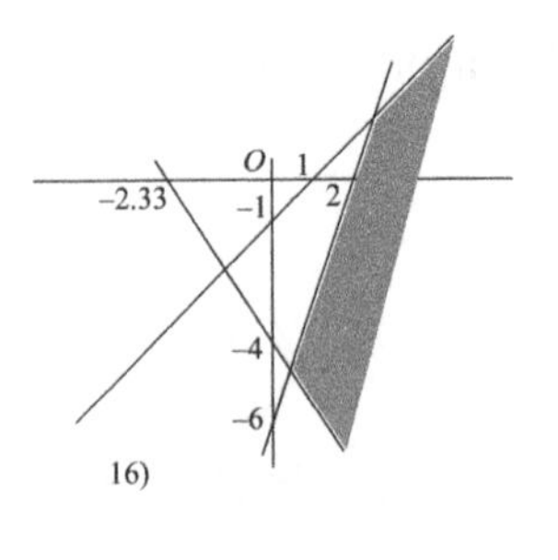

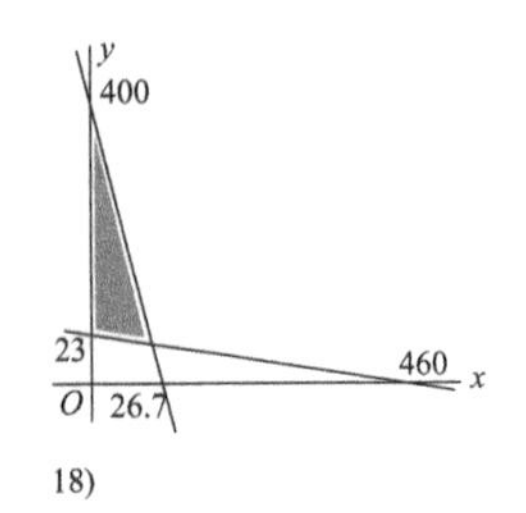

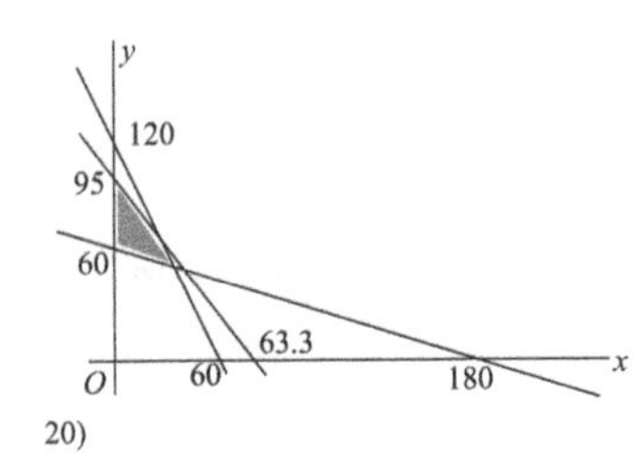

Exercises 5.2

1. 2, 101 2. 12, −28 3. 2, 92 4. −17, 23

5. 600 and 2400 lb in each case. 6. a, b) 600, 2400 c) 400, 2489
7. a) 1650 b) 1950 c) 3000

8. a) (40, 60) b, c) (75, 0)

9. a) 4000, 6000 b, c) 7000, 3000

10. a) 5000 to Zigzag, 6000 to Syzygy b, c) 6000 to Zigzag, 5000 to Syzygy

11. a, b) 80,000 and 40,000 c) 60,000 and 60,000
12. a, b) 5 radio and 10 TV ads c) 25 radio and 2 TV ads

13. a, b) 300 and 900 c) 0 and 1200

14. a) Use either no Tori and 15 Lou-Anne or 6 Tori and 3 Lou-Annes
    b) Use 6 Tori and 3 Lou-Annes c) Use 10.5 Tori and no Lou-Annes

15. Original question: Detroit 3 weeks and LA 6 weeks at a cost of $3,780,000.
    Second question: Detroit 6 weeks, LA 2 weeks at a cost of $4,000,000.

16. Schedule 7.5 days at each location. The minimum cost is $21,750.

Exercises 6.2

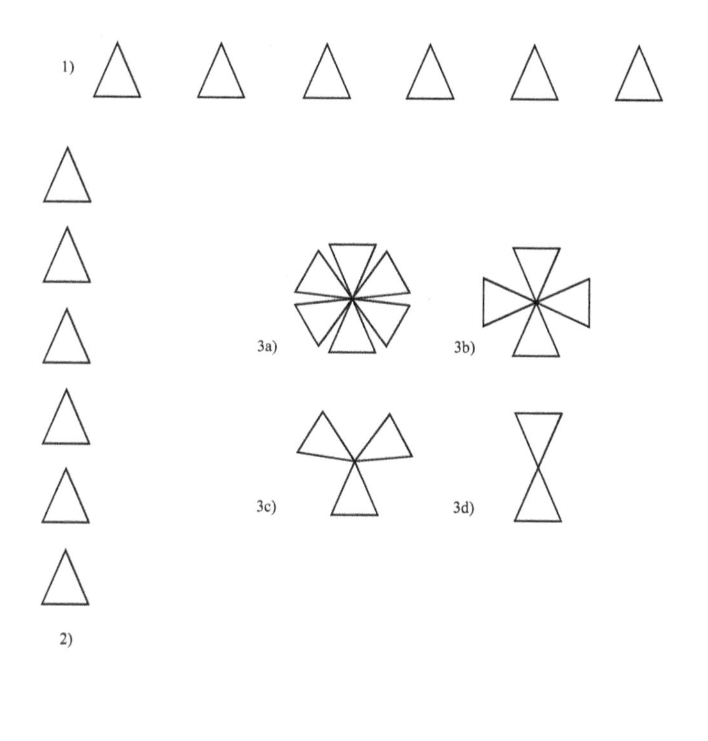

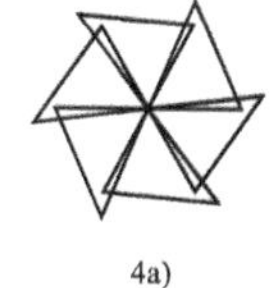
4a)

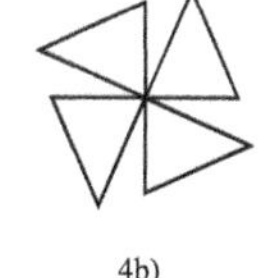
4b)

4c)

4d)

5a 5b

5c 5d

6a 6b

6c 6d

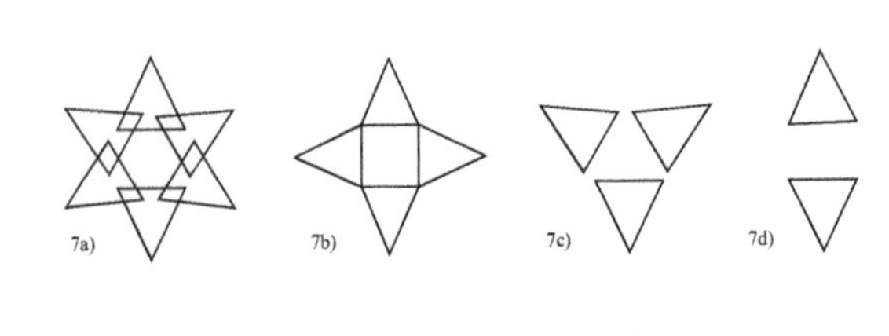

8) m

9) π

10) 11) 12)

13)

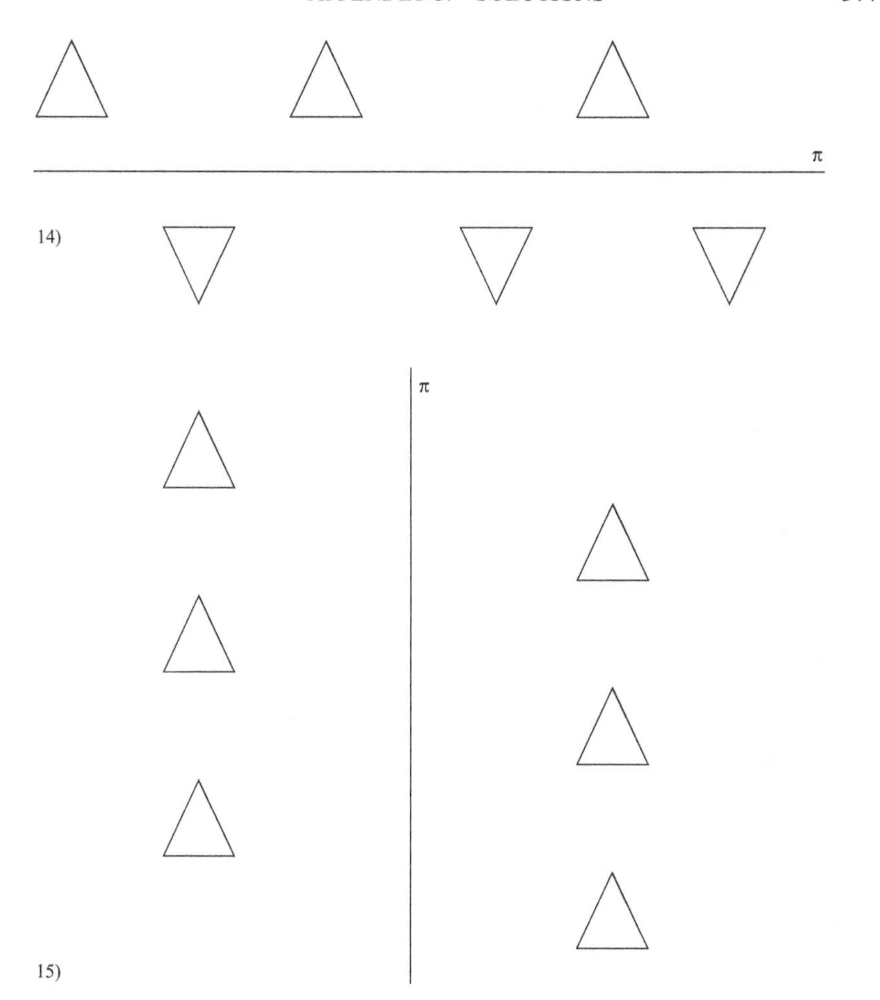

16) The effect of a 360° rotation is return each point to its original position.

Exercises 6.3

1. (See figure below)
   a) $\{\text{Id}, \rho_m, \rho_n, R_{C,180^\circ}\}$
   b) $\{\text{Id}, \rho_m, \rho_n, \rho_p, \rho_q, \rho_r, R_{C,72^\circ}, R_{C,144^\circ}, R_{C,216^\circ}, R_{C,288^\circ}\}$
   c) $\{\text{Id}, \rho_m, \rho_n, \rho_p, \rho_q, \rho_r, \rho_s, R_{C,60^\circ}, R_{C,120^\circ}, R_{C,180^\circ}, R_{C,240^\circ}, R_{C,300^\circ}\}$
   d) $\{\text{Id}, \rho_m, \rho_n, \rho_p, \rho_q, \rho_r, \rho_s, \rho_t, R_{C,\alpha}, R_{C,2\alpha}, R_{C,3\alpha} R_{C,4\alpha}, R_{C,5\alpha}, R_{C,6\alpha}\}$ where α is the angle of 360/7 degrees.
   e) $\{\text{Id}, \rho_m, \rho_n, \rho_p, \rho_q, \rho_r, \rho_s, \rho_t, \rho_u, R_{C,45^\circ}, R_{C,90^\circ}, R_{C,135^\circ}, R_{C,180^\circ}, R_{C,225^\circ}, R_{C,270^\circ}, R_{C,315^\circ}\}$

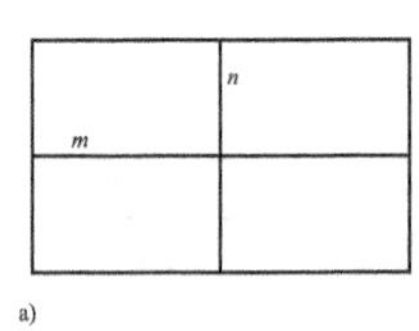

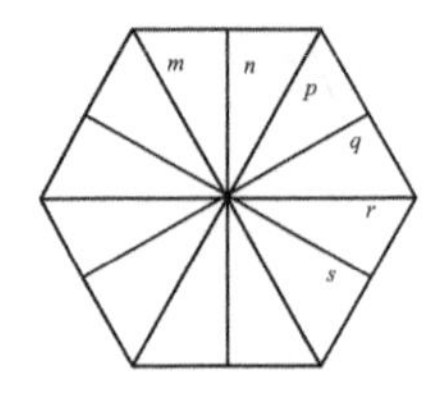

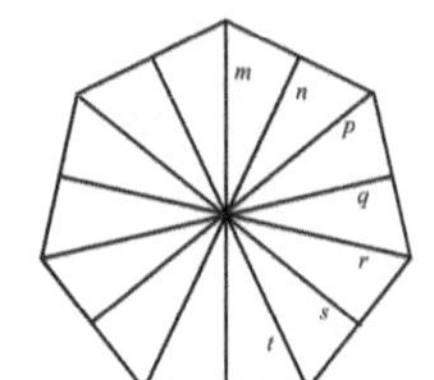

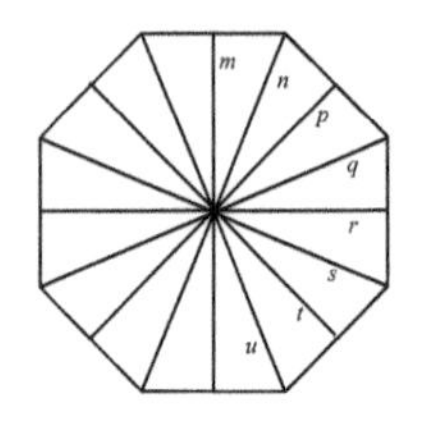

Exercises 6.4

1. Γ_1	2. Γ_2	3. Γ_4	4. Γ_1	5. Γ_2	6. Γ_4
7. Γ_3	8. Γ_7	9. Γ_5	10. Γ_3	11. Γ_3	12. Γ_7
13. Γ_5	14. Γ_3	15. Γ_7	16. Γ_6	17. Γ_6	18. Γ_5
19. Γ_4	20. Γ_1	21. Γ_6			

Exercises 6.5

1. pm	2. p2	3. p3m1	4. cmm	5. pgg	6. p6
7. p4g	8. cm	9. p3	10. p4m	11. pmg	12. p1
13. pmm	14. p31m	15. pmg	16. p4	17. p6m	18. p4
19. p4m	20. p31m	21. cm	22. pmg	23. p6	24. p6m
25. p3	26. pmm	27. p2	28. p1	29. cm	30. p4g
31. p3m1	32. pgg	33. pg	34. pmm		

Exercises 7.1

1. a) 24 b) 36 c) 6 squares and 8 hexagons
   d) c e) $24 - 36 + 14 = 2$

2. a) 12 b) 18 c) 4 triangles and 4 hexagons
   d) d e) $12 - 18 + 8 = 2$

3. a) 60 b) 90 c) 12 decagons and 20 triangles
   d) a e) $60 - 90 + 32 = 2$

4. a) 60 b) 90 c) 12 pentagons and 20 hexagons
   d) f e) $60 - 90 + 32 = 2$

5. a) 12 b) 24 c) 6 squares and 8 triangles
   d) g e) $12 - 24 + 14 = 2$

6. a) 6 b) 12 c) 8 triangles
   d) octahedron e) $6 - 12 + 8 = 2$

7. a) 30 b) 60 c) 12 pentagons and 20 triangles
   d) b = e e) $30 - 60 + 32 = 2$

8. a) 30 b) 60 c) 20 triangles and 12 pentagons
   d) b = e e) $30 - 60 + 32 = 2$

9. a) $72 - 108 + 38 = 2$, $48 - 72 + 26 = 2$
   b) $36 - 54 + 20 = 2$, $24 - 36 + 14 = 2$
   c) $72 - 108 + 38 = 2$, $48 - 72 + 26 = 2$
   d) $180 - 270 + 92 = 2$, $120 - 180 + 62 = 2$
   e) $180 - 270 + 92 = 2$, $120 - 180 + 62 = 2$
   f) $3v - (e + 3v) + (v + f) = v - e + f = 2$

10. a) $36 - 72 + 38 = 2$, $24 - 48 + 26 = 2$
    b) $18 - 36 + 20 = 2$, $12 - 24 + 14 = 2$
    c) $36 - 72 + 38 = 2$, $24 - 48 + 26 = 2$
    d) $90 - 180 + 92 = 2$, $60 - 120 + 62 = 2$
    e) $90 - 180 + 92 = 2$, $60 - 120 + 62 = 2$
    f) $e - 2e + (v + f) = 2$

11. $14 - 23 + 12 = 3$

12. $16 - 32 + 16 = 0$

13. $16 - 24 + 10 = 2$

14. $16 - 24 + 12 = 4$

Exercises 7.2

1. a) (1)(2)(3)(4)(5)(6)
   b) (1)(3)(2 5 4 6), (1)(3)(2 4)(5 6), (1)(3)(2 6 4 5),
   (5)(6)(1 2 3 4), (5)(6)(1 3)(2 4), (5)(6)(1 4 3 2),
   (2)(4)(1 5 3 6), (2)(4)(1 3)(5 6), (2)(4)(1 6 3 5)
   c) (1 2)(3 4)(5 6), (1 5)(3 6)(2 4), (1 6)(3 5)(2 4),
   (2 5)(4 6)(1 3), (2 3)(1 4)(5 6), (2 6)(4 5)(1 3)
   d) (1 5 2)(3 6 4), (1 2 5)(3 4 6)(2 5 3)(1 4 6),
   (2 3 5)(1 6 4), (1 2 6)(3 4 5), (1 6 2)(3 5 4)

2. a) Id   b) 20 of the type $R_{vertex,120^\circ}$ c) 15 of the type $R_{edge,180^\circ}$
   d) 12 of the type $R_{face,72^\circ}$, 12 of the type $R_{face,144^\circ}$

3. a) Id   b) 10 of the type $R_{vertex,72^\circ}$, 10 of the type $R_{vertex,144^\circ}$
   c) 15 of the type $R_{edge,180^\circ}$ d) 20 of the type $R_{face,120^\circ}$

4. a) (2 6)(4 8)(3 5)(1 7), (1)(7)(2 4 5)(6 3 8), (2 6 7 3)(1 5 8 4),
   (1 3)(2 4)(5 7)(8 6)
   b) i) Id   ii) $R_{23,180^\circ}$ iii) $R_{4,120^\circ}$ iv) $R_{15,180^\circ}$
   v) $R_{34,180^\circ}$ vi) $R_{7,120^\circ}$ vii) $R_{3487,-90^\circ}$
   viii) $R_{2,-120^\circ}$ ix) $R_{2,-120^\circ}$ x) $R_{1234,90^\circ}$
   xi) $R_{1584,180^\circ}$ xii) $R_{34,180^\circ}$ xiii) $R_{15,180^\circ}$
   xiv) $R_{4,-120^\circ}$ xv) $R_{12,180^\circ}$ xvi) Id

5. a) (1 5)(3 7)(2 8)(4 6), (2)(8)(1 3 6)(4 7 5), (1 5 6 2)(3 4 8 7),
   (1 3)(2 4)(5 7)(6 8)
   b) i) Id   ii) $R_{14,180^\circ}$ iii) $R_{3,120^\circ}$
   iv) $R_{26,180^\circ}$ v) $R_{34,180^\circ}$ vi) $R_{2,120^\circ}$
   vii) $R_{1234,270^\circ}$
   viii) $R_{7,240^\circ}$ ix) $R_{7,120^\circ}$ x) $R_{2376,270^\circ}$
   xi) $R_{1562,180^\circ}$ xii) $R_{23,180^\circ}$ xiii) $R_{26,180^\circ}$
   xiv) $R_{3,120^\circ}$ xv) $R_{14,180^\circ}$ xvi) Id

6. a) (1 3)(2 4), (1)(2 4 3), (2)(1 4 3), (1 4)(2 3)
   b) i) Id   ii) $R_{4,120^\circ}$ iii) $R_{1,240^\circ}$
   iv) $R_{12,180^\circ}$
   v) $R_{2,120^\circ}$ vi) $R_{1,240^\circ}$ vii) $R_{14,180^\circ}$
   viii) $R_{3,120^\circ}$ ix) $R_{3,240^\circ}$ x) $R_{13,180^\circ}$

xi) $R_{2,120°}$ xii) $R_{1,240°}$ xiii) $R_{12,180°}$
xiv) $R_{2,120°}$ xv) $R_{4,240°}$ xvi) Id

7. a) (1 4)(2 3), (3)(4 2 1), (3)(1 2 4), (1 2)(3 4)
b) i) Id ii) $R_{4,120}°$ iii) $R_{1,240}°$
iv) $R_{13,180}°$
v) $R_{1,120}°$ vi) $R_{3,240}°$ vii) Id
viii) $R_{2,120}°$ ix) $R_{4,240}°$ x) Id
xi) $R_{3,120}°$ xii) $R_{1,240}°$ xiii) $R_{13,180}°$
xiv) $R_{1,120}°$ xv) $R_{2,240}°$ xvi) Id

8. a) (1 3)(2 5)(4 6), (1)(3)(2 5 4 6), (1 4 6)(2 5 3), (1 3)(2)(4)(5 6)
b) i) Id ii) $R_{5,180}°$ iii) $R_{5,90}°$ iv) $R_{3,90}°$
v) $R_{2,180}°$ vi) $R_{1,180°}$ vii) $R_{23,180°}$
viii) $R_{26,180°}$ ix) $R2,90°$ x) $R_{16,180°}$
xi) $R_{235,120°}$ xii) $R_{263,120°}$ xiii) $R_{1,90°}$
xiv) $R_{25,180°}$ xv) $R_{125,120°}$ xvi) Id

9. a) (4 6)(2 5)(1 3), (1)(3)(2 4)(5 6), (3 4 5)(1 2 6), (1 4 3 2)(5)(6)
b) i) Id ii) $R_{26,180}°$ iii) $R_{14,180}°$
iv) $R_{125,120}°$
v) $R_{26,180}°$ vi) Id vii) $R_{125,120}°$ viii) $R23,180°$
ix) $R_{15,180}°$ x) $R_{235,240}°$ xi) $R_{162,120}°$
xii) $R_{3,90}°$ xiii) $R_{235,120}°$ xiv) $R_{12,180}°$
xv) $R_{4,90}°$ xvi) $R_{5,180}°$

10. a) (a c)(1 4)(2 3)(h i)(e k)(b 9)(5 7)(6 8)(f g)(d j),
(5)(7 d f)(g 1 b)(j 9 2)(a h 4)(3 e c)(i 6 k)(8),
(5 f b j 7)(6 8 i a e)(9 1 h k 3)(c 4 g d 2)
(1 f d 2 5)(i k 3 8 4)(j a c g 6)(b 9 h 7 e)
b) i) Id ii) $R_{5d9g7,72}°$ iii) $R_{5,240}°$ iv) $R_{9,240}°$ v) $R_{57jbf,-72}°$
vi) $R_{5,240}°$ vii) $R_{5f,180}°$ viii) $R_{d,240}°$ ix) $R_{7,240}°$ x) $R_{57,180}°$
xi) $R_{57jbf,144}°$ xii) $R_{5d,180}°$ xiii) $R_{a,120}°$ vix) $R_{f,240}°$
xv) $R_{2f,180}°$ xvi) $R_{1d5f2,-72}°$

11. a) (1)(2 e d)(4)(j k 3)(f 6 9)(h a 5)(7 c i)(8 g b)
(b j)(e a)(4 f)(1 i)(2 3)(7 c)(6 9)(h g)(5 k)(8 d)
(a 8 e i 6)(b 5 j f 7)(2 g c d 4)(9 k 1 3 h)
(2 h c b f)(a i 3 g 9)(1 6 k j 5)(d e 8 4 7)
b) i) $R_{1,120}°$ ii) $R_{7,240}°$ iii) $R_{5d9g7,216}°$ iv) $R_{j,120}°$ v) $R_{9,120}°$
vi) Id vii) $R_{3g9ai,-72}°$ viii) $R_{57jbf,216}°$ ix) $R_{g9ai3,144}°$
x) $R_{3g7j4,72}°$ xi) $R_{68iae,-72}°$ xii) $R_{5,240}°$ xiii) $R_{d,240}°$
xiv) $R_{1d9ae,144}°$ xv) $R_{j,240}°$ xvi) $R_{2fcbh,-72}°$

12. a) (1)(c)(7 b a 9 8)(5 4 3 2 6), (4 a)(2 7)(5 9)(6 8)(3 b)(1 c), (3 4 9)(a 8 1)(6 b 7)(2 5 c), (4)(7)(1 a 3 5 9)((6 c 2 b 8)

    b) i) $R_{1,144^\circ}$ ii) $R_{28,180^\circ}$ iii) $R_{4,144^\circ}$
    iv) $R_{54,180^\circ}$ v) $R_{67,180^\circ}$ vi) Id
    vii) $R_{a9c,240^\circ}$ viii) $R_{6,72^\circ}$ ix) $R_{b,216^\circ}$
    x) $R_{154,240^\circ}$ xi) $R_{349,120^\circ}$ xii) $R_{7cb,120^\circ}$
    xiii) $R_{34,180^\circ}$ xiv) $R_{5,-72^\circ}$ xv) $R_{49a,240^\circ}$
    xvi) $R_{4,-72^\circ}$

13. a) (5 3 6 4 2)(7 a 8 b 9)(1)(c), (8 9)(5 6)(a 2)(c 3)(1 b)(7 4), (1 2 6)(a 9 c)(8 b 4)(3 7 5), (9 5 c 4 b)(7 3 6 8 1)(a)(2)

    b) i) $R_{1,-72^\circ}$ ii) $R_{a94,120^\circ}$ iii) $R_{abc,120^\circ}$
    iv) $R_{56b,120^\circ}$ v) $R_{4a5,240^\circ}$ vi) Id
    vii) $R_{145,240^\circ}$ viii) $R_{34,180^\circ}$ ix) $R_{c89,120^\circ}$
    x) $R_{349,120^\circ}$ xi) $R_{126,240^\circ}$ xii) $R_{b,72^\circ}$
    xiii) $R_{349,120^\circ}$ xiv) $R_{14,180^\circ}$ xv) $R_{4,72^\circ}$
    xvi) $R_{a,-72^\circ}$

14. 4 15. 8 16. 6 17. 2 18. $2n$

Index